Polymer Supported Organic Catalysts

Editors

Narendra Pal Singh Chauhan

Department of Chemistry, Faculty of Science
Bhupal Nobles' University
Udaipur, Rajasthan, India

Sapana Jadoun

Departamento de Química, Facultad de Ciencias
Universidad de Tarapacá
Avda. General Velásquez, 1775, Arica, Chile

CRC Press
Taylor & Francis Group
Boca Raton London New York

CRC Press is an imprint of the
Taylor & Francis Group, an **informa** business

Cover credit: Image taken from Chapter 12. Reproduced by kind permission of Dr. Avinash Kumar Srivastava (corresponding author of the chapter).

First edition published 2025
by CRC Press
2385 NW Executive Center Drive, Suite 320, Boca Raton FL 33431

and by CRC Press
4 Park Square, Milton Park, Abingdon, Oxon, OX14 4RN

CRC Press is an imprint of Taylor & Francis Group, LLC

Library of Congress Cataloging-in-Publication Data (applied for)

ISBN: 978-0-367-48442-2 (hbk)
ISBN: 978-1-032-77096-3 (pbk)
ISBN: 978-1-003-03978-5 (ebk)

DOI: 10.1201/9781003039785

Typeset in Times New Roman
by Prime Publishing Services

Preface

Polymer-supported organic catalysts are largely insoluble in most reaction solvents, which allows for easy recovery and recycling of the catalysts. They are generally stable, readily available, and environmentally friendly, so they have attracted the interest of many synthetic chemists in the industrial and academic fields. The immobilization of achiral and chiral catalysts on polymer supports has drawn the interest of many researchers. To date, much effort has been put into the design and development of polymer-supported organic catalysts with the goal of achieving high catalytic performance while also being reusable. Some of these catalysts have recently been extended to continuous-flow processes, making them even more appealing tools in terms of operational simplicity, environmental compatibility, and cost efficiency. In this book, different types of polymer supported catalysts based on peptides, polystyrene, polyethers, poly(acrylic acid), poly(ethylene imine), poly(2-oxazoline), poly(isobutylene), poly(norbornene), etc., as well as metals such as Ruthenium, Iridium, Cobalt, Copper, and Rhodium, are included with their synthetic organic synthesis applications.

It is believed that this work will be of general interest to organic chemists, materials scientists, chemical engineers, polymer scientists and technologists. The editors would like to thank several scientists in this field at various institutes for their assistance in compiling unique knowledge of this class of materials.

Narendra Pal Singh Chauhan, Ph.D.
Sapana Jadoun, Ph.D.

Contents

Polymer-supported Organic Catalysts: An Introduction

Sapana Jadoun[1*] and Narendra Pal Singh Chauhan[2]

[1] Departamento de Química, Facultad de Ciencias, Universidad de Tarapacá,
   Avda. General Velásquez, 1775, Arica, Chile

[2] Department of Chemistry, Bhupal Nobles' University, Udaipur - 313002, Rajasthan, India

1. Background

Polymer-supported organic catalysts (PSOCs) are a class of heterogeneous catalysts that consist of an organic catalyst attached to a polymer matrix where organic catalysts can be defined by low molecular weight organic compounds able to boost a transformation in substoichiometric quantity (Shajahan et al. 2022, Benaglia et al. 2003). PSOCs have procured significance possessing inert, non-volatile, innocuous, reusable, and indissoluble astounding properties provided with the use of catalyst support which offers an easy synthetic path, materials availability, and higher efficiency (Li et al. 2023, Clapham et al. 2001). The use of PSOCs has gained significant attention in recent years as they offer several advantages over traditional catalysts such as ease of separation and recovery from reaction mixtures due to the immobilization on solid support resulting in easy filtration and removal (Liu et al. 2022, Ramey et al. 2023). These unique properties simplify the purification process by inhibiting the waste, making PSOCs more environmentally friendly than traditional catalysts (Wang et al. 2022a). These are highly versatile as the choice of polymers as well as catalysts can be tailored to suit a wide range of reactions (Silva et al. 2022). The polymer matrix can be compatible with various solvents, temperatures, and reaction conditions, making PSOCs suitable for use in a variety of industries, including pharmaceuticals, materials, and fine chemicals (Wang et al. 2022b, Chakravarthy et al. 2023).

*Corresponding author: sjadoun022@gmail.com; sjadoun@academicos.uta.cl

The application by suitable mixes of polymer-supported reagents, catalysts, and or scavengers is a strong strategy for the synthesis and simultaneous synthesis of small organic compounds (Siewniak et al. 2022). The use of polymer supports in organic synthesis began with solid-phase synthesis where the synthetic target is synthesized and attached to the polymer (Sherrington 2001). For a variety of reasons, including ease of reaction monitoring and product characterization, shorter method development time, etc., the role of the polymer has gradually shifted to supporting reagents for reacting with solution-phase substrates in what is sometimes referred to as polymer-assisted-synthesis (Hodge 2005).

The use of polymer supports in organic synthesis began with solid-phase synthesis where the synthetic target is synthesized and attached to the polymer (Kobayashi and Akiyama 2003). For a variety of reasons, including ease of reaction monitoring and product characterization, shorter method development time, etc., the role of the polymer has gradually shifted to supporting reagents for reacting with solution-phase substrates in what is sometimes referred to as polymer-assisted synthesis (Zhou et al. 2013). In this book, we have covered 13 chapters discussing various organic syntheses with support of various polymers such as peptides, polystyrene, alkaloids, polyisobutylene, poly (ethylene imine), polyacrylic acid, poly (2-oxazoline), polyethers, polynorbornenes, and polymers coordinating with transition metals like ruthenium, iridium, palladium, cobalt, copper, and rhodium.

In this context, chapter 2 discusses the peptides-supported catalysts including peptide catalysts in water and peptide catalysts in a mixture followed by their various synthetic approaches and applications. Chapter 3 includes polymer-supported phase transfer catalysts and discusses their types including quaternary ammonium and phosphonium salts, crown ethers, cryptands, and others. In addition, the chapter includes the principle and mechanism of phase transfer catalysts along with their numerous synthesis approaches. The characterization and applications of phase transfer catalysts have also been discussed in that chapter.

Chapter 4 comprises the achiral polymer-supported organic reagents in which the opening of epoxides using polymer-supported organic catalysts has been discussed along with the application of polymer-supported organic catalysts in the formation of cyclic products. This chapter has also been complimented with hydrogenation reactions using polymer-immobilized catalysts and the application of polymer-supported organic catalysts in aldol reactions. Chapter 5 discusses about alkaloid supported catalyst. The chapter includes the classification of alkaloids followed by pharmacological applications of alkaloids. The role of alkaloids has been discussed in the organic reactions as well as the discussion about platinum cinchona alkaloid catalyst for enantioselective α-Ketoester has also been given. The authors have summed up the chapter with conjugate addition reactions via cinchona alkaloid catalysts.

Chapter 6 includes cross-coupling reactions, organic synthesis and applications of polystyrene supported catalysts. Chapter 7 is a detailed discussion of polyacrylic acid and its derivatives-supported catalyst. The chapter discusses various derivatives of this catalyst using rhodium, platinum, palladium, TEMPO, and others. In a nutshell, applications of these catalysts have been discussed. Chapter 8 is a detailed discussion of polyether-supported catalysts comprising synthesis, properties such as selectivity, recyclability, stability, and applications in various sectors including organic synthesis, environmental remediations, pharmaceuticals, fuel cells, etc.

Chapter 9 is a sum up of poly(ethylene imine) and poly(2-oxazoline)-supported catalysts consisting of the general discussion about these followed by their synthetic approaches and applications in numerous sectors. Chapter 10 discusses the polyisobutylene and polynorbornenes-supported catalysts and includes a discussion of the mode of isobutylene and norbornene polymerization and a detailed discussion about these followed by the applications in numerous reactions such as coupling reactions, Heck coupling reactions, Suzuki coupling reactions, Sonogashira coupling reaction, epoxidation and some more. Chapter 11 tells us about different synthetic routes to prepare phosphine based polymer catalysts and their applications in organic functional group interconversion.

Chapter 12 gives a detailed discussion of the synthesis, mechanism, and applications of ruthenium and iridium-containing polymer-supported catalyst. Chapter 13 is about the palladium-containing polymer-supported catalyst and includes polystyrene-containing palladium catalysts. In addition the chapter describe about polystyrene supported palladium catalysts along with polymer support other than polystyrene. In a nutshell, chapter 14 includes cobalt, copper, and rhodium-containing polymer-supported catalysts. This chapter includes a detailed discussion about the chitosan-supported cobalt catalysts, chitosan-supported copper catalysts and chitosan-supported rhodium catalysts.

References

Benaglia, Maurizio, Alessandra Puglisi and Franco Cozzi. 2003. Polymer-supported organic catalysts. *Chemical Reviews* 103(9): 3401–3430. ACS Publications.

Chakravarthy, A.S. Jeevan, M.J. Madhura and V. Gayathri. 2023. A novel polymer supported copper (II) complex as reusable catalyst in oxidative esterification. *Catalysis Letters* 1–12. Springer.

Clapham, Bruce, Thomas S. Reger and Kim D. Janda. 2001. Polymer-supported catalysis in synthetic organic chemistry. *Tetrahedron* 57(22): 4637–4662. Pergamon.

Hodge, Philip. 2005. Synthesis of organic compounds using polymer-supported reagents, catalysts, and/or scavengers in benchtop flow systems. *Industrial & Engineering Chemistry Research* 44(23). 8542–8553. ACS Publications.

Kobayashi, Shū and Ryo Akiyama. 2003. Renaissance of immobilized catalysts. New types of polymer-supported catalysts, 'Microencapsulated Catalysts', which enable environmentally benign and powerful high-throughput organic synthesis. *Chemical Communications* 4: 449–460. Royal Society of Chemistry.

Li, Huadeng, Ke Zheng, Jiaxiang Qiu, Ruomeng Duan, Jianling Feng, Rui Wang, Zhimeng Liu, Guanqun Xie and Xiaoxia Wang. 2023. [HDBU] Br@ P-DD as porous organic polymer-supported ionic liquid catalysts for chemical fixation of CO_2 into cyclic carbonates. *ACS Sustainable Chemistry & Engineering* 11(10): 4248–4257. ACS Publications:.

Liu, Ruoyang, Shun-Cheung Cheng, Yelan Xiao, Kin-Cheung Chan, Ka-Ming Tong and Chi-Chiu Ko. 2022. Recyclable polymer-supported iridium-based photocatalysts for photoredox organic transformations. *Journal of Catalysis* 407: 206–212. Elsevier.

Ramey, Erin E., Elizabeth L Whitman, Cole E. Buller, James R. Tucker, Charles S. Jolly, Kjersti G. Oberle, Austin J. Becksvoort, Mark Turlington and Christopher R. Turlington. 2023. A biodegradable, polymer-supported oxygen atom transfer reagent. *Polymers* 15(9): 2052. MDPI.

Shajahan, Rubina, Rithwik Sarang and Anas Saithalavi. 2022. Polymer supported proline-based organocatalysts in asymmetric aldol reactions: A review. *Current Organocatalysis* 9(2): 124–146. Bentham Science Publishers.

Sherrington, David C. 2001. Polymer-supported reagents, catalysts, and sorbents: Evolution and exploitation—A personalized view. *Journal of Polymer Science Part A: Polymer Chemistry* 39(14): 2364–2377. Wiley Online Library.

Siewniak, Agnieszka, Edyta Monasterska, Ewa Pankalla and Anna Chrobok. 2022. Polymer-supported poly (ethylene glycol) as a phase-transfer catalyst for cross-aldol condensation of isobutyroaldehyde and formaldehyde. *Molecules* 27(19): 6459. MDPI.

Silva, Maria João, João Gomes, Paula Ferreira and Rui C. Martins. 2022. An overview of polymer-supported catalysts for wastewater treatment through light-driven processes. *Water* 14(5): 825. MDPI.

Wang, Bingyang, Jin Lin, Chungu Xia and Wei Sun. 2022a. Porous organic polymer-supported manganese catalysts with tunable wettability for efficient oxidation of secondary alcohols. *Journal of Catalysis* 406: 87–95. Elsevier.

Wang, Xiong, Wenqian Kang, Cuilan Zhang, Guangquan Li, Pingsheng Zhang and Yanqin Li. 2022b. Sulfonated porous organic polymer supported zeigler-natta polypropylene catalysts with high stereoregularity and broad molecular weight distribution. *Microporous and Mesoporous Materials* 343: 112151. Elsevier.

Zhou, You, Zhonghua Xiang, Dapeng Cao and Chang-Jun Liu. 2013. Covalent Organic Polymer Supported Palladium Catalysts for CO Oxidation. *Chemical Communications* 49(50): 5633–5635. Royal Society of Chemistry.

Peptide-supported Catalysts: Synthesis and Applications

Anupama Rajput[1], Prachika Rajput[2] and Srikanta Samanta[3*]

[1] Department of Applied Science, MRIIRS, Faridabad - 121001, Haryana, India
[2] Department of Chemistry, NSUT, Delhi - 110078, India
[3] Department of Chemistry, DNC College, University of Kalyani, West Bengal - 742201

1. Introduction

The peptides can be used in several organic syntheses as catalysts. Almost every peptide catalyst consists of few amino acids and is not that complex. The scope of modulation by slightly changing the sequences of amino acids creates an probability to serve use in specific reactions. Various strategies that have been put forward for obtaining specifically capable catalysts are the unique design of peptide structures, their screening, and modification of the peptides that are naturally obtained. Specific amino acid in the peptide catalysts provides a base to bind with substrate and sometimes benefit of an active catalyst in water (Arnold et al. 2003). In some cases, peptides are also subjected to be used as templates for noble metals, which can finally be used as catalysts.

Due to their asymmetric structure, they are highly preferred for selective reactions as well (Lieblich et al. 2017).

The extravagant nature of these special metal catalysts supported on a peptide template makes way for multiple applications, out of which a predominant one is of Au nanoparticle acting as a catalyst. The entire process - beginning from its synthesis to its use in real, or we can say practical, world - is enhanced and makes the Au NPs more biocompatible.

Further, the chapter includes an extended example of peptide-supported platinum and palladium catalysts with their applications in different fields and scope of modification by changing amino acid sequences in the peptide chain.

*Corresponding author: srikantachem81@gmail.com

2.   Peptide catalyst in water

The use of tripeptide with an alkyl chain as a base for the study of proceeding reaction can be taken into consideration.

For example: H-dPro-Pro-Glu-NHC12H151

The salient feature of this peptide catalyst is that it is stereo selective in nature. It is incorporated for the addition reaction including formation of nitro olefins from aldehydes in water (sometimes in other organic solvents too) (Lummis et al. 2005).

The first step of the reaction involves formation of intermediate (enamine). The factor affecting efficiency of the catalyst being used is the cis/trans ratio of amide bond in dPro-Pro. Higher the ratio, higher is the expected efficiency (Holzberger and Marx 2010).

The water media provides the facility to form emulsion and increases the efficacy of water insoluble substrates through the alkyl chain. Further addition of any additives like alcohols, carboxylic acids or metal complexes could affect the way the peptide (Figure 1) reacts as it can form intermediates. Drop in the rate of formation of certain by-product can also be observed as one of the effects.

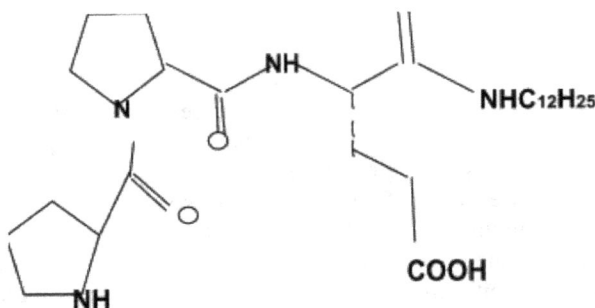

Figure 1. H-dPro-Pro-Glu-NHC$_{12}$H$_{25}$ (peptide 1)

Figure 1 shows high chemo selectivity even in the availability of abundant compounds and water as a medium. The functional groups present in the compounds too do not tend to interfere, thus a possible process in all.

3.   Peptide catalyst in mixture

The peptide1 catalysis also functioned well in a mixture of multi components of approximately related nature. Various sources of water were basically projected for the study.

Basically, they are alternative mixtures having water like solvents. The point where these mixtures differ is the varied pH, viscosity and polarity (and obviously composition). These variation factors help to provide a suitable environment for

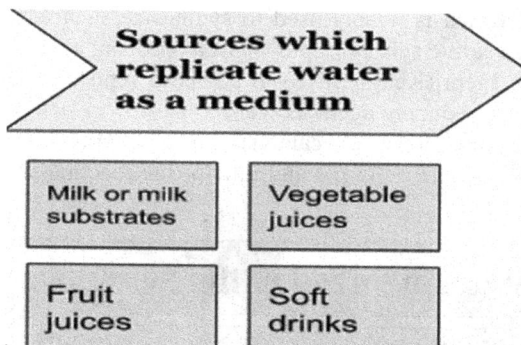

Figure 2. Diagram illustrates the sources replicating water in terms of medium

observing the effects each and every aspect can bring on the table (Schnitzer et al. 2022).

Various experiments showed that aldehydes and nitro olefins do not react at all when a peptide 1 catalyst was not present whereas they showed a good yield of products in its presence (Chatterjee et al. 2022). There was an expected difference in their stereo selectivity as in different types of solvents (Figure 2). The ratios of diastereomers in different type of solvents are observed as follows: 93:7 in vodka and 98:2 in green tea.

4. Synthesis of peptide supported catalyst

4.1 Peptide-supported noble metal nanoparticle

Attributed to the high activity and selectivity of the noble metal catalysts (Wang et al. 2017), as per the Wang et al 2017 their use in various fields. However, the sustained and in control syntheses of such types of nanoparticles is still a challenge (Slocik et al. 2006). Therefore, to overcome this problem some peptides are used as templates. The key points for their effectiveness are their ability to bind specifically to some specific binding sites of the metal and their property of self-assembly. On the plus side, the peptides as templates provide:

- Controlled size and shape
- Specific structure
- Desired composition

Enhancement in catalytic behavior is one of the perks too, in comparison with conventional catalysts. The factors that influence this increased activity are improvements in electron conductivity, metal dispersion and exposure to reactive sites.

The two types of variants which are studied extensively are:

1. Bimetallic
2. Monometallic

The method which is widely used to synthesize such catalyst is chemical reduction where ascorbic acid is employed as a reducing agent while the peptide acts as a capping agent (Kengo 2018). It has been reported that sometimes the peptide can act as a reducing agent as well.

Various types of nanoparticle catalysts can be obtained by just varying the peptide sequence, peptide ratio, pH and time of the reaction shown in Figure 3.

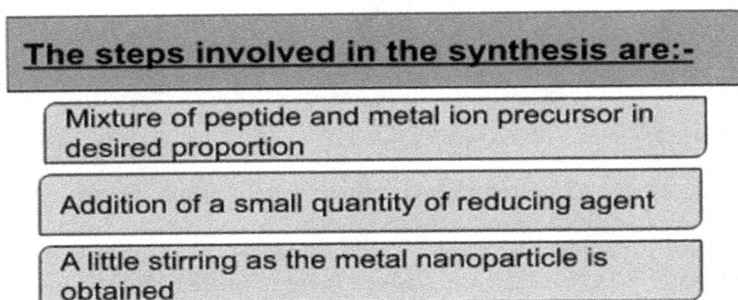

The steps involved in the synthesis are:-

Mixture of peptide and metal ion precursor in desired proportion

Addition of a small quantity of reducing agent

A little stirring as the metal nanoparticle is obtained

Figure 3. Flowchart illustrating the process of nanoparticle synthesis

Some notable examples are:

1. Peptide-supported Au catalyst
2. Pd and Pt catalysts (Briggs et al. 2015, Wang et al. 2016a)

The formation of 1D and 2D metal nanostructures can be attained by changing metal/peptide ratio as in case of Pd catalyst where morphological changes were observed by surging the Pd/ peptide ratio from 60 to 120 (change was from spherical structure to nanoparticle network) (Palafox-Hernandez et al. 2016).

4.2 Peptide-supported noble metal nanofilm

There are some noteworthy peptide templates which have the capability of forming nanofilms by sequential association of nanofibers (Choi et al. 2011, Vinod et al. 2013). The example supporting the statement is of a conductive Au nano film, which was synthesized by Jelinek and others.

- The nanofiber film was supported on a peptide monolayer.
- A TEM grid was employed to load the peptide, i.e. ((Phe-Lys) 5-Pro) nano film which was further floated over solution of Au (SCN) 41A. Here it is to be specified that the Au complex was negatively charged so it could easily bind to the nano film of peptide.
- Therefore, the complex eventually reduced into metallic Au.
- This process was followed by fabrication (Vinod et al. 2013) to a great extent; this type of peptide based synthesis is proven to be an effective way for fabrication of noble metal catalysts.

However, it's a time consuming procedure, which is a primary concern while adopting this. Hence, a more time efficient modification towards this approach is required.

Attributed to this, an advanced method of synthesizing nano film (peptide template noble metal) came into picture where argon glow discharge is used as an affordable electron source for electron reduction at room temperature (Li et al. 2016, Zhou et al. 2013a, Wang et al. 2014, 2016b). The method including glow discharge is also referred to as the cold plasma phenomenon. Also, this process is free from usage of any reducing agents (Pan et al. 2015, Yan et al. 2013). The fact that it takes place at room temperature leads to the formation of smaller nanoparticles.

Through many experimental studies, the researchers came to the conclusion that the electron reduction approach is time saving. The difference of the time span is due to the formation of hydrating electrons (Pan et al. 2012).

5. Applications

5.1 Application of peptide-supported gold-catalyst

Recently, more efforts have been made for the synthesis of peptides and proteins-supported gold composites, which can behave like stabilizing and reducing agents (Yu et al. 2017). The size, shape, composition, arrangement and self-organization of gold nanoparticles play a vital role on its catalytic activities. As the preparation of Au-NP with controlled size, required dimensions and morphology, preferred biocompatibility and random aggregation is a very challenging task, biomimetic approach (Khalil et al. 2022) is the best option for formation of gold nanoparticles with required biocompatibility.

Peptide functionalized gold-nanoparticles are found (Pengo et al. 2005) first to be hydrolytically active against carboxylate esters. Dipeptide-functionalized Au-nanoparticles (3) showed more efficiency over monomeric catalyst (2) on hydrolysis of 2,4-dinitrophenyl butanoate (DNPB) in lower pH. At pH<7, nanoparticle-supported dipeptide (3) initiates a cooperative hydrolytic mechanism with the formation of carboxylate and an imidazolium ion simultaneously which causes 300-fold rate acceleration of the hydrolytic process. This is not seen in case of an analogous monomeric dipeptide.

The scope of peptide-supported gold-NPs as esterase was further extended towards larger catalyst system like Au-PEP (Pengo et al. 2007). Among the number of ester compounds like 2,4-dinitrophenyl butyrate (DNPB) and the p-nitrophenyl esters of benzyloxycarbonyl N-protected leucine and glycine (Z-Leu-PNP and Z-Gly-PNP, respectively; PNP = p-nitrophenol), DNPB undergo hydrolysis at low pH value than Z-Leu-PNP and Z-Gly-PNP.

When the position of histidine amino acid is changed in the peptide chain linked with gold nanoparticles, a new set of catalyst system is generated, which can be denoted as Au@ExHy (x = heptad repeat, y = position of histidine residue)

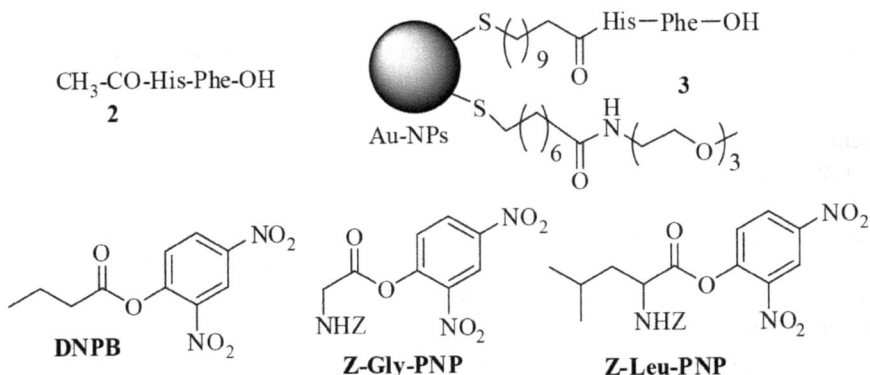

Figure 4. Different functionalised peptide and ester molecules

(Mikolajczak et al. 2018), e.g. E3H8, E3H15 and E3H22. During the ester hydrolysis of 4-nitrophenylacetate (4-NPA), it was observed that the hydrophilic esters are cleaved most rapidly by Au@E3H15 and more hydrophobic esters are efficiently cleaved by Au@E3H8 than Au@E3H22.

Gold-peptide nanoparticle (GPNP) composites Au@Ac-IVFK-NH$_2$ have been synthesized using an ultrashort peptide through a photochemical reduction method in absence of any reducing agents. The peptide nanofibers and Au@ NP are attached with each other through the noncovalent type interactions with the amino group of the lysine amino acid in the sequence. In presence of UV irradiation on the aromatic residue, the photoionization of the peptide causes the reduction of gold ions. At the same time, the peptide takes the role of a capping and stabilizing, reducing agent. Furthermore, the pollutant p-nitrophenol can be reduced using these gold-peptide composites within a very short time. However, the rate of reaction depends on the peptide-gold nanocomposites' concentration and the rate decreases on dilution of catalyst. This green synthetic method will establish new approaches in biocatalysis and environmental applications (Abbas et al. 2022).

Peptide-supported gold nanoparticles (pep-Au-NPs) are combined inorganic-organic non-materials, where catalytically active peptides are immobilized onto the metal surface. However, Au-NPs are well known as cascade catalysts used for hydrogenation reactions (Corma and Serna 2006) or oxidative transformations (Jv et al. 2010). In presence of Pep-Au-NPs catalyst, two sequential transformations occur under mild conditions in one pot. As an example of Cascade Catalysis, Au@ E1H8 can be used to hydrolyse 4-nitrophenylacetate (4-NPA) to 4-nitrophenol (4-NP) which will be further reduced to 4-aminophenol via NaBH4-mediated hydrogenation in the same reaction medium in one pot reaction (Mikolajczak et al. 2018). Here in a single reaction system, peptide monolayer function as esterase mimic and Au-NP takes part in reduction process efficiently.

Figure 5. Reduction of 4-nitrophenol catalysed by peptide-supported gold catalyst

Another example of cascade catalysis is (AuNP@CDs-Azo-GFGH) where azobenzene terminated peptides (Azo-GFGH) are linked with β-cyclodextrin-coated gold-nanoparticles (AuNP@CDs) via photo-switchable host-guest like interaction. The imidazole moiety of histidine present in Azo-GFGH is also the active site of many natural hydrolases like protease, esterase, lipase and glycosyl hydrolase (Nothling et al. 2019). High load of peptide (Azo-GFGH) on Au-NP enhances the ester hydrolysis. The azobenzene molecule, on UV irradiation, undergoes photoisomerization, and leads to the disassembly of peptide (Tan et al. 2021). The hydrolysis product 4-nitrophenol will be further reduced to 4-aminophenol in presence of catalyst (AuNP@CDs) and reducing agent (NaBH4) under same reaction medium.

It has been found that β-sheet-forming hepta-peptide IHIHIQI (IHQ)-supported gold nanoparticles (Au@IHQ-NQ) can mimic the *in-vivo* activities of esterase and Carbonic Anhydrase (CA) which contains zinc-based metalloenzyme that can reversibly and efficiently convert CO_2 into hydrogen carbonate (HCO_3^-). In carbonic anhydrase, zinc (Zn(II)) ion is coordinated by three histidine group

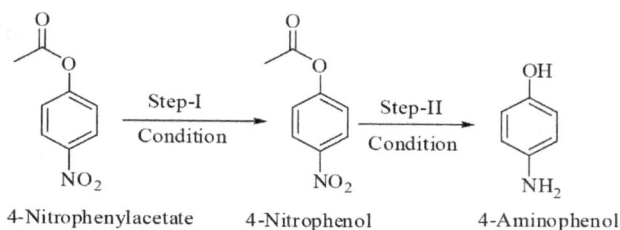

Entry	Step-I Condition	Step-II Condition	Reference
1	Au@E_1H_8, H_2O	Au@E_1H_8, $NaBH_4$	Mikolajczak et al. 2018
2	AuNP@CDs-Azo-GFGH, H_2O	AuNP@CDs, $NaBH_4$, UV	Tan et al. 2021

Figure 6. One-pot ester hydrolysis and reduction reaction by peptide-templated gold catalyst.

leaving a vacant space for water molecule association. We have already discussed that Peptide-Au-NP composite is a very efficient catalyst for ester hydrolysis. It has been found that Zn(II)-associated peptide-gold nanoparticle conjugates (Zn(II)-Au@IHQ-NP) show more efficiency towards ester hydrolysis and CO_2 hydration in comparison with un-conjugated peptide variants. The efficiency of Au@IHQ-NP remains almost intact (94%) even after five time run (Mikolajczak and Koksch 2019).

5.2 Application of peptide-supported platinum-catalyst

The peptide (Eosin Y) self-assembled biofilm supported platinum catalyst is found (Pan et al. 2015) to show very efficient visible-light-driven photo catalysis for the water splitting and CO_2 reduction. The rate of reaction is found to be enhanced in presence of both visible light as well as peptide-Pt-biofilm composite catalyst. The result is not satisfactory if we remove any component of the catalyst system. In the visible-light driven photocatalytic water splitting and CO_2 reduction with water on EY/Pt/Film, the highly conducive biofilm allows the transition of the photo-excited electrons from EY to Pt-nanoparticles, to supply more photo-induced electrons to participate in the reaction with H_2O (or CO_2 + H_2O). This could be the main reason for the significantly improved performance of EY/Pt/Film. The Tri-ethanolamine (TEOA) is used as electron donor to reduce back EY.

An attractive method (Zhou et al. 2013b) has been seen for the immobilization of platinum nanoparticle (Pt-NPs) on the surface of aniline-GGAAKLVFF peptide (AFP) fibrils to form novel amyloid fibril platinum-nanoparticle composites (Pt-AFP fibrils). The interactions between Pt-NPs and AFP fibrils are totally electrostatic in nature. The excellent electro catalytic activities of Pt–AFP fibrils toward oxygen reduction make it attractive for its applications in batteries, polymer electrolyte fuel cells (PEFCS) and other electrodes. Electro-catalytic activity of Pt-AFP fibrils and Pt NPs on oxygen reduction was further studied and it was seen that the electro catalytic efficiency of Pt-AFP fibrils for oxygen reduction is better than Pt-NPs and bulk Pt electrodes. Further scope for metal-AFP fibrils can be studied with several nanoparticles (Ru, Pd, Au, CeO_2 and others) and bio molecules via electrostatic assembly to generate new catalytic behaviors.

5.3 Application of peptide-supported palladium-catalyst

Self-assembling peptide template (Coppage et al. 2010, Bedford et al. 2015) can be used for the generation of non-spherical Pd-nanostructures. On the basis of Pd/peptide ratio, different catalytically reactive inorganic morphologies can be developed within the peptide scaffolds using linear nanoribbons, nanoparticles and complex nanoparticle networks (NPNs). The catalytic effects of bioinspired peptide based system have been discussed on two different reactions e.g. 4-nitrophenol reduction and Stille C-C coupling. Enhanced reactivity was seen for the peptide catalyst with Pd-nanoparticles and NPNs than the nanoribbons for all type of reactions (Bhandari and Knecht 2011).

Another type of peptide-supported palladium nano-materials was developed (Bedford et al. 2014) using water-soluble hydroxyl-terminated fourth generation poly-amidoamine (G4-PAMAM-OH) dendrimers framework covalently attached with R5-peptide and palladium nanomaterials. The R5-peptide/dendrimer/Pd-nano-materials show multiple morphologies which were absent for free peptide template. The calculated turn over frequency (TOF) for the hydrogenation of allyl alcohol was found to be higher for R5-peptide-dendrimer template Pd-nano-materials in comparison to the catalyst generated using the dendrimer-free R5 scaffold. After characteristics study, it was observed that R5-peptide/dendrimer/Pd-nanomaterials have a more porous macrostructure in solution with more catalytic surface area of Pd-component in comparison to structure with native R5-template.

6. Conclusion

Ongoing research on the peptide-supported noble metal catalysts have shown that the peptide can act as excellent template for synthesis of various biocatalysts with required criteria by just controlling the metal/peptide ratios, peptide sequences and metal precursors. Therefore, the exposed metal surface area and the penetration depth of noble metal catalysts can be both tuned. In spite of few limitations of peptide-templated approach, the benefits have proven it as an effective and novel route for noble catalysts' syntheses and applications. The field could be further explored via association with proteolytic enzyme activity which will enhance catalytic activity of noble metal nanoparticles after cleavage from peptides and new research directions may be opened up for the exploration of these opportunities.

References

Abbas, M., H.H. Susapto and C.A.E. Hauser. 2022. Synthesis and organization of gold-peptide nanoparticles for catalytic activities. *ACS Omega*, 7: 2082-2090. https://doi.org/10.1021/acs omega.1c05546

Arnold, U., M.P. Hinderaker, J. Köditz, R. Golbik, R. Ulbrich-Hofmann and R.T. Raines. 2003. Protein prosthesis: 1,5-disubstituted[1,2,3]triazoles as cis-peptide bond surrogates. *Journal of American Chemical Society*, 125: 7500-7501. https://doi.org/10.1021/ja0351239

Bedford, N.M., R. Bhandari, J.M. Slocik, S. Seifert, R.R. Naik and M.R. Knecht. 2014. Peptide-modified dendrimers as templates for the production of highly reactive catalytic nanomaterials. *Chemistry of Materials* 26: 4082-4091. https://doi.org/10.1021/acsami.6b11651

Bedford, N.M., H. Ramezani-Dakhel, J.M. Slocik, B.D. Briggs, Y. Ren, A.I. Frenkel, V. Petkov, H. Heintz, R.R. Naik and M.R. Knecht. 2015. Elucidation of peptide-directed palladium surface structure for biologically-tunable nanocatalysts. *ACS Nano* 9(5): 5082-5092. https://doi.org/10.1021/acsnano.5b00168

Bhandari, R. and M.R. Knecht. 2011. Effects of the material structure on the catalytic activity of peptide-templated Pd-nanomaterials. *ACS Catalysis* 1: 89-98. https://doi. org/10.1021/cs100100k.

Briggs, B.D., Y. Li, M.T. Swihart and M.R. Knecht. 2015. Reductant and sequence effects on the morphology and catalytic activity of peptide-capped Au nanoparticles. *ACS Applied Materials & Interfaces* 7(16): 8843-8851. doi: 10.1021/acsami.5b01461.

Chatterjee, A., A. Reja, S. Pal and D. Das. 2022. Systems chemistry of peptide-assemblies for biochemical transformations. *Chemical Society Review* 51: 3047-3070. https:/doi. org/10.1039/DICSO1178B

Choi, B.G., M.H. Yang, T.J. Park, Y.S. Huh, S.Y. Lee, W.H. Hong and H. Park. 2011. Programmable peptide-directed two dimensional arrays of various nanoparticles on grapheme sheets. *Nanoscale* 3: 3208-3213. https://doi.org/10.1039/C1NR10276A

Coppage, R., J.M. Slocik, M. Sethi, D.B. Pacardo, R.R. Naik and M.R. Knecht. 2010. Elucidation of peptide effects that control the activity of nanoparticles. *Angew. Chem. Int. Ed.* 49: 3767-3770. https://doi.org/10.1002/anie.200906949

Corma, A. and P. Serna. 2006. Chemoselective hydrogenation of nitro compounds with supported gold catalysts. *Science* 313: 332–334. https://doi.org/10.1126/ science.1128383.

Holzberger, B. and A. Marx. 2010. Replacing 32 proline residues by a noncanonical amino acid results in a highly active DNA polymerase. *Journal of American Chemical Society* 44: 15708-15713. https://doi.org/10.1021/ja106525y

Jv, Y., B. Li and R. Cao. 2010. Positively-charged gold nanoparticles as peroxidiase mimic and their application in hydrogen peroxide and glucose detection. *Chemical Communication* 46: 8017-8019. https://doi.org/10.1039/C0CC02698K

Kengo, A. 2018. Catalysis by peptides. *Peptide Applications in Biomedicine* 513-564. (Book) Doi:10.1016/B978-0-08-100736-5.00021-1

Khalil, A.T., M. Ovais, J. Iqbal, A. Ali, M. Ayaz, M. Abbas, I. Ahmad and H.P. Devkota. 2022. Microbes-mediated synthesis strategies of metal nanoparticles and their potential role in cancer therapeutics. *Semin. Cancer Biol.* 86: 693-705. DOI: 10.1016/j. semcancer.2021.06.006.

Li, M.Y., Q.D. Sun and C.-J. Liu. 2016. Self-healing, reshaping, and recycling of vulcanized chloroprene rubber. *ACS Sustainable Chemistry & Engineering* 4: 3255-3260. https:// doi.org/10.1021/acssuschemeng.6b00224

Lieblich, S.A., K.Y. Fang, J.K.B. Cahn, J. Rawson, J. LeBon, H.T. Ku and D.A. Tirrell. 2017. 4S-hydroxylation of insulin at ProB28 accelerates hexamer dissociation and delays fibrillation. *Journal of American Chemical Society* 139: 8384-8387. 10.1021/ jacs.7b00794.

Lummis, S.C.R., D.L. Beene, L.W. Lee, H.A. Lester, R.W. Broadhurst and D.A. Dougherty. 2005. Cis-trans isomerization at a proline opens the pore of a neurotransmitter-gated ion channel. *Nature* 438: 248-252. http://dx.doi.org/10.1038/nature04130

Mikolajczak, D.J. and B. Koksch. 2018. Peptide-gold nanoparticle conjugates as sequential cascade catalysts. *ChemCatChem* 10(19): 4324-4328. http://dx.doi.org/10.1002/ cctc.201800961.

Mikolajczak, D.J., J. Scholz and B. Koksch. 2018. Tuning the catalytic activity and substrate specificity of peptide-nanoparticle conjugates. *ChemCatChem* 10(19): 5665-5668. https://doi.org/10.1002/cctc.201801521

Mikolajczak, D.J. and B. Koksch. 2019. Peptide–gold nanoparticle conjugates as artificial carbonic anhydrase mimics. *Catalysts* 9(11): 903. https://doi.org/10.3390/ catal9110903.

Nothling, M.D., Z. Xiao, A. Bhaskaran, M.T. Blyth, C.W. Bennett, M. L. Coote and L.A. Connal. 2019. Synthetic catalysts inspired by hydrolytic enzymes. *ACS Catalysis* 9(1): 168-187. https://doi.org/10.1021/acscatal.8b03326

Pan, Y.X., C.-J. Liu, S. Zhang, Y. Yu and M. Dong. 2012. 2D-oriented self-assembly of peptides induced by hydrated electrons. *Chem. Eur. J.* 18(46): 14614-14617. https://doi.org/10.1002/chem.201200745

Pan, Y.-X., H.-P. Cong, Y.-L. Men, S. Xin, Z.-Q. Sun, C.-J. Liu and S.-H .Yu. 2015. Peptide self-assembled biofilm with unique electron transfer flexibility for highly efficient visible-light-driven photocatalysis. *ACS Nano* 9(11): 11258-11265. https://doi.org/10.1021/acsnano.5b04884

Palafox-Hernandez., J.P., C.-K. Lim, Z. Tang, K.L.M. Drew, Z.E. Hughes, Y. Li, M.T. Swihart, P.N. Prasad, M.R. Knecht and T.R. Walsh. 2016. Optical actuation of inorganic/organic interfaces: Comparing peptide-azo benzene ligand reconfiguration on gold and silver nanoparticles. *ACS Applied Materials & Interfaces* 8(1): 1050-1060. https://doi.org/10.1021/acsami.5b11989

Pengo, P., S. Polizzi, L. Pasquato and P. Scrimin. 2005. Carboxylate-imidazole cooperativity in dipeptide-functionalized gold nanoparticles with esterase-like activity. *Journal of American Chemical Society* 127(6): 1616-1617. https://doi.org/10.1021/ja043547c

Pengo, P., L. Baltzer, L. Pasuato and P. Scrimin. 2007. Substrate modulation of the activity of an artificial nanoesterase made of peptide-functionalized gold nanoparticles. *Angew Chemie International Edition* 46: 400-404. https://doi.org/10.1002/anie.200602581

Schnitzer, T., J.W. Rackl and H. Wennemers. 2022. Stereo selective peptide catalysis in complex environments – From river water to cell lysates. *Chemical Science* 13: 8963-8967. DOI: https://doi.org/10.1039/d2sc02044k

Slocik., J.M. and R.R. Naik. 2006. Biologically programmed synthesis of bimetallic nanostructures. *Adv. Mater.* 18(15): 1988-1992. https://doi.org/10.1002/adma.200600327

Tan, X., Y. Xu, S. Lin, G. Dai, S. Zhang, F. Xia and Y. Dai. 2021. Peptide-anchored gold nanoparticles with bicatalytic sites for photo-switchable cascade catalysis. *Journal of Catalysis* 402: 125-129. https://doi.org/10.1016/j.jcat.2021.08.023.

Vinod, T.P., S. Zarzhitsky, A. Morag, L. Zeiri, Y. Levi-Kalisman, H. Rapaport and R. Jelinek. 2013. Transparent, conductive, and SERS-active Au nanofiber films assembled on an amphiphilic peptide template. *Nanoscale* 5: 10487-10493. https://doi.org/10.1039/C3NR03348A

Wang, Q., Z. Tang, L. Wang, H. Yang, W. Yan and S. Chen. 2016a. Morphology control and electro catalytic activity towards oxygen reduction of peptide-templated metal nanomaterials: A comparison between Au and Pt. *Chemistry Select* 1(18): 6044-6052. https://doi.org/10.1002/slct.201601362

Wang, W., M.M. Yang, Z.Y. Wang, J.M. Yan and C.-J. Liu. 2014. Silver nanoparticle aggregates by room temperature electron reduction: Preparation and characterization. *RSC Advance* 4: 63079-63084. https://doi.org/10.1039/C4RA11803K

Wang, Z.Y., M.Y. Li, W. Wang, M. Fang, Q.D. Sun and C.-J. Liu. 2016b. Floating silver film: A flexible surface-enhanced Raman spectroscopy substrate for direct liquid phase detection at gas–liquid interfaces. *Nano Research* 9: 1148-1158.

Wang, W., C.F. Anderson, Z. Wang, W. Wu, H. Cui and C.-J. Liu. 2017. Peptide-templated noble metal catalysts: Syntheses and applications. *Chemical Science* 8: 3310-3324. doi: 10.1039/c7sc00069c.

Yan, J.M. , Y.X. Pan, A.G. Cheetham, Y.A. Lin, W. Wang, H. Cui and C.-J. Liu. 2013. One-step fabrication of self-assembled peptide thin films with highly dispersed noble metal nanoparticles. *Langmuir* 29(52): 16051-16057. https://doi.org/10.1021/la4036908

Yu, X., Z. Wang, Z. Su and G. Wei. 2017. Design, fabrication, and biomedical applications of bioinspired peptide–inorganic nanomaterial hybrids. *Journal of Material Chemistry B* 5: 1130-1142. https://doi.org/10.1039/C6TB02659A.

Zhou, B., Z. Sun, D. Li, T. Zhang, L. Deng and Y.-N. Liu. 2013a. Platinum nanostructures via self-assembly of an amyloid-like peptide: A novel electrocatalyst for the oxygen reduction. *Nanoscale* 5: 2669-2673. https://doi.org/10.1039/C3NR33998J.

Zhou, Y., Z.H. Xiang, D.P. Cao and C.-J. Liu. 2013b. Covalent organic polymer supported palladium catalysts for CO oxidation. *Chemical Communication* 49(50): 5633-5635. https://doi.org/10.1039/C3CC00287J

Polymer-supported Phase Transfer Catalyst

Nisha Tewatia[1,2*], Shagufta Jabin[1], Radhamanohar Aepuru[3], Manda Sathish[4] and Manjinder Kour[5]

[1] Department of Applied Sciences, Faculty of Engineering and Technology, Manav Rachna International Institute of Research and Studies, Faridabad - 121001, Haryana, India

[2] Department of Chemistry, Pt. J.L.N. Government P.G. College, Faridabad - 121002, Haryana, India

[3] Departamento de Ingeniería Mecánica, Facultad de Ciencias Físicas y Matemáticas, Universidad de Chile, Santiago, Chile

[4] Centro de Investigación de Estudios Avanzados del Maule (CIEAM), Vicerrectoría de Investigación y Postgrado, Universidad Católica del Maule, 3460000, Talca, Chile

[2] Microbiology and Cell Biology Department, Montana State University, Bozeman, MT, 59715, USA

1. Introduction

Most of the useful chemical reactions are not easy to carry out because of the unavailability of the reactants towards each other. The addition of a solvent that has both water-like and organic-like properties, such as ethanol, which derives its hydrophilic nature from its hydroxyl group and its lipophilicity from the ethyl group, traditionally solves the crucial complexity of bringing a water soluble nucleophilic reagent and an organic, water insoluble electrophilic reagent together to carry out the required chemical reaction (Wang and Weng 1988). Now-a-days, phase transfer catalysis (PTC) is recognized in organic chemistry as a flexible and significant synthetic approach. In both liquid-liquid and solid-liquid systems, phase transfer catalysis (PTC) allows interactions between reagents in two immiscible phases by using catalytic quantities of phase transfer

*Corresponding author: nisha.tewatia13@gmail.com

agents that enhance interphase transfer of species (Jensen et al. 2002). Many reviews on phase transfer catalysts (PTCs) as synthetic methods have also been published before (Brändström 1977, Dehmlow and Fastabend 1993, Gokel and Weber 1978, Starks et al. 1994, Freedman 1986). PTCs come in a variety of forms, including quaternary ammonium and phosphonium salts, crown ethers, cryptands and others. The quaternary ammonium salts are among the most frequently used PTCs in the market since they are the cheapest of all (Wang 2019). Although phase transfer catalysts (PTCs) are widely used in the industry, polymer-supported (PS) phase transfer catalysts, commonly referred as triphase catalysts, have not yet attracted extensive industrial recognition (Fiamegos and Stalikas 2005).

The reaction between benzyl chloride and sodium bromide is previously studied by using two different phase transfer catalysts. One was water soluble phase transfer catalyst i.e. tetrabutylammonium bromide (QBr), which is slightly soluble in toluene also, hence the reaction system is two-phase (liquid-liquid), and another was a polymer-supported phase transfer catalyst, which was prepared by immobilizing tributylamine in chloromethyl polystyrene (Wang and Weng 1988). The reaction was carried out in triphase mixtures of organic liquid, solid or aqueous inorganic salt; and solid catalyst, hence the reaction system is three-phase (triphase). The authors have concluded that the rate of the reaction is higher in the latter case than that of the former one (when the same milliequivalents of quaternary ammonium ions are used) due to lower activation energy of the triphase system than that of the two-phase system. Triphase catalysis has been introduced as an immobilized phase transfer catalysis, where a polymer is the most suitable support for immobilizing the phase transfer catalysts (Marken et al. 2019). The method of triphase catalysis was initially introduced by Regen (Regen 1975). Initially, crown ether catalysts and polymer-bound quaternary onium ions in triphase were reported independently in the literature (Regen 1975, Brown and Jenkins 1976, Cinouini et al. 1976). Polymer-supported (PS) phase transfer catalysts are heterogeneous types of catalysts which are insoluble in nature. They have covalently bound functional groups (FGs) which are active as catalysts for reactions between anions and neutral organic substrates. The active functional groups include quaternary ammonium or phosphonium ions, cryptands, crown ethers, grafted poly(ethylene glycols), analogues of dipolar aprotic solvents etc. (Srivastava et al. 2020). The two most often utilised polymer-supports are poly(ethylene glycol) and the copolymer of polystyrene crosslinked with divinylbenzene (Lemanowicz et al. 2021). The 1% and 2% cross-linked polystyrenes have been used as supports for various catalysts in different organic reactions. Osada and Chiba (1979) used quaternized poly(4-vinylpyridine) gels as catalysts in aspirin hydrolysis; Yamazaki et al. (1980) used quaternary ammonium ion substituted polystyrene gels as catalysts for decarboxylation of 6-nitrobenzisoxazole-3-carboxylate anion in aqueous buffer solutions. Organic or organometallic functional groups exhibiting higher specificity and lower temperature stability are typically used as active sites in polymer-supported catalysts. In the design of catalysts for specific reaction processes, the plethora of polymers and functionalization methods available provide controlled swelling properties, balanced hydrophilic/lipophilic ratio and the pore structure of the support (Zhao et al. 2019). The synthesis of alpha-amino esters, esterification reactions, amidation reactions, polymerisation reactions and other organic chemical reactions use polymer-supported (PS) phase transfer catalysts as intermediates (Kaur et al. 2011, Thierry et al. 2001). Classical soluble phase-transfer catalysts offer many disadvantages when compared with the analogous insoluble catalysts, which involve slow reaction rates, difficulty in separating the catalyst from the reaction mixture, etc. (Mąkosza and Fedoryński 2020). Soluble phase transfer catalysts like quaternary ammonium and phosphonium salts possess modest surfactant properties, which can cause emulsions resulting in difficulty for isolation of the product from organic phase (Polarz et al. 2018). Using insoluble polymer-supported (PS) phase-transfer catalysts, usually known as triphase catalysts, is a solution to all the problems related to the classical soluble phase transfer catalysts. Triphase catalysts can be easily removed from the reaction mixture by filtration or centrifugation (Yang et al. 2019). The lipophilicity of the triphase catalyst facilitates transport of the organic phase into the catalyst, thereby increasing the reaction rate (Nakade et al. 2020). Polymer-supported phase transfer catalysts offer an appealing method of reusing the catalyst after the reaction. Even though they are more expensive than classical soluble phase-transfer catalysts, the recovery of the catalyst by its extensive recycling makes it available for continuous industrial phase transfer catalysis (Katole and Yadav 2019). This chapter presents the synthesis of polymer-supported phase transfer catalysts, their multifarious applications as intermediates in various organic chemical reactions, their reaction mechanisms, reactivity, selectivity and characterizations.

1.1 Principle of phase transfer catalysis (PTC)

Reuben and Sjoberg (1981) discussed the principle of phase transfer catalysis (PTC) (Reuben and Sjoberg 1981). The basis of PTC is the ability of specific "phase-transfer agents" or "phase transfer catalysts" to make it easier for one reagent to move from one phase into an opposite (immiscible) phase where the other reagent is present. Thus, the reagents that were initially in various stages and were not previously in close proximity to one another are combined to enable the reaction. However, for the phase transfer catalytic action to be efficient, it is also necessary that the transferred species be in an active state and that it be regenerated during the organic process (Brändström 1977).

1.2 General mechanism

The mechanism of PTC reaction was initially proposed by Starks and Halper (Starks and Halper 2012). He illustrated the complete mechanism by taking a classic example of a nucleophilic substitution reaction between 1-chlorooctane (R-X; R=C_8H_{17}) and aqueous sodium cyanide (M-Y). No desired product was obtained without the use of a catalyst irrespective of refluxing this two-phase mixture with vigorous stirring for 1 or 2 days. Although if 1% (w/v) of a quaternary ammonium salt $(C_6H_{13})_4N^+Cl^-$; Q^+X) is added to the reaction system, the desired product, 1-cyanooctane (R-Y), was obtained with 100% selectivity and 100% conversion in 2-3 hrs.

The author marked the following plausible mechanism for the above reaction:

1. A quaternary ammonium salt (Q^+X^-), dissolved in the aqueous phase, transfers cyanide ion into the organic phase (Q^+Y^-).
2. In the organic phase, the cyanide ion (Y^-) being quite nucleophilic undergoes a nucleophilic substitution reaction with 1-chlorooctane (R-X; $R=C_8H_{17}$) forming the desired product (R-Y).
3. The catalyst subsequently returns to the aqueous phase by transferring displaced chloride ion (X^-) back to the aqueous phase (Q^+X^-) and the cycle continues.

Due to its lipophilic nature, the ion-pair formed (Q^+Y) can cross the liquid-liquid/solid-liquid interface and diffuses into the organic phase - this is the "phase-transfer" step (Scheme 1a).

2. Synthesis of polymer-supported phase transfer catalysts

2.1 Synthesis of methyltributylammonium chloride catalyst

The triphase catalyst, methyltributylammonium chloride, was prepared (Desikan and Doraiswamy 2000). In a 200 ml round-bottom flask, 50 ml of toluene was taken, then 5 g of chloromethylated polystyrene precursor, having 4.3 mmol of chlorine per gram of solid, was added to the flask. After purging the reaction mixture with nitrogen for 5 minutes, 40.13 g of 0.216 mol tributylamine was added into the flask and with the help of a magnetic stirrer, continuously stirred. The reaction mixture was left undisturbed for 36 hours at 90 °C (Scheme 1b). It was vacuum-filtered before being washed with toluene and then water, thrice. Using a soxhlet continuous extractor and refluxing toluene, the filtered solids were extracted. After extraction, solids were filtered to eliminate excessive toluene. The sample was dried for 24 hours at 90 °C under vacuum.

2.1.1 Characterization

Characterization is one of the important pieces of evidence that clearly indicates the surface morphology of the polymer matrix where the reaction occurs. Optical microscope, scanning electron microscope (SEM), electron spectroscopy for chemical analysis (ESCA), secondary ion mass spectroscopy (SIMS), ion scattering spectroscopy (ISS), Auger electron spectroscopy (AES), electron probe microanalyzer (EPMA), transmission electron microscope (TEM), and electron

$$Q^+ = \quad R_4N^+$$

R-Cl + QCN $\xrightarrow{\text{Organic Phase Reaction}}$ R—CN + QCl

NaCl + QCN $\rightleftharpoons$ NaCN + QCl

Aqueous Phase Reaction

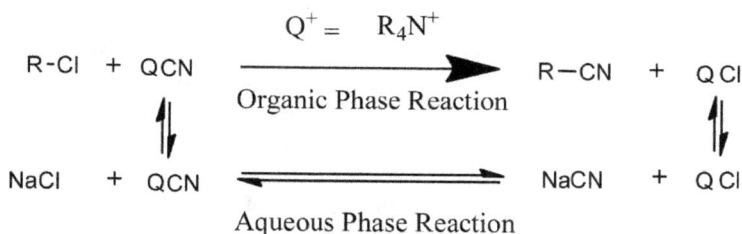

Scheme 1a. Schematic representation of cyanide displacement on 1-chlorooctane via. PTC

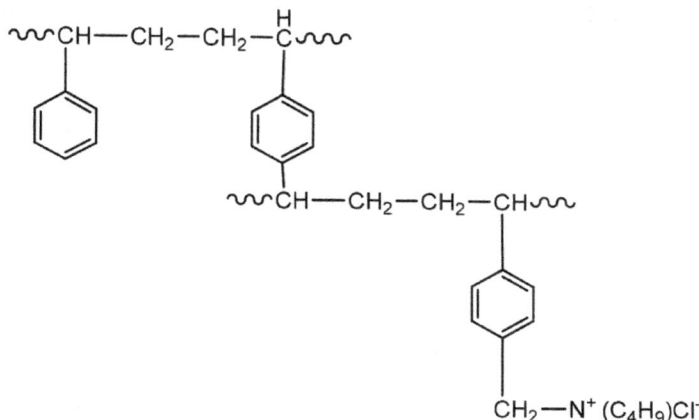

Scheme 1b. Methyltributylammonium chloride catalyst

paramagnetic resonance (EPR) are the several characterization techniques used for the analysis of the heterogeneous catalysts (Chou and Weng 1990).

BET (Brunauer-Emmett-Teller) analysis was used to measure the surface area of the polymer precursor with the nitrogen adsorption method. The catalyst support was found to have a very low surface area ($0.40 \ m^2/g$ approx.), which indicates its nonporous nature. Volhard titration method and energy dissipation spectroscopy (EDS) were used for chloride analysis and elemental analysis of the polymer-supported phase transfer catalyst, respectively (Desikan and Doraiswamy 2000).

2.2 Synthesis of polyethylene-g-quaternary ammonium salt (PE-g-Q_N^+)

A polymer-supported (PS) phase transfer catalyst, polyethylene-g-quaternary ammonium salt (PE-g-Q_N^+), was prepared through a three-step graft copolymerization (Kaur et al. 2011) (Scheme 2) through the following steps:

1. In the first step, maleic anhydride (MAn), in 5 ml of acetone, was reacted with 100 mg polyethylene (PE), which was already suspended in 10 ml

of xylene, by photochemical method using 1% benzophenone (Bz) as photosensitizer. Excess methanol was added to the reaction mixture and the product, polyethylene-g-maleic anhydride (PE-g-MAn), was filtered, washed with water and dried.

2. Acid hydrolysis of polyethylene-g-maleic anhydride (PE-g-MAn) was carried out by refluxing the mixture of 0.100 g PE-g-MAn (Resin I) with water (20 ml) and concentrated H_2SO_4 (0.5 ml) for 4 hours. The product, PE-g-succinic acid (Resin II), was filtered, then washed with water, and dried finally.

3. Resin II on further refluxing with 0.200 g of tetrabutylammonium bromide (TBAB) and 20 ml of tetrahydrofuran (THF), upto the boiling temperature of THF, under basic conditions to obtain PE-g-Q_N^+ (Resin III).

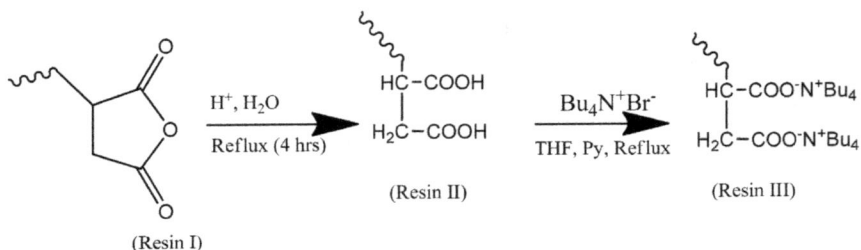

Scheme 2. Synthesis of polyethylene-g-quaternary ammonium salt (PE-g-QN⁺)

The structural analysis for the polymer-supported phase transfer catalyst (Resin III) was done by using spectroscopic and conductometric techniques. The FTIR spectrum of PE-g-Q_N^+ (Figure 1) shows a characteristic peak of ammonium salt at 2367.1 cm⁻¹. The characteristic bands at 1628.9 cm⁻¹ and 1378.5 cm⁻¹

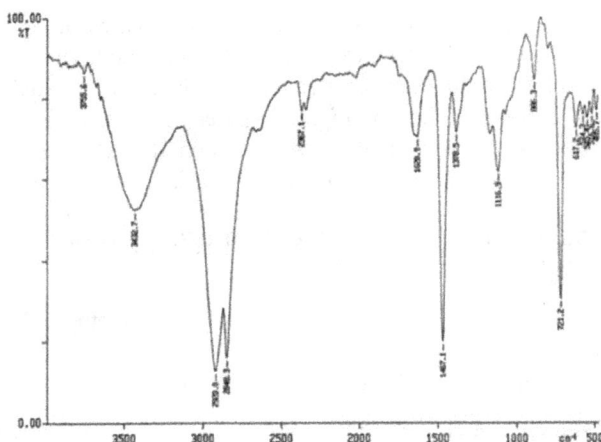

Figure 1. FTIR Spectrum of PE-g-QN⁺ (Kaur et al. 2011)

are observed due to COO⁻ asymmetric and symmetric stretching, respectively. The similar IR spectral pattern during the synthesis of quaternary salt (Sarkar et al. 2021).

Conductance measurements were made on the shedlosky conductance cell and conductivity bridge to confirm the ionic nature of quaternary salt, Resin III. The specific conductance of the aqueous solution of PE-g-Q_N^+ was observed at 47×10^{-6} ohm⁻¹, which was higher than that of PE-g-MAn (Resin I) i.e. 37×10^{-6} ohm⁻¹ indicating the ionic nature of the former one.

3. Organic reactions

On the polymer surface or within the polymer matrix, chemical reactions occur. More lightly cross-linked polystyrenes were developed as support in peptide synthesis (Erickson 1976, Merrifield 1963) and synthesis of polynucleotides (Letsinger and Kornet 1963), following the introduction of ion exchange resins based on polystyrene. These polymers expand and contract more than ordinary ion exchange resins, which are typically packed into columns for water treatment, even with just 1% or 2% divinylbenzene. Although more fragile, compressible, and difficult to filter than commercial ion exchange resins, the more swollen gels with increased solvent concentration have shown outstanding yields in peptide synthesis on a small scale. The lightly cross-linked polymers, as supports for general organic synthesis (Akelah and Sherrington 1981, Fréchet 1981, Sherrington 2001) and for immobilized "homogeneous" transition metal catalysts (Kaneko and Tsuchida 1981, Ciardelli et al. 1982, Whitehurst 1980), were used in earlier times. Various organic reactions were carried out using polymer-supported phase transfer catalysts. Few of them are discussed below.

3.1 Polymerization reaction

There are various polymerization reactions obtained with PE-g-Q_N^+ as a polymer-supported phase transfer catalyst reported in literature (Kaur et al. 2011, Vajjiravel and Umapathy 2008). Different monomers like styrene, methyl methacrylate, acrylonitrile etc. are polymerized by refluxing them with the suitable amount of the catalyst under mild acidic condition to obtain polymers polystyrene (PS), poly(methyl methacrylate) (PMMA) and poly(acrylonitrile) (PAN). The residue obtained is treated with suitable solvent depending upon the polymer obtained; for instance, benzene was used for PS, acetone for PMMA and dimethyl formamide for PAN.

3.2 Amidation reaction

Amidation reaction of ethyl bromide was carried out by refluxing its suitable amount with the catalyst at boiling temperature of benzene under basic conditions (Kaur et al. 2011). The reaction mixture was filtered and the residue was washed with ethanol. The organic phase of the filtrate was separated and the solution obtained was decanted and evaporated to give a crystalline solid product, N-ethyl benzamide.

3.3 Esterification reaction

A solution of sodium benzoate in water was refluxed with an appropriate amount of the catalyst, PE-g-Q_N^+, and a benzene solution of ethyl bromide (Kaur et al. 2011). The reaction mixture was cooled at rt. (room temperature), diluted with water and filtered to separate out the polymer-support. After separating the organic layer, it was washed with water and dried over anhydrous $MgSO_4$. The organic layer was decanted off and evaporated to give the product, that is, ethyl benzoate with a fruity smell.

3.4 Synthesis of dialkyl peroxides

Dialkyl peroxides are widely used as initiators of free radical reactions, cross-linking agents or bleaching and oxidizing agents (Van Deurzen et al. 1997). The authors investigated the synthesis of dialkyl peroxides from alkyl hydroperoxides and alkyl bromides with the help of polymer-supported phase-transfer catalysts. In his work, Cumyl hydroperoxide was made to react with 1-bromobutane (Scheme 3) in the presence of aqueous solution of sodium hydroxide (Baj et al. 2006).

Scheme 3. Synthesis of dialkyl peroxide using polymer-supported PTC

This process involves the following six steps as shown in scheme 4.

1. Sodium hydroxide, in aqueous phase, undergoes mass transfer and intraparticle diffusion in the presence of the catalyst.
2. Exchange of OH⁻ with Cl⁻ takes place at the active sites of the catalyst.
3. Generation of organic anion, $PhC(CH_3)_2OO^-$ during mass transfer and intraparticle diffusion of cumyl hydroperoxide ($PhC(CH_3)_2OOH$).
4. Mass transfer and intraparticle diffusion of 1-bromobutane from the organic phase to the active sites of the catalyst.
5. Organic ion, $PhC(CH_3)_2OO^-$ and 1-bromobutane are now in close proximity and easily undergo chemical reaction at the active sites.
6. Diffusion of the product obtained, $PhC(CH_3)_2OOC_4H_9$, into the organic phase.

Reactivity of polymer-supported phase transfer catalysts

Earlier, it was assumed that due to external and intraparticle diffusion resistances, the heterogeneous catalysts as compared to homogeneous catalyst were likely to show a lower reactivity (Ford and Tomoi 1984). But, literature provides a number of references where the scenario is totally opposite. Supported polyethylene

$C_4H_9Br_{(org)}$ (4)

$PhC(CH_3)_2OOH_{(org)}$ (3)

$NaOH_{(aq.)}$ (1)

$PhC(CH_3)_2OO(C_4H_9)_{(org)}$

$NaBr_{(aq.)}$

(6)

Q^+Cl^-

$Q^+Cl^- + OH^-_{(aq.)} \rightleftharpoons Q^+OH^- + Cl^-_{(aq.)}$

Q^+Cl^-

(5)

(2)

$PhC(CH_3)_2OO^-Q^+ + BrC_4H_{9(org)} \longrightarrow PhC(CH_3)_2OO(C_4H_9)_{(org)} + Q^+Br^-$

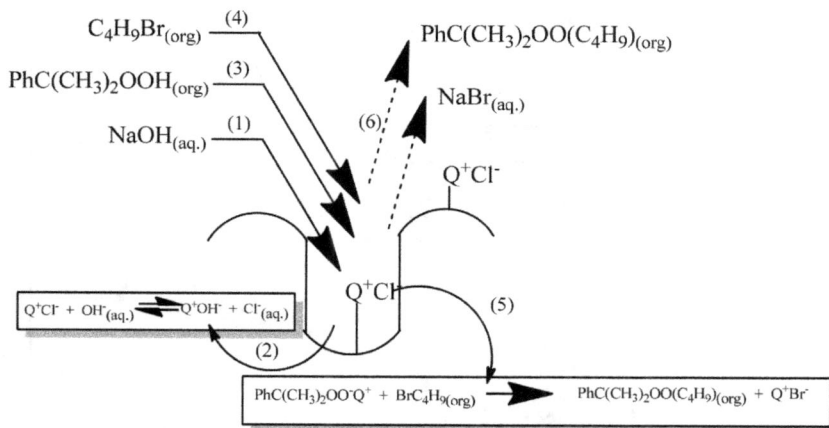

Scheme 4. Proposed mechanism in the synthesis of dialkyl peroxide using polymer-supported PTC. $Q^+ =$ quaternary onium cation

glycols (PEG) and crown ethers have been shown to have higher reactivity than soluble ones (MacKenzie and Sherrington 1981, Regen and Nigam 1978, Kimura and Regen 1983, Hradil and Švec 1984).

An esterification reaction between benzyl chloride in the organic phase and sodium acetate in the aqueous phase to form benzyl acetate was chosen as the model reaction (Desikan and Doraiswamy 2000). The reaction was preceded by "phase transfer catalysis" where different types of phase transfer catalysts were used to carry out the reaction. Different concentrations of polymer-supported catalysts were used. A comparative study was done by taking their soluble correspondents of equal concentration.

When 0.1 mol of benzyl chloride is reacted with 1 mol of sodium acetate by using 2 mmol of polymer-supported methyltributylammonium chloride at the agitation rate of 700 rpm, it has been found that a polymer bound phase transfer catalyst is actually more reactive than its soluble correspondents (Figure 2). Desikan and Doraiswamy (2000) explained this rate enhancement very well in their work. The lipophilicity of the catalyst is increased by introducing the polymer-support to the catalyst, which results in the easier passage of the organic phase into the active sites of the catalyst, hence the reaction rate is increased. Now, the support is not at all inert as in the case of many other heterogeneous catalysts. This approach clearly concludes that the manipulations in the support system can result in the maximum conversions for a given reaction system.

In another case, where the reaction conditions were same but this time polymer-supported phosphonium catalyst and its soluble analogous (2 mmol) were used, in this case also the rate enhancement effect due to the polymer-bound catalyst is indeed significant (Figure 3).

In both the cases studied, the enhancement of the reactivity of the reaction is due to the polymer backbone, which is used as a support to the phase transfer

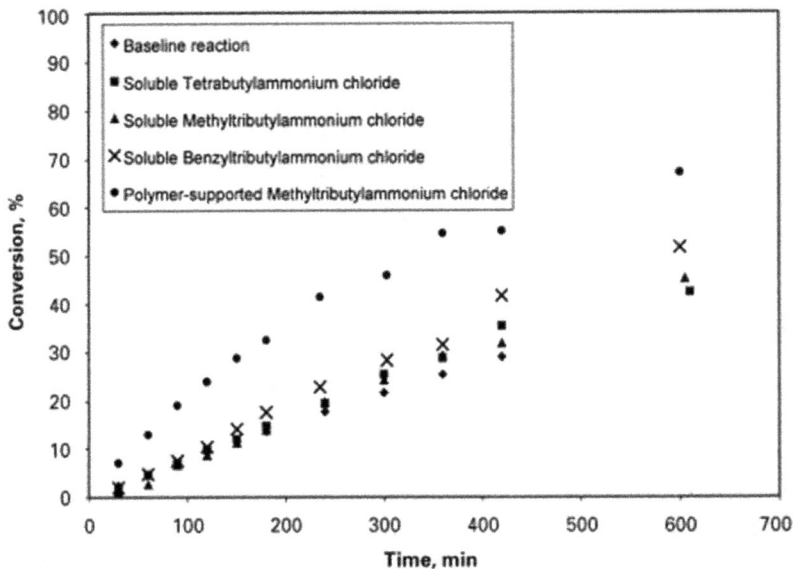

Figure 2. Comparison between soluble and polymer-supported ammonium catalysts (equal concentration of 2 mmol Eq. of catalysts was used) (Desikan and Doraiswamy 2000)

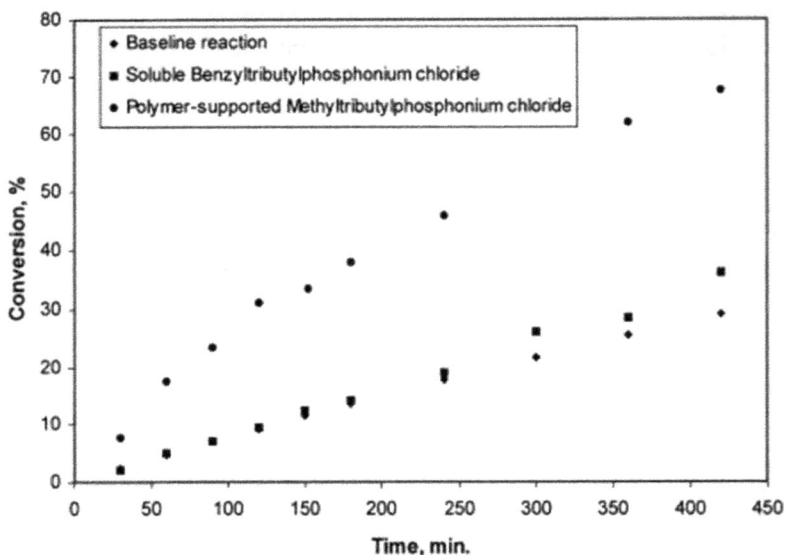

Figure 3. Comparison between soluble and polymer-supported phosphonium catalysts (equal concentration of 2 mmol Eq. of catalysts was used) (Desikan and Doraiswamy 2000)

catalyst. This was again proven by comparing the effects of soluble and polymer-supported catalysts with silica-supported tributylmethylammonium chloride (1 mmol) during the same chemical reaction at the same reaction condition. Figure 4 clearly shows that the silica-supported catalyst has lower reactivity than the soluble catalyst. This is again due to the increased lipophilicity of the polymer backbone, which is not the case of silica as a support to the catalyst. This is known as the "polymer effect".

The reduction of external mass transfer effects of catalysts can also enhance the rate of some reactions. This is famously known as "microphase effect of the catalyst" (Mehra 1990). In his studies, reaction was carried out with chloromethylated polymer-support but without the phase transfer catalyst on it. It resulted in the same reaction rate as that in the case of a baseline reaction (where no catalyst was used), clearly indicating that there is no microphase effect without the catalyst.

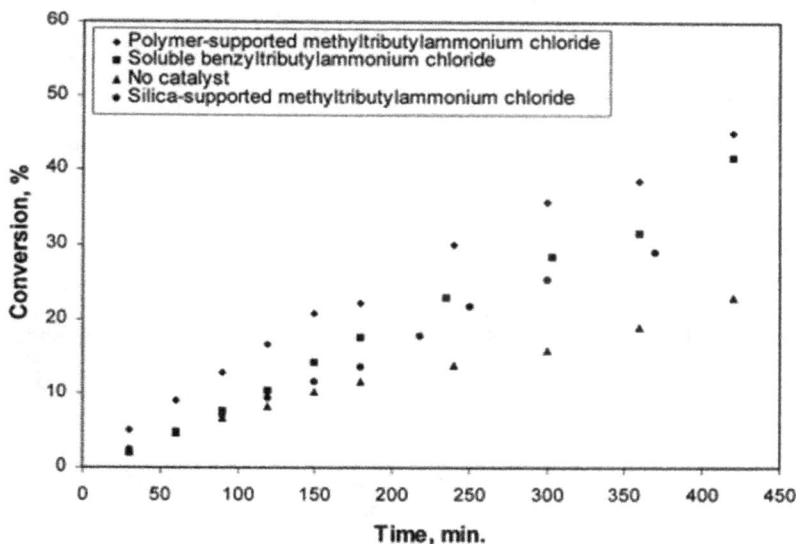

Figure 4. Comparison between soluble, polymer-supported, and silica-supported methyltributylammonium chloride catalysts (equal concentration of 1 mmol Eq. of catalysts was used) (Desikan and Doriswamy 2000)

4. Conclusion

The triphase reactions involve three phases as a reaction system, viz. organic liquid, aqueous salt solution, and solid polymer-supported catalyst. Hence, the catalyst is named as a "triphase catalyst" although, with other traditional heterogeneous catalysts, the experimental reactivity of some chemical reactions is generally retarded due to mass transfer and intra particle diffusional restrictions.

However, the polymer-supported phase transfer catalysts are remarkably showing a trend of increased reactivity when they are compared with their soluble analogous. Polymer-supported quaternary ammonium and phosphonium catalysts are the examples that have been discussed within the chapter. They are successful to overcome such restriction by the lipophilic nature of the polymer backbone, with vigorous stirring. The use of lightly cross-linked polymer-supports enhance the particle durability. The polymer-supported catalyst has reduced activation energy compared to the soluble form and to the baseline reaction where no catalyst is used. Polymer-supported phase-transfer catalysts are active in the synthesis of various chemical precursors like dialkyl peroxides, polystyrene, poly (acrylonitrile), alpha-amino acids, etc. Triphase catalysts are easy to separate from the reaction mixture. Without a significant loss of activity, they can be reused many times after the regeneration from the reaction mixture. The advantages of polymer-supported catalysts create the perception of using them for future manufacturing demands and needs of laboratories and research industries.

Acknowledgement

Manda Sathish is thankful for ANID FONDECYT INCIACION (#11200555).

References

Akelah, A. and D.C. Sherrington. 1981. Application of functionalized polymers in organic synthesis. *Chemical Reviews* 81(6): 557-587.

Baj, S., A. Siewniak and B. Socha. 2006. Synthesis of dialkyl peroxides in the presence of polymer-supported phase-transfer catalysts. *Applied Catalysis A: General* 309(1): 85-90.

Brändström, A. 1977. Principles of phase-transfer catalysis by quaternary ammonium salts. *Advances in Physical Organic Chemistry* 15: 267-330.

Brown, J.M. and J.A. Jenkins. 1976. Micelle-related heterogeneous catalysis. Anion-activation by polymer-linked cationic surfactants. *Journal of the Chemical Society, Chemical Communications* (12): 458-459.

Chou, S.C. and H.S. Weng. 1990. Characterization of the polymer-supported phase transfer catalyst. *Journal of Applied Polymer Science* 39(8): 1665-1679.

Ciardelli, F., G. Braca, C. Carlini, G. Sbrana and G. Valentini. 1982. Polymer-supported transition metal catalysts: Established results, limitations and potential developments. *J. Mol. Catal.*, 14(1).

Cinouini, M., S. Colonna, H. Molinari, F. Montanari and P. Tundo. 1976. Heterogeneous phase-transfer catalysts: Onium salts, crown ethers, and cryptands immobilized on polymer supports. *Journal of the Chemical Society, Chemical Communications*, 11: 394-396.

Dehmlow, E.V. and U. Fastabend. 1993. Do the structures of phase-transfer catalysts influence dihalogenocarbene–carbenoid selectivities? *Journal of the Chemical Society, Chemical Communications* 16: 1241-1242.

Desikan, S. and L.K. Doraiswamy. 2000. Enhanced activity of polymer-supported phase transfer catalysts. *Chemical Engineering Science*, 55(24): 6119-6127.

Erickson, B.W. and R.B. Merrifield. 1976. *The Proteins*. Academic Press, New York. 3: 255-426.

Fiamegos, Y.C. and C.D. Stalikas. 2005. Phase-transfer catalysis in analytical chemistry. *Analytica Chimica Acta*, 550(1-2): 1-12.

Ford, W.T. and M. Tomoi. 1984. Polymer-supported phase transfer catalysts: Reaction mechanisms. *Solar Energy - Phase Transfer Catalysis - Transport Processes* 49-104.

Fréchet, J.M. 1981. Tetrahedron report number 103: Synthesis and applications of organic polymers as supports and protecting groups. *Tetrahedron* 37(4): 663-683.

Freedman, H.H. 1986. Industrial applications of phase transfer catalysis (PTC): Past, present and future. *Pure and Applied Chemistry*, 58(6): 857-868.

Gokel, G.W. and W.P. Weber. 1978. Phase transfer catalysis. Part II: Synthetic applications. *Journal of Chemical Education* 55(7): 429.

Hradil, J. and F. Švec. 1984. Phase transfer catalysis. *Polymer Bulletin* 11(2): 159-164.

Jensen, M.P., J.A. Dzielawa, P. Rickert and M.L. Dietz. 2002. EXAFS investigations of the mechanism of facilitated ion transfer into a room-temperature ionic liquid. *Journal of the American Chemical Society* 124(36): 10664-10665.

Kaneko, M. and E. Tsuchida. 1981. Formation, characterization, and catalytic activities of polymer-metal complexes. *Journal of Polymer Science: Macromolecular Reviews* 16(1): 397-522.

Katole, D.O. and G.D. Yadav. 2019. Process intensification and waste minimization using liquid-liquid-liquid tri-phase transfer catalysis for the synthesis of 2-((benzyloxy) methyl) furan. *Molecular Catalysis* 466: 112-121.

Kaur, I., V. Kumari and P.K. Dhiman. 2011. Synthesis, characterization and use of polymer-supported phase transfer catalyst in organic reactions. *Journal of Applied Polymer Science* 121(6): 3185-3191.

Kimura, Y. and S.L. Regen. 1983. Poly (ethylene glycols) and poly (ethylene glycol)-grafted copolymers are extraordinary catalysts for dehydrohalogenation under two-phase and three-phase conditions. *The Journal of Organic Chemistry* 48(2): 195-198.

Lemanowicz, M., A. Mielańczyk, T. Walica, M. Kotek and A. Gierczycki. 2021. Application of polymers as a tool in crystallization—A review. *Polymers* 13(16): 2695.

Letsinger, R.L. and M.J. Kornet. 1963. Popcorn polymer as a support in multistep syntheses. *Journal of the American Chemical Society* 85(19): 3045-3046.

MacKenzie, W.M. and D.C. Sherrington. 1981. Substrate selectivity effects involving polymer-supported phase transfer catalysts. *Polymer*, 22(4): 431-433.

Mąkosza, M. and M. Fedoryński. 2020. Interfacial processes—The key steps of phase transfer catalyzed reactions. *Catalysts* 10(12): 1436.

Marken, F., E. Madrid, Y. Zhao, M. Carta and N.B. McKeown. 2019. Polymers of intrinsic microporosity in triphasic electrochemistry: Perspectives. *ChemElectroChem* 6(17): 4332-4342.

Mehra, A. 1990. An overview of microphase catalysis. *Current Science* 970-979.

Merrifield, R.B. 1963. Solid phase peptide synthesis. I: The synthesis of a tetrapeptide. *Journal of the American Chemical Society* 85(14): 2149-2154.

Nakade, P.G., G. Singh and S. Sen. 2020. Tri-liquid phase transfer catalysis: A green reaction technology. *Green Sustainable Process for Chemical and Environmental Engineering and Science* 453-480.

Osada, Y. and T. Chiba. 1979. Effects of polymeric cations and their gels on aspirin hydrolysis. *Die Makromolekulare Chemie: Macromolecular Chemistry and Physics* 180(6): 1617-1621.

Polarz, S., M. Kunkel, A. Donner and M. Schlötter. 2018. Added-value surfactants. *Chemistry – A European Journal* 24(71): 18842-18856.

Regen, S.L. 1975. Triphase catalysis. *Journal of the American Chemical Society* 97(20): 5956-5957.

Regen, S.L. and A. Nigam. 1978. Selectivity features of polystyrene-based triphase catalysts. *Journal of the American Chemical Society* 100(24): 7773-7775.

Reuben, B. and K. Sjoberg. 1981. Phase-transfer catalysis in industry. *Chemtech* 11(5): 315-320.

Sarkar, R., A. Pal, A. Rakshit and B. Saha. 2021. Properties and applications of amphoteric surfactant: A concise review. *Journal of Surfactants and Detergents* 24(5): 709-730.

Sherrington, D.C. 2001. Polymer-supported reagents, catalysts, and sorbents: Evolution and exploitation—A personalized view. *Journal of Polymer Science Part A: Polymer Chemistry* 39(14): 2364-2377.

Srivastava, A., B. Sharma and P. Mandal. 2020. Polymers used as catalyst. *Fundamentals and Prospects of Catalysis* 1: 169-193.

Starks, C.M. and M. Halper. 2012. Phase-transfer catalysis: Fundamentals, applications, and industrial perspectives. *Springer Science & Business Media* 23-47.

Starks, C.M., C.L. Liotta and M.E. Halpern. 1994. Phase-transfer catalysis: Fundamentals I. *Phase-Transfer Catalysis* 23-47.

Thierry, B., J.C. Plaquevent and D. Cahard 2001. New polymer-supported chiral phase-transfer catalysts in the asymmetric synthesis of α-amino acids: The role of a spacer. *Tetrahedron: Asymmetry* 12(7): 983-986.

Vajjiravel, M. and M.J. Umapathy. 2008. Free radical polymerisation of methyl methacrylate initiated by multi-site phase transfer catalyst—A kinetic study. *Colloid and Polymer Science* 286(6): 729-738.

Van Deurzen, M.P., F. van Rantwijk and R.A. Sheldon. 1997. Selective oxidations catalyzed by peroxidases. *Tetrahedron* 53(39): 13183-13220.

Wang, D.H. and H.S. Weng. 1988. Preliminary study on the role played by the third liquid phase in phase transfer catalysis. *Tenth International Symposium on Chemical Reaction Engineering* 2019-2024.

Wang, H. 2019. Chiral phase-transfer catalysts with hydrogen bond: A powerful tool in the asymmetric synthesis. *Catalysts* 9(3): 244.

Whitehurst, D.D. 1980. Catalysis by heterogenized transition-metal complexes. *Chemischer Informationsdienst* 11(26).

Yamazaki, N., S. Nakahama, A. Hirao and J. Kawabata. 1980. Catalysis by cross-linked cationic polymers. I. The decarboxylation of 6-nitrobenzisoxazole-3-carboxylate anion catalyzed by cross-linked polystyrene resins having quaternized ammonium chloride. *Polymer Journal* 12(4): 231-238.

Yang, X., J. Zhai, T. Xu, B. Xue, J. Zhu and Y. Li. 2019. Grafted polyethylene glycol–graphene oxide as a novel triphase catalyst for carbenes and nucleophilic substitution reactions. *Catalysis Letters* 149(10): 2767-2775.

Zhao, W., H. Zuo, Y. Guo, K. Liu, S. Wang, L. He, X. Jiang, G. Xiang and S. Zhang. 2019. Porous covalent triazine-terphenyl polymer as hydrophilic–lipophilic balanced sorbent for solid phase extraction of tetracyclines in animal derived foods. *Talanta* 201: 426-432.

Achiral Polymer-supported Organic Reagents

Rajesh Kumar Meena[1*], Pragati Fageria[2], Aruna Sharma[3*], Aprajita Gaur[1], Sudha Gulati[4] and Roopa Kumari[5]

[1] Department of Chemistry, Kalindi College, University of Delhi, New Delhi, India
[2] Department of Chemistry, Rajasthan University, Jaipur, Rajasthan, India
[3] Department of Chemistry, JECRC University, Jaipur, Rajasthan, India
[4] Department of Physics, Kalindi College, University of Delhi, New Delhi, India
[5] Department of Chemistry, University of Kota, Kota, Rajasthan, India

1. Introduction

Polymers with catalytically active moieties are called polymer supported organic catalysts. Synthetic polymers like polystyrene, poly (ethylene glycol), polyethylene, polyisobutylene (PIB), polyester, alkylated polystyrene, etc. typically have an organic catalytic moiety connected to the side chain. Organic catalysts supported by polymers are utilised as catalysts in a variety of organic synthesis processes. Due to their insolubility, cross-linked polymers are frequently used as polymer support material. It is simple to remove insoluble polymer catalysts from the reaction mixture and recycle them. The polymer's main chain has some organic catalysts inserted into it. Catalysts are typically needed in chemical reactions with green chemical processes in order to speed up the reactions under mild reaction conditions. When the reaction is finished, the catalysts must be removed from the reaction mixture. We can only hope that they are retrieved, renewed, and used again. However, the majority of the catalysts are difficult to recycle. It is straightforward to remove the cross-linked polymer catalyst from the reaction mixture when an organic molecule with catalytic activity is immobilized to it. Precipitation methods can also be used

*Corresponding authors: rajeshkumarmeena@kalindi.du.ac.in;
aruna.sharma@jecrcu.edu in

to separate supports made of linear polymers. The use of polymer catalysts can lessen the toxicity and the odour because polymers are not volatile. Polymer catalysts are used in a wide variety of organic synthesis processes (Ding and Uozumi 2008, Itsuno et al. 2011, Shinichi et al. 2008, Yang and Bergbreiter 2013). To demonstrate great selectivity and specificity, several organometallic complexes, for instance, have been utilized as catalysts for a variety of chemical processes.

The polymer immobilisation approach is a crucial alternative since it is typically challenging to retrieve these complicated catalysts. Organic polymers like polystyrene, poly (ethylene glycol), polyethylene, etc. are used as polymer supports for the organic catalysts in order to use these catalysts in organic solvents. To limit the use of hazardous chemical solvents, however, using water as a solvent is equally crucial in greener organic synthesis (Chiwara et al. 2009, Hashiguchi et al. 1995, Terhalle et al. 1997). In several chemical reactions involved in the production of organic compounds, polymeric reagents have proven helpful. Easy setup, reuse of the supported reagent after regeneration, and sterile reaction conditions are a few benefits of polymer supported reagents (Arakawa et al. 2006, Haraguchi et al. 2010a, Li et al. 2004). The literature on using polymer-bound organic and organometallic reagents in organic synthesis is covered in this chapter. In the last ten years, both polymer-supported chemistry and synthesis have experienced enormous progress. Recently, the ability to recycle the catalyst has received a lot of attention in the field of polymer supported organometallic and organic reagents, which is crucial from the perspective of green chemistry.

A review of recent advancements in the use of organometallic and organic reagents supported by polymers in organic synthesis is given, with a particular emphasis on the process for forming carbon-carbon bonds. The traditional Suzuki, Sonogashira, and Heck couplings, as well as aryl amination, epoxide opening, rearrangements, metathesis, and cyclopropanation, are a few examples of the reactions addressed. Also explored are applications in the area of asymmetric synthesis (Ley et al. 2000, Clapham et al. 2001, Bergbreiter et al. 2002, Leadbeater et al. 2002).

2. Polymer-supported organic catalysts for various reactions

Numerous significant organic reagents and catalysts have been created during the past few decades to meet the demands of synthetic chemists for reagents and catalysts for bond formation, functional group transformation, and the creation of complex organic frameworks.

Therefore, research on the recovery and repurposing of these organocatalysts/ reagents has received a lot of attention. Despite the fact that many organocatalysts are air- and moisture-stable, column chromatography is frequently required to separate these catalysts from the end products. Since immobilisation can more effectively separate catalysts and can be used as a way to reuse these catalysts, which are frequently used at relatively high mol percent loadings, immobilisation of organocatalysts on polymer supports to facilitate recovery and reuse of these catalysts is a desirable strategy. Numerous recent papers describing separable insoluble polymer-supported organocatalysts have been published as a result of the potential economic and green chemical benefits of this method. Selective illustrations of these catalysts can be found in Figure 1 (Benaglia et al. 2006, 2003, Zhang et al. 2011, Itsuno et al. 2011, Kristensen et al. 2010, Cozzi et al. 2006, Gruttadauria et al. 2008, Lu 2009, Kristensen et al. 2010, Hara et al. 2010, Alza et al. 2011, Arakawa et al. 2011, Gleeson et al. 2011, Demuynck et al. 2011, Haraguchi et al. 2010b, Wennemers et al. 2011, Ohtani et al. 1994, Font et al. 2008, Kudo et al. 2011, Yang and Bergbreiter 2013).

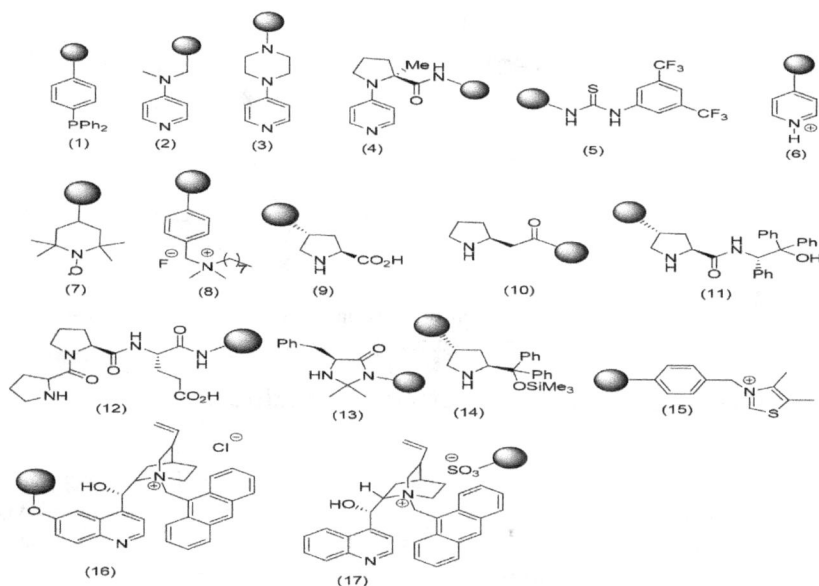

Figure 1. Examples of insoluble polymer-supported organocatalysts

3. Opening of epoxides using polymer-supported organic catalysts

3.1 Co catalyst attached to a dendronized polystyrene support (Salen)

According to Weck and colleagues, a (salen) Co catalyst connected to a dendronized polystyrene support has proven successful in achieving kinetic resolution of epoxides by hydrolysis. Three units of the cobalt-salen catalyst can be attached to the dendron linker, improving cooperative contacts and raising the local catalyst concentration, enabling the use of a far lower catalytic loading than was previously possible with a catalyst supported by a polymer for this process (Scheme 1) (Goyal et al. 2008). Depending on the epoxide's substitution pattern, pre-activating the catalyst by oxidising Co(II) to Co(III) in the presence of air and acetic acid produced basically enantiomerically pure epoxide with yields between 40-47% (the maximum theoretical yield was 50%).

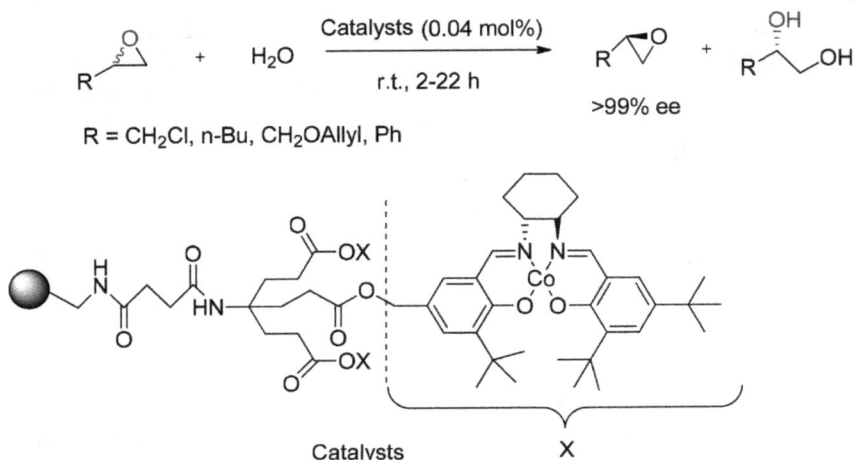

R = CH$_2$Cl, n-Bu, CH$_2$OAllyl, Ph

Catalysts   X

Scheme 1. Epoxides are resolved hydrolytically and kinetically with a (salen)-cobalt catalyst supported by a polymer

3.2 Amberlite IRA-400 supported phenoxide and naphthoxide anions

Phenoxide and naphthoxide anions are readily produced on the IRA-400 supported by amberlite. These very air stable polymer supported reagents are employed for regioselective ring opening reactions of various epoxides to produce aryl ether alcohols in large quantities under accommodating reaction conditions. The reaction of Amberlite IRA-400 (Cl$_2$ form) with an aqueous sodium salt of various substituted phenols or naphthols produced the Amberlite IRA-400 phenoxide or naphthoxide anion resin quickly and easily. These polymer-supported counterparts to conventional phenoxides or naphthoxides are highly air stable and can be stored for months without any oxidation. Ordinary phenoxides or naphthoxides are air sensitive and should be made only prior to their usage. The regioselective ring opening of several epoxides was accomplished using anions supported by polymers. Investigated were the effects of the solvent and polymer's molar ratio on the ring opening process of epoxides. The reactions were conducted in ethanol, dichloromethane, acetonitrile, and benzene. The outcomes of the experiment demonstrated that ethanol was the best solvent. It was discovered that a 3.1 molar ratio between the polymer-supported phenoxide or naphthoxide and the epoxide worked best. The regenerated polymeric reagents can be used again without losing any of their reactivity. Additionally, it is equally simple to employ polymer-supported phenoxides bearing electron-withdrawing groups (Table 1). The results for opening the rings of various substituted epoxides using phenoxide and naphthoxide ions supported by polymers are displayed in Tables 1 and 2. The benefits of using these air stable polymer-supported phenoxides or naphthoxides for the ring opening of epoxides over other methods described in the literature include nonaqueous and neutral reaction conditions, no catalyst used, easy reaction set-up, and the regeneration and reuse of the polymeric reagent (Akelah and Sherrington 1981, Akelah and Moet 1990, Tamami et al. 1993, Tamami and Mahdavi 2001, Tamami and Mahdavi 2002, Chen et al. 1995) (Figure 2).

3.3 Polystyrene-supported catalysts

The widespread use of phase-transfer catalysis in organic synthesis has significantly aided in the development of organic catalysts as practical preparative materials. It has also been reported (Regen 1975, 1976, 1977, 1979, Margaret et al. 1980, Nishikubo et al. 1990) to immobilise quaternary ammonium salts on polystyrene cross-linked with 2% divinylbenzene (DVB) and to employ a polymer-supported catalyst to catalyse the nucleophilic ring opening of dioxiranes (Scheme 2).

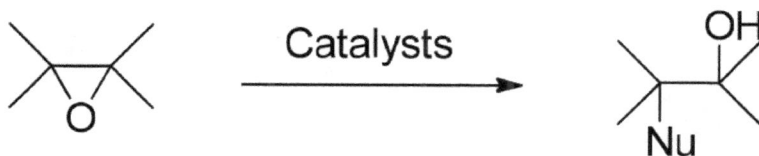

Scheme 2. Reaction catalyzed by insoluble polymer-supported phase-transfer catalysts (Regen et al. 1975)

3.4 Ruthenium (III) complex was anchored to a polymer-bound bis(2-picolyl)amine

Kim and Lee have demonstrated assisted ruthenium catalysis for ring opening of epoxides. Both aromatic and aliphatic epoxides were reacted with either methanol or water using a ruthenium (III) complex that was tethered to a polymer-bound bis(2-picolyl)amine ligand with nearly full conversion (Scheme 3). Styrene reacted quickly (1 h), whereas other epoxides took a little longer (2.5-35 h), and 1,2-epoxy-hexane required 200 h to completely convert. With no loss of catalytic activity, the polymer-bound catalyst may be recycled up to ten times. Although significant leaching was noted in the case of hydrolysis, the same kinds of reactions could also be carried out with a supported iron catalyst (Lee et al. 2007).

Table 1. Epoxides reacting in ethanol using Amberlite IRA-400 supported $X-C_6H_5O_2$. Products were identified by comparing their IR and NMR spectra and/or physical data with the original samples. Reactions were carried out at 50°C, and yield refers to an isolated product

S. No.	Epoxide	Time (hr.)			Product	Yield (%)		
		X=H	X=Cl	X=CH$_2$		X=H	X=Cl	X=CH$_2$
1.		1	2	1.5		57	55	60
2.		1.5	2	1.5		90	80	93
3.		1.5	2	1.5		93	82	95
4.		1	2	1.5		87	80	90
5.		1	2	1.5		90	85	92
6.		1	2	1.5		95	85	96

Table 2. Epoxides react with ethanol-supported Amberlite IRA-400 b-naphthoxide. Products were identified by comparing their IR and NMR spectra and/or physical data with the original samples. Yield denotes an isolated product. Reactions were conducted at 50°C

S. No.	Epoxide	Time (hr.)	Product	Yield (%)
1.	Ph—epoxide	2	Ph—CH(OH)—CH₂—ONaphtyl	85
2.	Ph—O—CH₂—epoxide	2	PhO—CH₂—CH(OH)—CH₂—ONaphtyl	80
3.	acetyl ester—CH₂—epoxide	2	ester—CH₂—CH(OH)—CH₂—ONaphtyl	82
4.	cyclohexene oxide	2	cyclohexane OH / ONaphtyl	85
5.	HO—CH₂—epoxide	2	HO—CH₂—CH(OH)—CH₂—ONaphtyl	80
6.	allyl ether—CH₂—epoxide	2	allyl—O—CH₂—CH(OH)—CH₂—ONaphtyl	85

4. Application of polymer-supported organic catalysts in formation of cyclic product

4.1 Cycloaddition of an imine to N-phenylmaleimide

In the past, Carretero and colleagues created a family of ferrocenes that could be employed as ligands in metal-catalyzed asymmetric transformations by substituting them with tert-butylsulfide and other phosphines (Priego et al. 2002). In order to permit attachment to Wang resin and Merrifield resin, one of these, the so-called Fesulphos ligand, was subsequently given two alternative handles and used in the copper-catalyzed 1,3-dipolar cycloaddition of azomethine ylides (Scheme 4) (Martin et al. 2007)

It was discovered that copper coordination to the polymer-bound ligands was slower than the equivalent preparation in solution, taking an hour as opposed to a few seconds. The cycloaddition reaction of different imines with

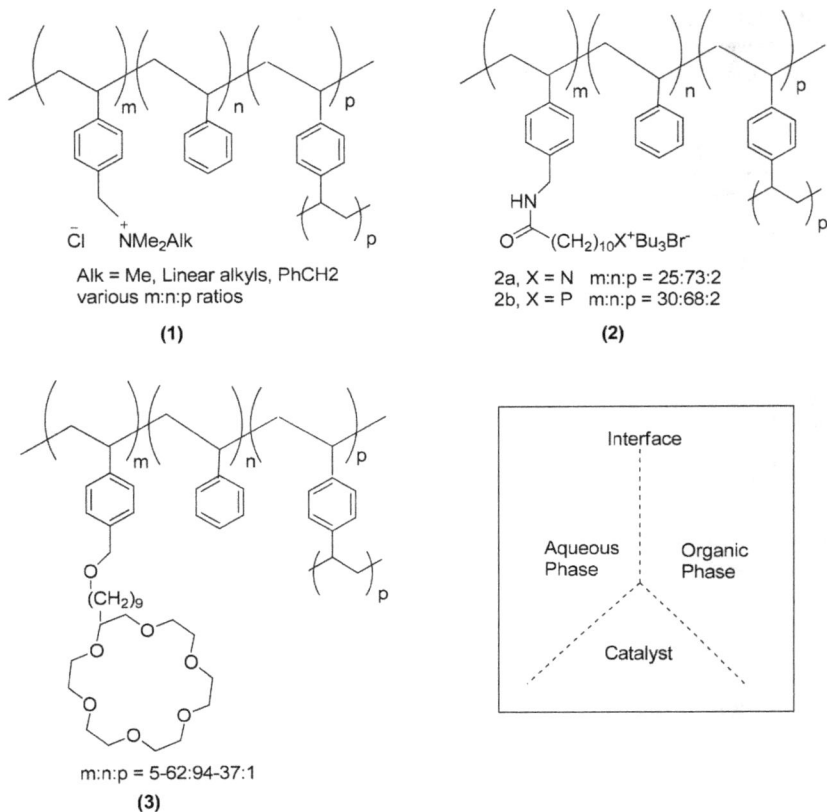

Alk = Me, Linear alkyls, PhCH2
various m:n:p ratios

(1)

2a, X = N m:n:p = 25:73:2
2b, X = P m:n:p = 30:68:2

(2)

m:n:p = 5-62:94-37:1

(3)

Interface

Aqueous Phase | Organic Phase

Catalyst

Figure 2. Crown ethers and ammonium and phosphonium salts on polystyrene support (an outline of triphase catalysis is reported in the box)

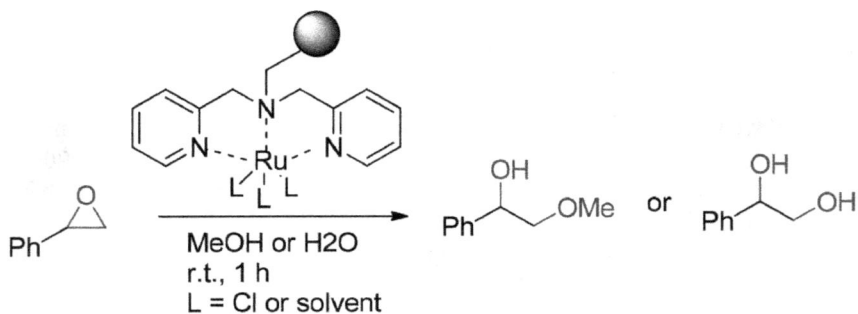

MeOH or H2O
r.t., 1 h
L = Cl or solvent

Scheme 3. Epoxide ring opening with methanol or water with ruthenium as the catalyst, using styrene oxide as the substrate

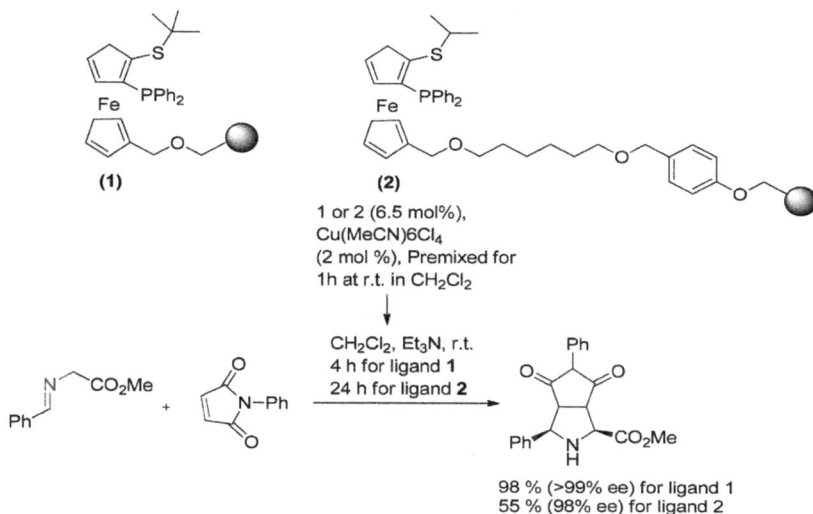

Scheme 4. In the copper-catalyzed 1,3-dipolar cycloaddition of an imine to N-phenylmaleimide, Fesulphos ligands attached to polymers are used

N-phenylmaleimide and methyl fumarate, on the other hand, produced the cyclized products in high yield and with outstanding enantioselectivity thanks to the polymer-supported catalyst synthesized from ligand 1. Surprisingly, the same excellent stereoselectivity was achieved using Fesulphos ligand 2 linked to a Wang resin via a spacer, but the yields were reduced (Hein et al. 2010).

4.2 Copper-catalyzed azide-alkyne cycloaddition

Due to its ease of usage and high yields, the copper-catalyzed azide-alkyne cycloaddition (also known as CuAAC) to produce triazole compounds is a popular technique for joining two molecules. Chan and Fokin have created a click-reaction-produced tristriazole ligand that stabilises Cu (I) and acts as a catalyst for CuAAC-type reactions. The ligand was attached to a NovaSynr TG amino resin (a polystyrene-polyethylene glycol support) to reduce copper contamination of the product, and it was tested in the click reaction between phenyl acetylene and benzyl azide (Scheme 5). By washing the resin-tethered ligand with a copper salt solution, copper (I) was preloaded onto it. When utilising a variety of solvents, the click reaction produced yield that was practically quantitative. The polymer-bound ligand could be recycled up to ten times with only a slight reduction in efficiency, and there was barely any copper leaching (Chan et al. 2004, 2007).

4.3 Multi-Step organic synthesis

In the presence of magnesium sulphate or molecular sieves, the readily accessible aldehyde 1 was treated with methyl hydrazine at room temperature to

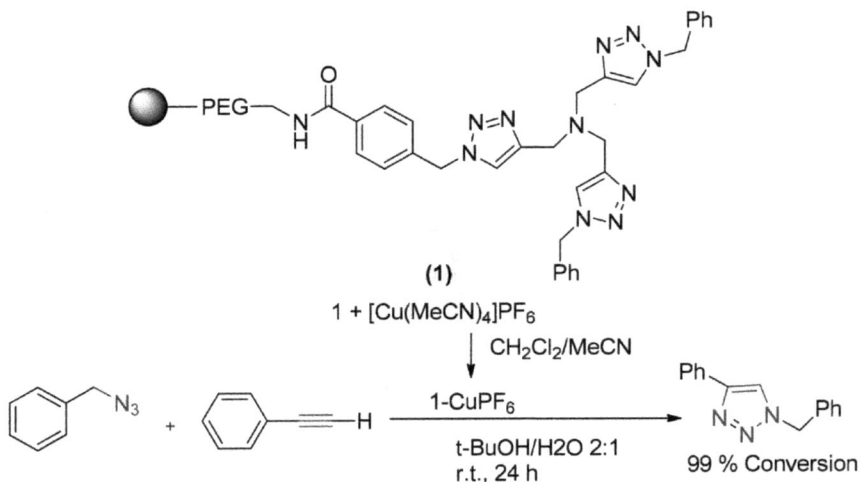

(1)

1 + [Cu(MeCN)₄]PF₆

CH₂Cl₂/MeCN

1-CuPF₆

t-BuOH/H2O 2:1
r.t., 24 h

99 % Conversion

Scheme 5. A copper-tris(triazolyl) complex supported by a polymer serves as a catalyst
for the cycloaddition of an azide to an alkyne

produce the hydrazone 2. The polymer-supported base BEMP (PS BEMP) was
used to N-alkylate the compound 2 with a small excess of ethyl bromoacetate.
Excellent yields were obtained, and any unreacted bromoester was removed
with aminomethyl polystyrene (Hill et al. 1998, Ley et al. 1999a, 2000, Dale et
al. 2000, Ryan et al. 1999). Trisamine that is supported by a polymer could also
be employed for the sequestering process without affecting the achievable yields
or purity. Then, using an ion exchange cyanide resin in a modified Strecker
process, it was possible to convert 3 to 4. Two equivalents of the resin at reflux
in ethanol containing a catalytic quantity of acetic acid were necessary for the
ideal circumstances. With no unreacted starting material 36 found by either LC-
MS or 1H NMR, this reaction proved to be exceptionally efficient, producing
solely the required hydrazine product, nitrile 4. For the oxidation of chemical 4
to the matching hydrazone 5, many processes were looked into.

Despite mostly producing the desired product, the polymer-supported
pyridinium bromide perbromide also produced a number of unidentified by-
products that could not be eliminated by traditional sequestering methods.
When compared to other "clean" oxidation methods, using Pd/C/cyclopentene
or manganese dioxide produced only the oxidation product, though not in a fully
converted state. However, treatment of the mixture with an isocyanate resin
supported by a polymer made it easier to remove the unreacted starting material
4 and produce a pure solution of 5. It was discovered that the phosphazene base
PS-BEMP is appropriate for the quick deprotonation and concurrent cyclization
of 5 to the tetra-substituted pyrazole 6. The ester 6 was then converted to the
necessary amide 7 by dissolving it in a saturated solution of ammonia in methanol.
Initial detection of two products - the amide 7 and the methyl ester produced by

the transesterification of 6 - led to their identification. This substance underwent multiple transformations before becoming amide 7 (Kim et al. 1996, Pop et al. 1997, Tronchet et al. 1993, Ian et al. 2000) (Scheme 6).

4.4 Fixation of carbon dioxide in the form of a cyclic carbonate

Jiang et al. reported on an intriguing derivatization of propargylic alcohols using supercritical carbon dioxide on a solid substrate (Jiang et al. 2008). They did this by using copper(I) iodide bonded to a dimethylamino-polystyrene resin to create alkylidene cyclic carbonates. The reaction was shown to have an optimal CO_2-pressure range of 14-18 MPa, resulting in a variety of cyclized compounds being produced in good yields. No product was produced when the substrate 2-phenyl-3-butyn-2-ol or primary or internal alkynes were utilised in the reaction; the reaction was restricted to terminal secondary propargylic alcohols with aliphatic substitutents. An article by Cai and colleagues that demonstrates how copper (II) coordinated to a polymerbound proline can catalyse the cross-coupling of oximes with arylboronic acids in respectable yields should also be briefly mentioned (Wang et al. 2010) (Scheme 7).

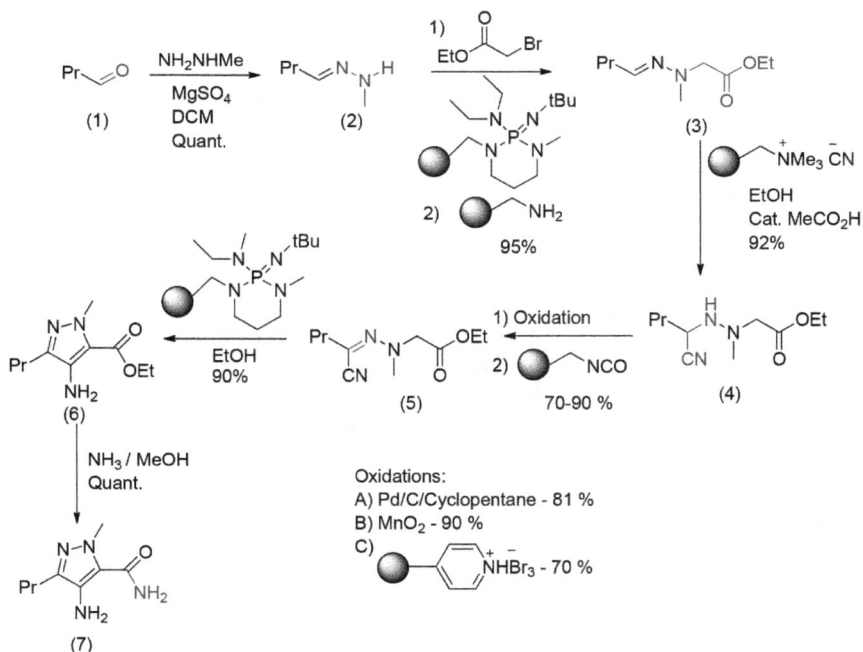

Scheme 6. The use of reagents supported by polymers in the sequential, multi-step production of heterocycle 3 (Ian et al. 2000)

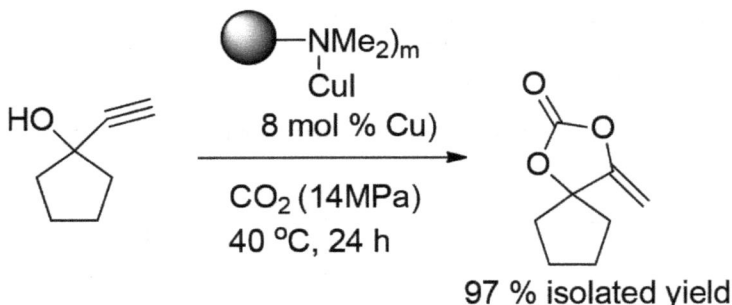

97 % isolated yield

Scheme 7. Fixation of carbon dioxide using a cyclic carbonate, facilitated by a Cu(I)-amine complex supported by a polymer (Wang et al. 2010)

4.5 Dichlorocyclopropane synthesis

Numerous preparative uses for insoluble phase-transfer polystyrene-supported ammonium and phosphonium salts have been documented, including the synthesis of dichlorocyclopropane from substituted alkenes (Scheme 8). (Margaret et al. 1980, Nishikubo et al. 1990).

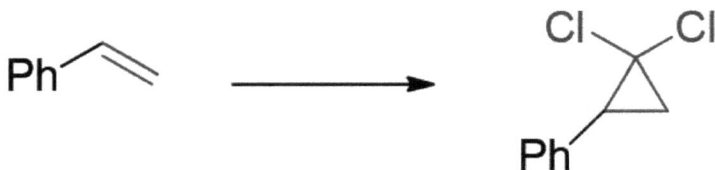

Scheme 8. Ammonium and phosphonium salts supported on polystyrene promote insoluble phase-transfer reactions

4.6 Stereoselective reactions catalyzed by Cinchona alkaloid-derived polymer-supported catalysts

Colonna and colleagues found that the epoxide product of the darzens reaction between 4-methylphenyl chloromethyl sulfone and 2-butanone (equation-1, Scheme 9) occurred at a maximum ee of 23% when it was conducted at room temperature in a two-phase system of acetonitrile and 50% aqueous sodium hydroxide. The more rigid structure of cinchona alkaloids, whose capacity to catalyse a range of activities had already been discovered at the time, was sought after for improvement in stereoselectivity. It's intriguing how the alkaloid skeleton provides various locations for polymer attachment. Kobayashi and Iwai chose the more often used bridge-head nitrogen for the immobilisation of quinine by quaternization with regular and modified DVB cross-linked polystyrene. As a result, catalysts 2 (Y = Cl) and (Y = Br) (Figure-4) were made and used as catalysts (0.2 mol equiv) in the Weitz-Scheffer epoxidation of chalcone (Scheme

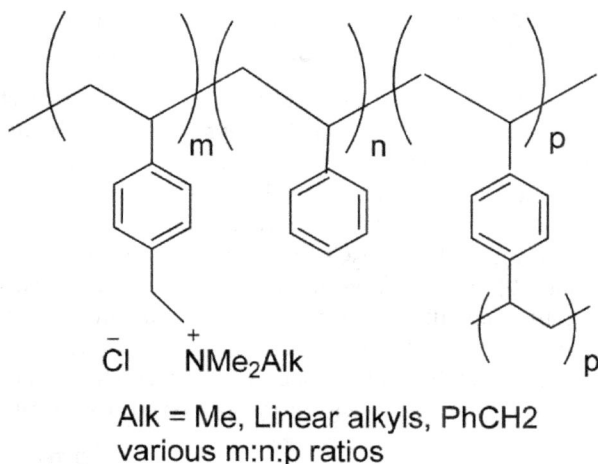

Alk = Me, Linear alkyls, PhCH2
various m:n:p ratios

(1)

Figure 3. Structure of ammonium and phosphonium salts supported by polystyrene

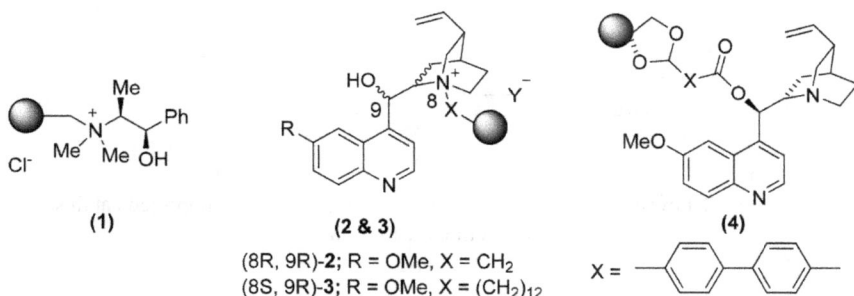

(1)   **(2 & 3)**   **(4)**

(8R, 9R)-**2**; R = OMe, X = CH$_2$
(8S, 9R)-**3**; R = OMe, X = (CH$_2$)$_{12}$ X =

Figure 4. a: (1, 2 and 3) structure of chiral phase-transfer catalysts supported on polystyrene and poly(ethylene glycol) (shaded circle = polystyrene with different degree of divinylbenzene cross-linking), b: (4) structure of supported cinchona alkaloids catalysts (shaded circle = Wang resin)

9, equation-2) using sodium hydroxide and a solvent mixture of 30% hydrogen peroxide and toluene. Extremely low ee (up to 4%) were seen, with catalyst 3 working better than catalyst 2. The catalyst 4 was employed in the cycloaddition of phenyl ketene with the N-tosylimine of ethyl glyoxalate to produce 1-p-tolylsulfonyl-3-phenyl-4-carbethoxy â-lactam (Scheme 9, equation-3). The product, which is a 93:7 mixture of cis and trans-azetidinones, was produced with a 62% yield (Colonna et al. 1978, Kacprzak and Gawronski 2001, Hodge et al. 1983). The predominant diastereoisomer has a % ee.

4.7 New soluble-polymer bound ruthenium carbene catalysts and application in ring-closing metathesis

Since this strategy should make it easier to both recycle the catalyst and purify the produced product, polymer supported catalysis has recently attracted more attention. A few synthetic and recycling investigations on ruthenium precatalysts supported by polymers have been published for this purpose.

On a manual synthesizer, the reactions were carried out in parallel by reacting multiple diene systems 1a-f in CH_2Cl_2 in the presence of 5 mol% of catalysts 3a or 3b (Scheme 10). The reaction mixture was then transferred to ether for precipitation and filtration after it had finished. After the filtrate was evaporated and the little quantity of Ru that had leached into the vessel was removed, column chromatography produced the cyclic products 2a-f in excellent yields (Table 3). Catalyst 3b was active enough to ensure cyclization of a substrate containing

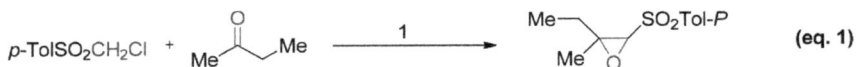

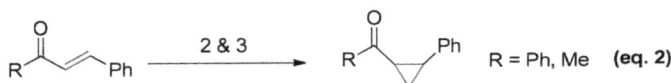

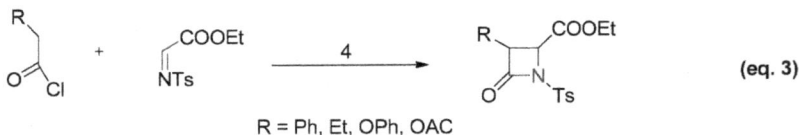

$$p\text{-TolSO}_2\text{CH}_2\text{Cl} \quad + \qquad \xrightarrow{\quad 1 \quad} \qquad \qquad \text{(eq. 1)}$$

$$\xrightarrow{\quad 2 \ \& \ 3 \quad} \qquad \qquad R = Ph, Me \quad \text{(eq. 2)}$$

$$\xrightarrow{\quad 4 \quad} \qquad \qquad \text{(eq. 3)}$$

R = Ph, Et, OPh, OAC

Scheme 9. Catalysing stereo-selective processes with polymer-supported catalysts generated from Cinchona alkaloids

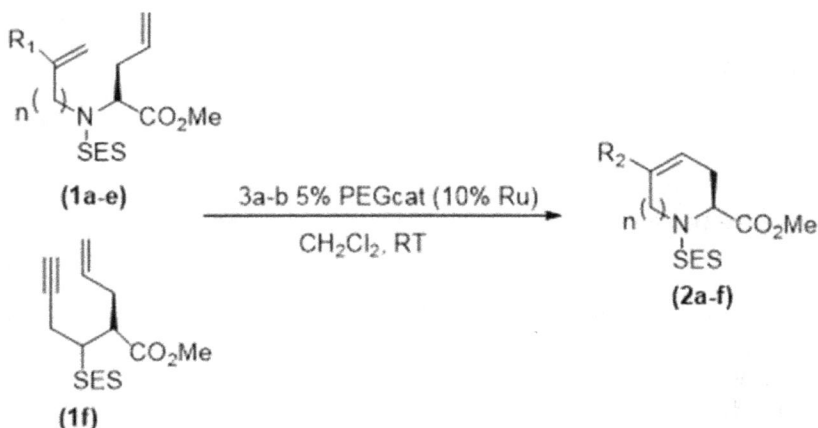

(1a-e)

$$\xrightarrow[\text{CH}_2\text{Cl}_2, \ RT]{\text{3a-b 5\% PEGcat (10\% Ru)}}$$

(2a-f)

(1f)

Scheme 10. Synthesis of cyclic α-aminoesters

substituted electronpoor olefins such as 4. This method provided an original access to the carboxy pyrroline 5 (Scheme 11) (Clapham et al. 2001, Miguel 2000, Leadbeater et al. 2002, McNamara et al. 2002, Barrett et al. 2002, Dickerson et al. 2002, Bergbreiter et al. 2002, Yao 2000, Nguyen and Grubbs 1995, Schurer et al. 2000, Randl et al. 2001, Connon et al. 2002, Stephane et al. 2003).

Table 3. Ring-closing metathesis of α-aminoesters by supported catalysis

Entry	R_1	(1)	n	R_2	Yield of 2%
1	H	a	1	H	91
2	H	b	2	H	95
3	H	c	3	H	93
4	CH_3	d	1	CH_3	87
5[a]	CO_2Me	e	1	CO_2Me	86
6		f	1	$CH=CH_2$	89

[a] Catalyst 3b was used (in Figure 5)

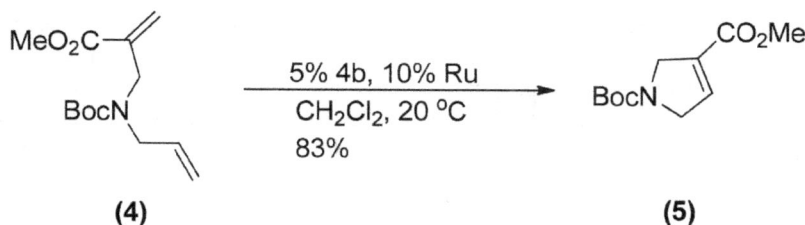

Scheme 11. Synthesis of carboxy pyrroline

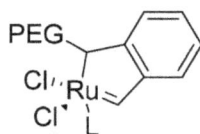

L = PCy_3 **3a**
L = H_2IMes **3b**

PEG-OH = H-$(O-CH_2-CH_2)_n$-OH with anaverage MW = 3400

Figure 5. Structure of polyoxygenated catalysts

4.8 Cyclization of Triketone

(S)-Prolinamide was linked to polystyrene using a sulfonamide linkage to create 2, which acted as a catalyst for the asymmetric cyclization of triketone 1 to create 3 and 4 (Scheme 12) (Pedrosa et al. 2013).

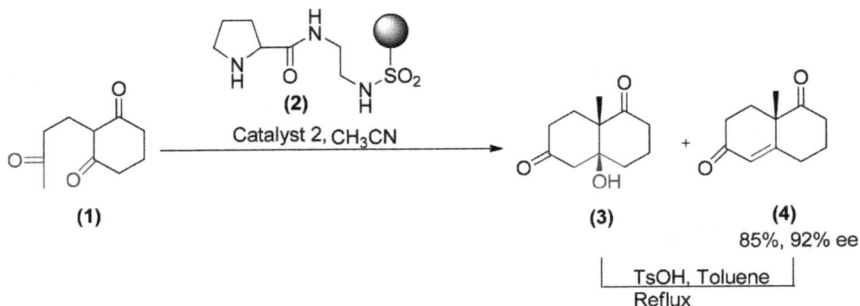

Scheme 12. Asymmetric cyclization of triketone

5. Hydrogenation reactions using polymer-immobilized catalyst

5.1 Asymmetric hydrogenation of acetophenone

One example of the most effective catalysts is the asymmetric hydrogenation catalysts created for ketone reduction, complexes generated from chiral 1, 2-diamines, and RuCl$_2$/diphosphines. For the asymmetric hydrogenation, a number of polymer catalysts that include the chiral Ru complex have been developed and employed. As chiral polymeric ligands of Ru catalyst, chiral 1,2-diamines immobilised on polystyrene (PSt) and ones immobilised on PEG are both successfully employed. The polymer catalysts and their application in the enantioselective hydrogenation of acetophenone 1 are shown in (Scheme 13). In the asymmetric hydrogenation of aromatic ketones, the chiral Ru complex made from 3, XylBINAP ((S)-(-)-2,20-bis[di(3,5-xylyl) phosphino]-1,10-binaphthyl) and RuCl$_2$ has good catalytic activity. Numerous recycling cycles of these polymer catalysts are possible without losing any of their catalytic activity (Chiwara et al. 2009).

Asymmetric transfer hydrogenation, which is a promising catalytic approach for the synthesis of chiral alcohols and amines, is a notable method of asymmetric reduction without hydrogen gas. TsDPEN transition metal complexes (TsDPEN14 p-toluenesulfonyl 1,2-diphenylethylenediamine), created by Noyori and Ikariya (Hashiguchi et al. 1995), are one of the potent asymmetric catalysts for the enantioselective reduction of carbonyl and imine compounds under the transfer hydrogenation reaction condition. The first instance of immobilising monosulfonylated 1,2-diphenylethylenediamine into cross-linked polystyrene was described by Lemaire et al. (Ter et al. 1997)

Lemaire's cross-linked polymer catalyst has decreased catalytic activity with 84% ee in the enantioselective reduction of acetophenone. 94% of ee is produced by the comparable original low-molecular-weight catalyst in solution system. When quaternary ammonium sulfonate structure is added to cross-linked

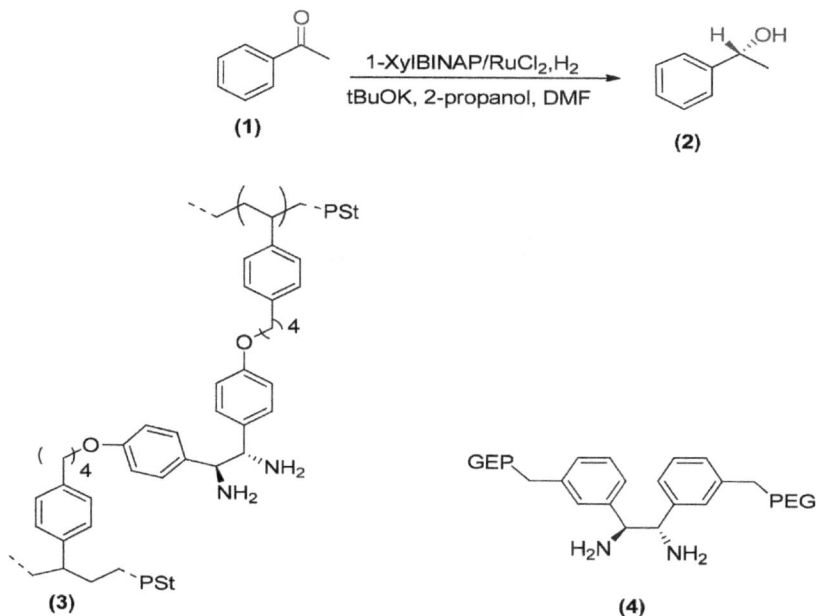

Scheme 13. Asymmetric hydrogenation of acetophenone with polymer-immobilized chiral diamine-Ru complex

polystyrene, significant improvement is seen. By terpolymerizing 4, 5, and 6, the polymeric chiral ligand 3 is easily created (Scheme 14).

In the reduction of acetophenone, the polymer catalyst containing quaternary ammonium sulfonate provides a quantitative conversion with a % ee (Arakawa et al. 2006). This illustration demonstrates the significance of changing the structure of support polymers in order to increase the catalytic activity and stereoselectivity of polymer catalysts. Utilizing microspheres as a support polymer for the polymeric catalyst is another intriguing strategy. High catalyst loading is achievable since the microspheres' surface area is rather large. Precipitation polymerization is used to create polymer microspheres that are functionalized with chiral TsDPEN ligand (Haraguchi et al. 2010b).

Precipitation polymerization is successfully used to create monodisperse, cross-linked poly (divinylbenzene) and poly(methacrylic acid-co-ethylene glycol dimethacrylate) microspheres with (R,R)-TsDPEN moiety. By altering the order in which the matching monomers are added, the site of the introduction of the (R, R)-TsDPEN moiety into polymer microspheres can be controlled. Other polymer supports, such as PEG (Li et al. 2004b), PE (Xiao et al. 2012), and silica (El-Shehawy and Itsuno 2005), are utilised in addition to polystyrene (Itsuno et al. 2013).

A productive technique for the asymmetric hydrogenation of imines utilising a hydrophobic polymer-supported N-toluenesulfonyl-1,2-diphenylethylenediamine

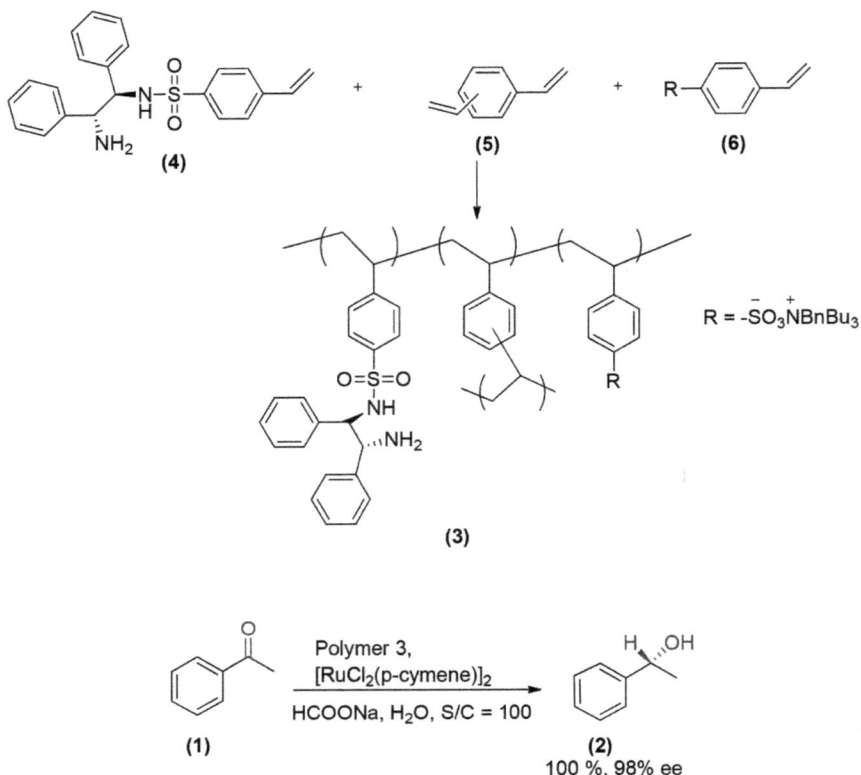

Scheme 14. Asymmetric transfer hydrogenation of acetophenone in water using polymer-immobilized catalyst

(TsDPEN) ruthenium catalyst has been described by Haraguchi, Itsuno, and colleagues (Haraguchi et al. 2009). The polymer was made hydrophilic by adding pendant quaternary ammonium sulphate groups to the polymer backbone, allowing for the hydrogenation of ketones like acetophenone in water (Scheme 15). The level of cross-linking was also looked at; however, it was discovered to be less significant as even using a highly cross-linked polymer (20% DVB) as the catalyst support resulted in a high yield of product (Arakawa et al. 2008a).

Chiral alcohol synthesis can be accomplished easily and safely using the catalytic ATH of ketones. For transfer hydrogenation, a variety of chiral catalysts based on complexes of Ti, Ru, Rh, and Ir have been created. The compound of Ru(II) and optically active N1-p-toluenesulfonyl-1,2-diphenylethylene-1,2-diamine (TsDPEN) created by the research team of Ikariya and Noyori is now the most notable among them [6]. The complex was immobilised on a variety of supports, including silica, sulfonated polystyrene (Arakawa et al. 2008), PEG (Li et al. 2004b) and polystyrene copolymers that include phosphonates. These

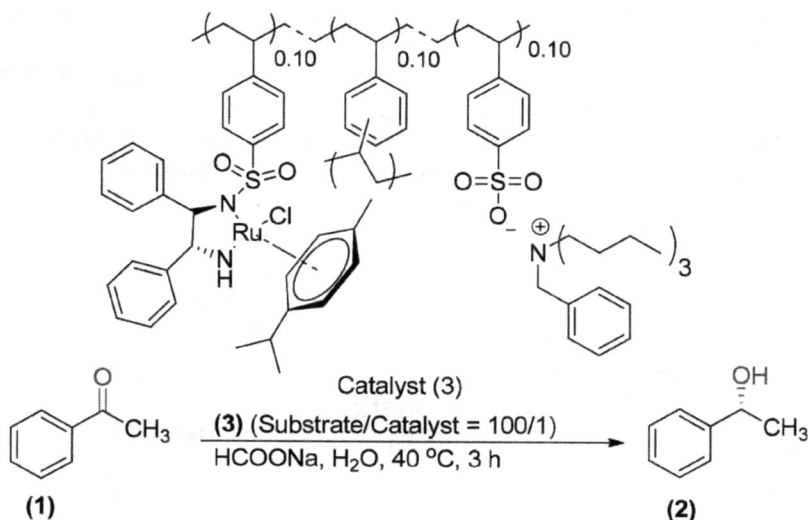

Scheme 15. Utilizing complexes of 1,2-diamine monosulfonamide and ruthenium, chiral polymer supports asymmetric transfer hydrogenation

catalytic complexes that were immobilised worked well in watery media. Narrowly dispersed polymer microspheres 3 were created by precipitation polymerizing styrene, the crosslinking agent, and the TsDPEN monomer. For the reduction of ketone 1 to produce 2 with high enantioselectivity (Scheme 16), core-shell type

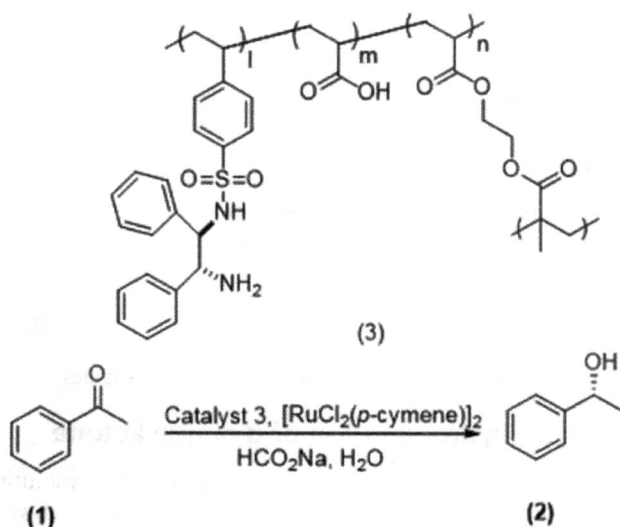

Scheme 16. Enantioselective reduction of ketone to form chiral alcohol

microspheres demonstrated stronger catalytic activity than other microsphere catalysts (Liu et al. 2013, 2004, Xu et al. 2013).

It has become possible to create synthetic-amino amides 3, which are produced from actual amino acids. In a high yield and with enantioselectivity of 95%, their chiral Zn(II) complexes catalysed the enantioselective addition reaction between ZnEt$_2$ and aldehyde 1 to form chiral secondary alcohol 2 (Scheme 17) (Escorihuela et al. 2013, Wang et al. 2014).

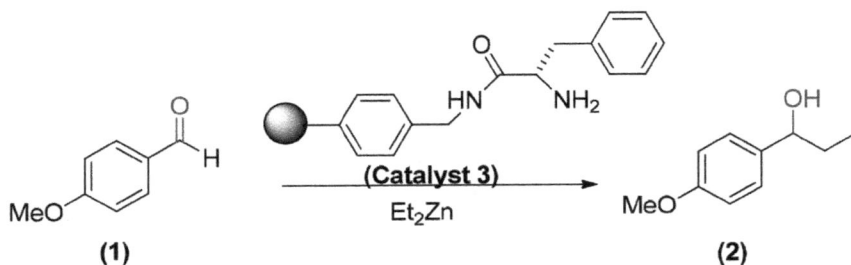

Scheme 17. Enantioselective addition reaction between ZnEt2 and aldehyde 1 to form chiral secondary alcohol

On a polymethacrylate support, borane reduction of ketone was carried out under the catalysis of chiral oxazaborolidine. On crosslinked polymethacrylate, 4-hydroxy,-diphenyl-(S)-prolinol was immobilised. The asymmetric borane reduction of ketones, such as the reaction of 1 to create 2, was carried out using the polymeric chiral amino alcohol 3 (Scheme 18). (Itsuno et al. 1998, Thvedt et al. 2011) obtained high catalytic activity and enantioselectivities.

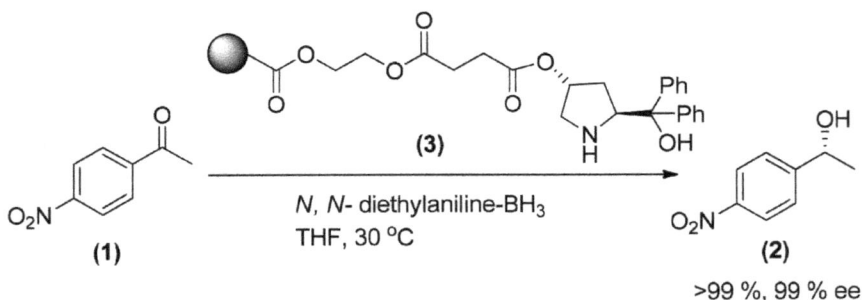

Scheme 18. Asymmetric borane reduction of ketones

5.2 Asymmetric hydrogenation of α-amide ketone

The asymmetric hydrogenation under dynamic kinetic resolution (DKR) conditions is another intriguing application of polymeric catalyst containing chiral 1,2-diamine ligand. A-(N-benzoyl-N-methylamino) propiophenone 1 is asymmetrically hydrogenated through DKR utilising the polymer catalyst made from polymer 3, (R)-BINAP, and RuCl$_2$ to generate syn-b-amide alcohol 2 solely with nearly complete enantioselectivity (Scheme 19). In a homogenous solution system, the polymer catalyst exhibits better stereoselectivity than the corresponding low-molecular-weight catalyst. This example demonstrates how the stereoselective performance of the catalyst is significantly influenced by the microenvironment produced in the polymer network. The hydroxyethyl methacrylate unit and methacrylate cross-linkage in polymer 3 offer the catalytic moiety a favourable micro environment (Kocienski et al. 1978).

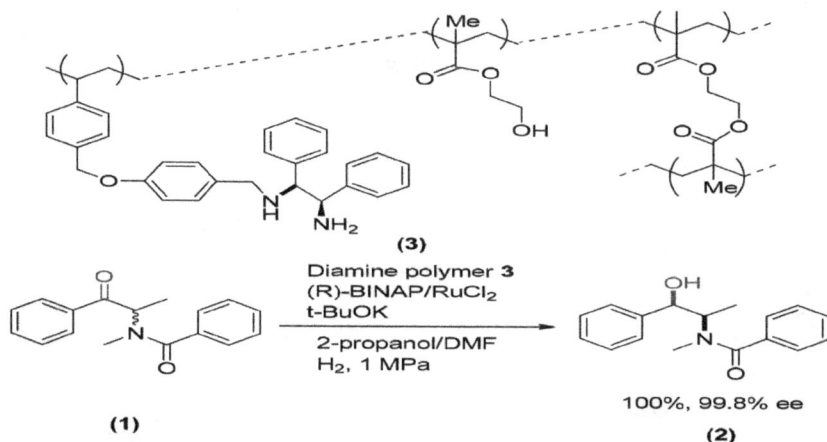

Scheme 19. Asymmetric hydrogenation of α-amide ketone

5.3 Haloform reaction of benzaldehyde

The haloform reaction of benzaldehyde (Scheme 20) was carried out in the presence of 0.05 mol equiv of compound 3 in acetonitrile/50% aqueous sodium hydroxide two phase system at room temperature with good chemical yields and distinct stereoselectivity. The more rigid structure of Cinchona alkaloids, whose capacity to catalyse a range of activities had already been discovered at the time, was sought after for improvement in stereoselectivity. The alkaloid skeleton, it turns out, provides a variety of locations for polymer attachment (Stefane et al. 2014).

5.4 Asymmetric hydrogenation of keto ester

BINAP 3 was produced by copolymerizing divinylbenzene with a chiral BINAP monomer. In the asymmetric hydrogenation of the keto ester 1 to produce the -hydroxyester 2 with strong enantioselectivity and quantitative conversion (Scheme 21), the Ru complex of 3 demonstrated good catalytic activity (Du et al. 2014, Sun et al. 2012).

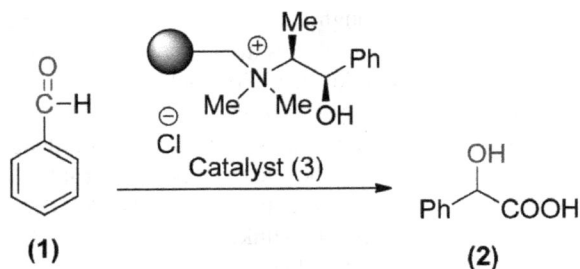

Scheme 20. Cinchona alkaloid-derived polymer-based catalysts for stereoselective processes and the structure of chiral phase-transfer catalysts supported on polystyrene and poly (ethylene glycol)

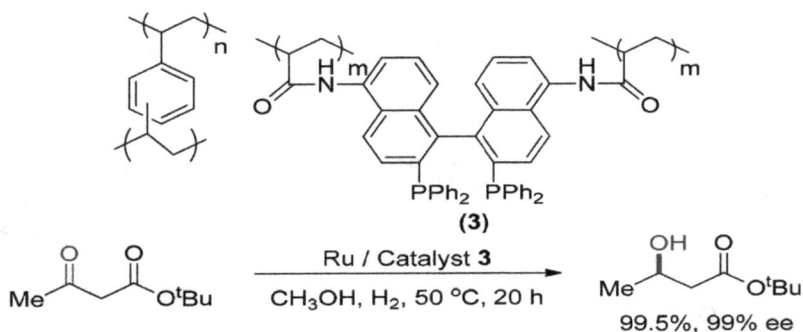

Scheme 21. Asymmetric hydrogenation of keto ester

5.5 Asymmetric hydrogenation of imines

A polymeric chiral monosulfonamide ruthenium complex was used for the ATH of cyclic sulfonimine 1. The ATH reaction in the organic solvent CH_2Cl_2 performed best on polystyrene crosslinked with divinylbenzene. But because of its great hydrophobicity, the polystyrene-based support did not react with the aqueous system. The facilitation of this reaction in water depends on the quaternary ammonium pendant group. Therefore, 95% ee was achieved for the conversion of 1 to 2 in water by employing the polymeric chiral ligand 3 (Scheme 22) (Sugie et al. 2014).

6. Application of polymer-supported organic catalyst in aldol reactions

6.1 Aldol reactions between cyclic ketone and aryl aldehyde

Quite efficient organocatalysts for a number of asymmetric reactions include proline and its derivatives. Proline derivatives immobilised on polymers have

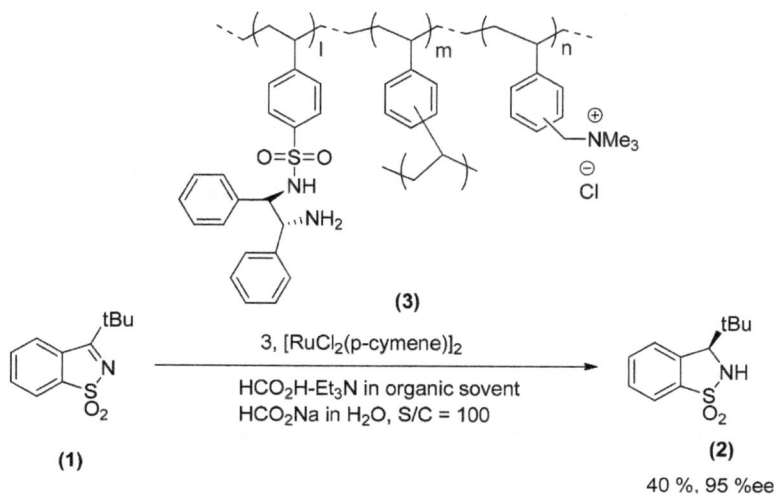

Scheme 22. Asymmetric hydrogenation of imines

been created and are effective catalysts for the asymmetric processes listed below. The anchoring of (S)-proline moieties on block copolymers 4 made of thermoresponsive poly (N-isopropylacrylamide) and PEG-grafted polyacrylate blocks were recently described by Suzuki et al. Then, in aqueous solution, 12 aldol reactions between ketone 1 and aryl aldehyde 2 were carried out. With a high yield and high diastereo- and enantioselectivity (96% ee) (Scheme 23), the aldol product 3 was produced (Suzuki et al. 2013).

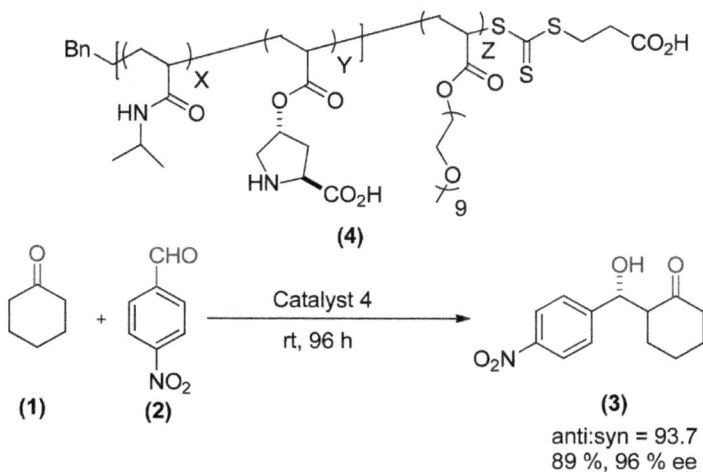

Scheme 23. Aldol reactions between ketone and aryl aldehyde

6.2 Aldol reactions between acetone and aryl aldehyde

A polymer was linked to a (S)-proline functionalized chiral amide alcohol to produce 4, which served as a catalyst for the asymmetric aldol reaction of 1 and 2 to produce 3 (Heidlindemann et al. 2014) (Scheme 24).

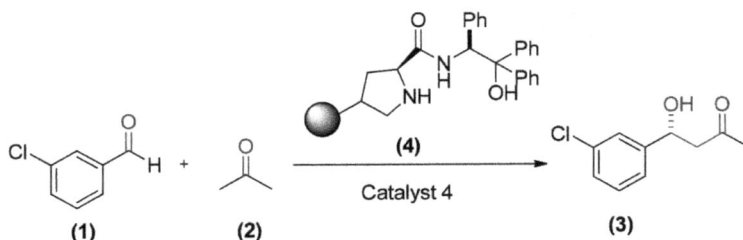

Scheme 24. As a catalyst for the asymmetric aldol reaction, a polymer was joined to a (S)-proline functionalized chiral amide alcohol

6.3 Asymmetric aldol reaction of aldehyde and acetone

A prolinamide structure was attached to a crosslinked polymer using an ester linkage and the hydroxy group of hydroxyl proline. In the asymmetric direct aldol synthesis of aldehyde 1 and acetone 2, the resultant polymeric prolinamide alcohol 4 was utilized as a catalyst to produce 3 (Scheme 25) (Rulli et al. 2013).

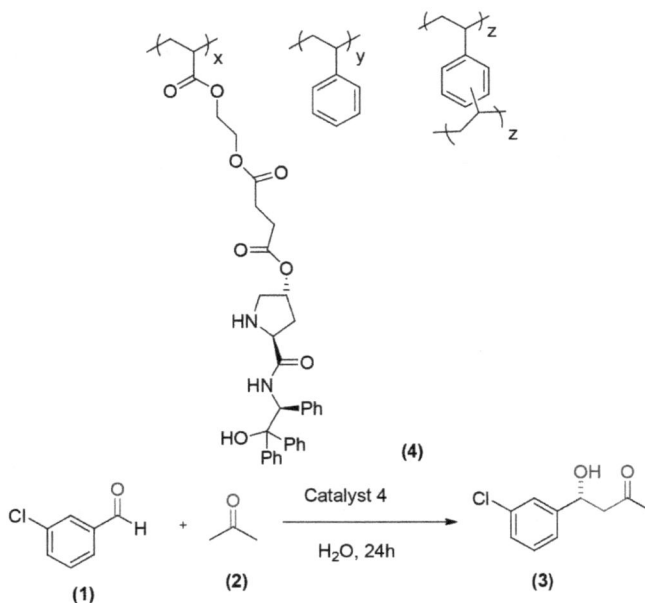

Scheme 25. Asymmetric direct aldol reaction of aldehyde and acetone

6.4 Asymmetric aldol reaction between cyclohexanone and aromatic aldehyde

The asymmetric aldol reaction between cyclohexanone and aromatic aldehyde in water was successfully driven by (S)-proline moieties attached to a thermoresponsive polymer nanoreactor with outstanding yields and enantioselectivities. An azide-acetylene click reaction was used to join a crosslinked polymer to a (S)-proline structure. In a flow system (Scheme 26), the resultant polymeric proline 4 was utilized as a catalyst in the asymmetric aldol reaction between aromatic aldehydes 1 and 2 to generate 3 (Ayats et al. 2012).

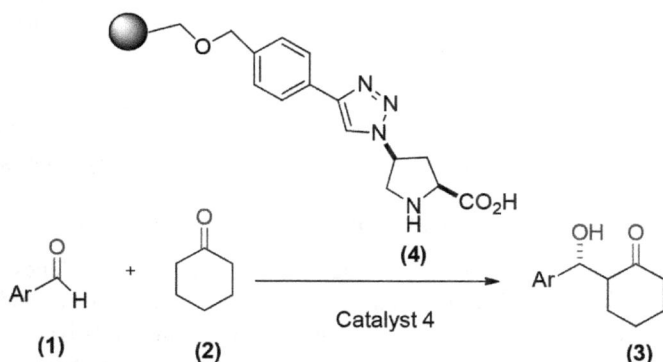

Scheme 26. Asymmetric aldol reaction between aromatic aldehyde and cyclohexanone

6.5 Aldol reaction with high diastereoselectivity

The direct aldol reaction of 1 and 2 was catalysed by polymer-immobilized prolinamide 4 to produce the aldol addition product 3 in an excellent yield and with strong diastereoselectivity (Scheme 27) (Qu et al. 2014).

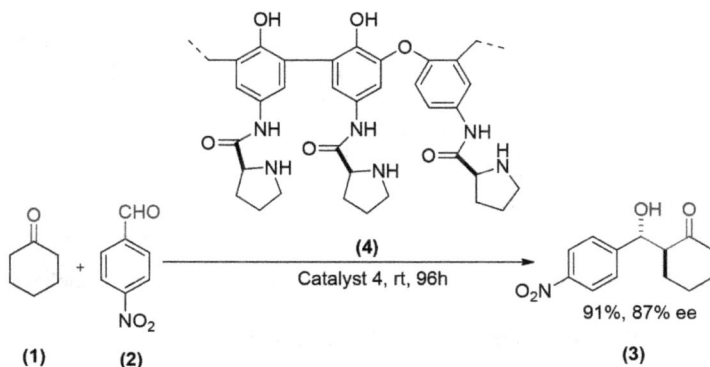

Scheme 27. Polymer-immobilized prolinamide catalyst was used as a direct aldol reaction

6.6 Stereoselective aldol reaction of aromatic aldehyde with ketone

Other amino acids, besides proline and its derivatives, have been employed as chiral organocatalysts. L-threonine, whose hydroxy group facilitated its attachment to the polymer side chain, is one recent example. When aromatic aldehyde 1 reacts with ketone 2 to generate 3, the resulting polymeric L-threonine 4 works as an easily recyclable, highly reactive, and stereoselective (up to 99% ee) catalyst (Scheme 28) (Henseler et al. 2014).

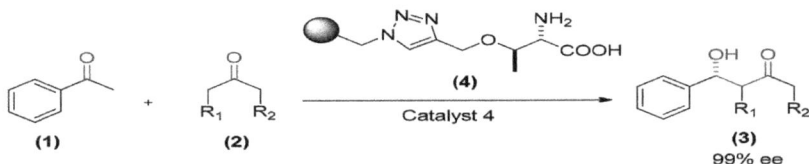

Scheme 28. Stereo selective aldol reaction of aromatic aldehyde with ketone

6.7 Aldol reaction of p-nitrobenzaldehyde and cyclohexanone

Prolinamide pendant group-containing substituted polyacetylenes 4 were created. In order to produce the chiral aldol adduct 3 in an 80% yield with an 80%t efficiency (Scheme 29), the prolinamide moiety of the polymer catalysed the asymmetric aldol reaction of p-nitrobenzaldehyde 1 and cyclohexanone 2 (Zhang et al. 2012).

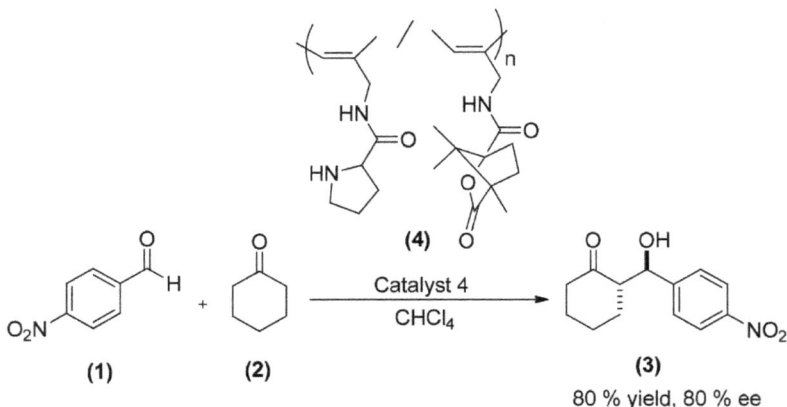

Scheme 29. Asymmetric aldol reaction of p-nitrobenzaldehyde and cyclohexanone

6.8 Aldol and Michael addition reaction

Wennemers et al. announced the development of the most reactive peptidic organocatalysts for the Aldol and Michael addition processes. In order to synthesise 3 in a continuous flow procedure with >99% yield and 80% efficiency from p-nitrobenzaldehyde 1 and aliphatic ketone 2 (Scheme 30), Wennemers' tripeptide 4 was polymer-immobilized (Otvos et al. 2012, Itsuno et al. 2014).

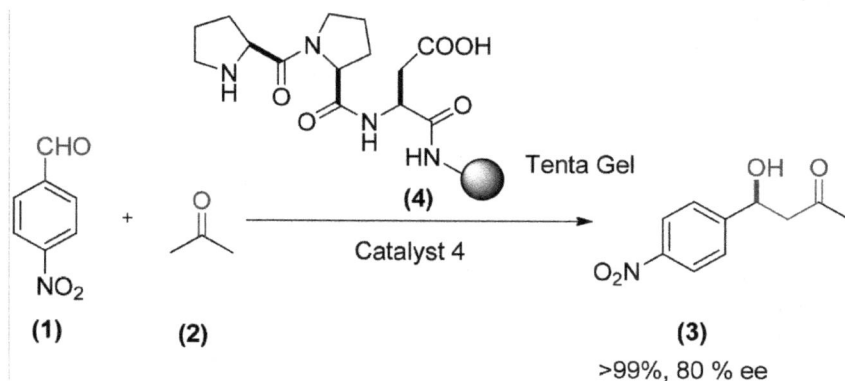

Scheme 30. Aldol and Michael addition reaction of p-nitrobenzaldehyde with aliphatic ketone

References

Akelah, A. and D.C. Sherrington. 1981. Application of functionalized polymers in organic synthesis. *Chemical Reviews* 81(6): 557-587.

Akelah A. and A. Moet. 1990. Functionalized Polymers and Their Applications. Chapman and Hall, London. pp. 345.

Alza, S. Sayalero, X.C. Cambeiro, R. Martín-Rapún, P.O. Miranda and M.A. Pericàs, 2011. Catalytic batch and continuous flow production of highly enantioenriched cyclohexane derivatives with polymer-supported diarylprolinol silyl ethers. *Synlett* 464-468.

Arakawa, Y., N. Haraguchi and S. Itsuno. 2006. Design of novel polymer-supported chiral catalyst for asymmetric transfer hydrogenation in water. *Tetrahedron Letters* 47: 3239-3243.

Arakawa, Y., A. Chiba, N. Haraguchi, S. Itsuno. 2008a. Asymmetric transfer hydrogenation of aromatic ketones in water using a polymer-supported chiral catalyst containing a hydrophilic pendant group. *Advanced Synthesis and Catalysis* 350: 2295-2304.

Arakawa, Y., N. Haraguchi, S. Itsuno. 2008b. An immobilization method of chiral quaternary ammonium salts onto polymer supports. *Angewandte Chemie International Edition* 47: 8232-8235.

Arakawa, Y., M. Wiesner and H. Wennemers. 2011. Efficient recovery and reuse of an immobilized peptidic organocatalyst. *Advanced Synthesis & Catalysis* 353(8): 1201-1206.

Ayats, C., A.H. Helseler and M.A. Pericàs. 2012. A solid-supported organocatalyst for continuous-flow enantioselective Aldol reactions. *Chem Sus Chem* 5: 320-325.

Barrett, A.G.M., B.T. Hopkins and J. Kobberling. 2002. ROMP gel reagents in parallel synthesis. *Chemical Reviews* 10(102): 3301-3324.

Benaglia, M., A. Puglisi and F. Cozzi. 2003. Polymer-supported organic catalysts. *Chemical Reviews* 103: 3401-3430.

Benaglia, M. 2006. Recoverable and recyclable chiral organic catalysts. *New Journal of Chemistry* 30: 1525-1533.

Bergbreiter, D.E. 2002. Using soluble polymers to recover catalysts and ligands. *Chemical Reviews* 102: 3345-3383.

Chan, T.R., R. Hilgraf, K.B. Sharpless and V.V. Fokin, 2004. Polytriazoles as copper(I)-stabilizing ligands in catalysis. *Organic Letters* 6: 2853-2855.

Chan, T.R. and V.V. Fokin. 2007. Polymer-supported copper (I) catalysts for the experimentally simplified azide-alkyne cycloaddition. *QSAR & Combinational Science* 26(11-12): 1274-1279.

Chen, J. and W. Shum. 1995. A practical synthetic route to enantiopure 3-aryloxy-1,2-propanediols from chiral glycidol. *Tetrahedron Letters* 36(14): 2379-2380.

Chiwara, V.I., N. Haraguchi and S. Itsuno. 2009. Polymer-immobilized catalyst for asymmetric hydrogenation of racemic a-(N-Benzoyl-N-methylamino) propiophenone. *Journal of Organic Chemistry* 74(3): 1391-1393.

Clapham, B., T.S. Reger and K.D. Janda. 2001. Polymer-supported catalysis in synthetic organic chemistry. *Tetrahedron* 57: 4637-4662.

Colonna, S., R. Fornasier and U. Pfeiffer. 1978. Asymmetric induction in the darzens reaction by means of chiral phase-transfer in a two-phase system. The effect of binding the catalyst to a solid polymeric support. *Journal of Chemical Society, Perkin Transactions 1*, 1: 8-11.

Connon, S.J., A.M. Dunne and S. Blechert. 2002. A self-generating, highly active, and recyclable olefinmetathesis catalyst. *Angewandte Chemie International Edition* 41: 3835-3838.

Cozzi, F. 2006. Immobilization of organic catalysts: When, why, and how. *Advanced Synthesis & Catalysis* 348: 1367-1390.

Dale, D.J., P.J. Dunn, C. Golightly, M.L. Huges, A.K. Pearce, P.M. Searle, G. Ward and A.S. Wood. 2000. The chemical development of the commercial route to sildenafil: A case history. *Organic Process Research & Development* 4(1): 17-22.

Demuynck, A.L.W., L. Peng, F. de Clippel, J. Vanderleyden, P.A. Jacobs and B.F. Sels. 2011. Solid acids as heterogeneous support for primary amino acid derived diamines in direct asymmetric Aldol reactions. *Advanced Synthesis & Catalysis* 353(5): 725-732.

Dickerson, T.J., N.N. Reed and K.D. Janda. 2002. Soluble polymers as scaffolds for recoverable catalysts and reagents. *Chemical Reviews* 102(10): 3325-3344.

Ding, K. and Y. Uozumi (eds). 2008. Handbook of Asymmetric Heterogeneous Catalysis. Wiley-VCH: Weinheim. ISBN: 978-3-527-62302-0.

Du, Y., D. Feng. J. Wan and X. Ma. 2014. The enhanced asymmetric hydrogenation of unsymmetrical benzils to hydrobenzoin catalyzed by organosoluble zirconium phosphonate-immobilized ruthenium catalyst. *Applied Catalysis A: General* 479: 49-58.

El-Shehawy, A.A. and S. Itsuno. 2005. *Current Topics in Polymer Research.* R.K. Bregg (ed.). Chapter 1, pp. 1-69. Nova Science Publishers, New York.

Escorihuela, J., L. González, B. Altava, M.I. Burguete and S.V. Luis. 2013. Chiral catalysts immobilized on achiral polymers: Effect of the polymer support on the performance of the catalyst. *Applied Catalysis A: General* 462-463: 23-30.

Font, D., S. Sayalero, A. Bastero, C. Jimeno and M.A. Pericas. 2008. Toward an artificial aldolase. *Organic Letters* 10(2): 337-340.

Gleeson, O., G.-L. Davies, A. Peschiulli, R. Tekoriute, Y.K. Gun'ko and S.J. Connon. 2011. The immobilisation of chiral organocatalysts on magnetic nanoparticles: The support particle cannot always be considered inert. *Organic & Biomolecular Chemistry* 9(22): 7929-7940.

Goyal, P., X.L. Zheng and M. Weck. 2008. Enhanced cooperativity in hydrolytic kinetic resolution of epoxides using poly(styrene) resin-supported dendronized Co-(salen) catalysts. *Advanced Synthesis and Catalysis* 350: 1816-1822.

Gruttadauria, M., F. Giacalone and R. Noto. 2008. Supported proline and proline-derivatives as recyclable organocatalysts. *Chemical Society Reviews* 37(8): 1666-1688.

Hara, N., S. Nakamura, N. Shibata and T. Toru. 2010. Enantioselective Aldol reaction using recyclable montmorillonite entrapped N-(2-thiophenesulfonyl) prolinamide. *Advanced Synthesis & Catalysis* 352(10): 1621-1624.

Hashiguchi, S., A. Fujii, J. Takehara, T. Ikariya and R. Noyori. 1995. Asymmetric transfer hydrogenation of aromatic ketones catalyzed by chiral ruthenium (II) complexes. *Journal of the American Chemical Society* 117(28): 7562-7563.

Haraguchi, N., K. Tsuru, Y. Arakawa and S. Itsuno. 2009. Asymmetric transfer hydrogenation of imines catalyzed by a polymer-immobilized chiral catalyst. *Organic and Biomolecular Chemistry* 7: 69-75.

Haraguchi, N., A. Nishiyama and S. Itsuno. 2010a. Synthesis of polymer microspheres functionalized with chiral ligand by precipitation polymerization and their application to asymmetric transfer hydrogenation. *Journal of Polymer Science Part A: Polymer Chemistry* 48: 3340-3349.

Haraguchi, N., Y. Takemura and S. Itsuno. 2010b. Novel polymer-supported organocatalyst via ion exchange reaction: Facile immobilization of chiral imidazolidin-4-one and its application to Diels-Alder reaction. *Tetrahedron Letters* 51(8): 1205-1208.

Heidlindemann, M., G. Rulli, A. Berkessel, W. Hummel and H. Gröger. 2014. Combination of asymmetric organo and biocatalytic reactions in organic media using immobilized catalysts in different compartments. *ACS Catalysis* 4: 1099-1103.

Hein, J.E. and V.V. Fokin. 2010. Copper-catalyzed azide-alkyne cycloaddition (CuAAC) and beyond: New reactivity of copper (I) acetylides. *Chemical Society Reviews* 39: 1302-1315.

Henseler, A.H., C. Ayats and M.A. Pericàs. 2014. An enantioselective recyclable polystyrene-supported threonine-derived organocatalyst for aldol reactions. *Advanced Synthesis and Catalysis* 356: 1795-1802.

Hill, D.C. 1998. Trends in development of high-throughput screening technologies for rapid discovery of novel drugs. *Current Opinion in Drug Discovery Development* 1: 92-97.

Hodge, P., E. Khoshdel and J. Waterhouse. 1983. Michael reactions catalysed by polymer-supported quaternary ammonium salts derived from cinchona and ephedra alkaloids. *Journal of Chemical Society, Perkin Transactions 1* 2205-2209.

Ian, R. Baxendale and S.V. Ley. 2000. Polymer-supported reagents for multi-step organic synthesis: Application to the synthesis of sildenafil. *Bioorganic & Medicinal Chemistry Letters* 10: 1983-1986.

Itsuno, S. 1998. Enantioselective reduction of ketones. pp. 395-576. *In:* Paquette, L.A. (ed.), Organic Reactions, Vol. 52: 395-576. Wiley.

Itsuno, S. (ed). 2011. Polymeric Chiral Catalyst Design and Chiral Polymer Synthesis. Wiley: Hoboken. ISBN: 978-0-470-56820-0.

Itsuno, S., M.M. Parvez and N. Haraguchi. 2011. Polymeric chiral organocatalysts. *Polymer Chemistry* 2(9): 1942-1949.

Itsuno, S. 2013. Encyclopedia of polymeric nanomaterials. *Polymer Catalysts* 1-9.

Itsuno, S. and M.M. Hassan. 2014. Polymer-immobilized chiral catalysts. *RSC Advances* 4: 52023-52043.

Jiang, H.F., A.Z. Wang, H.L. Liu and C.R. Qi. 2008. Reusable polymer-supported amine-copper catalyst for the formation of alpha-alkylidene cyclic carbonates in supercritical carbon dioxide. *European Journal of Organic Chemistry* 13: 2309-2312.

Kacprzak, K. and J.R. Gawronski. 2001. Cinchona alkaloids and their derivatives: Versatile catalysts and ligands in asymmetric synthesis. *Synthesis* 07: 0961-0998.

Kim, Y.H. and J.Y. Choi. 1996. Novel palladium catalyzed dehydrogenation of α-hydrazinonitriles to hydrazonoyl cyanides using cyclopentene. Synthesis of 1H-pyrazole-4-carboxylate. *Tetrahedron Letters* 37(48): 8771-8774.

Kocienski, P.J., B. Lythgoe and S. Ruston. 1978. Scope and stereochemistry of an olefin synthesis from β-hydroxysulphones. *Journal of Chemical Society, Perkin Transactions 1* 8: 829-834.

Kristensen, T.E. and T. Hansen. 2010. Polymer-supported chiral organocatalysts: Synthetic strategies for the road towards affordable polymeric immobilization. *European Journal of Organic Chemistry* 17: 3179-3204.

Kristensen, T.E., K. Vestli, M.G. Jakobsen, F.K. Hansen and T. Hansen. 2010. A general approach for preparation of polymer-supported chiral organocatalysts via acrylic copolymerization. *Journal of Organic Chemistry* 75(5): 1620-1629.

Kudo, K., K. Akagawa and S. Itsuno (eds.). 2011. Polymeric Chiral Catalyst Design and Chiral Polymer Synthesis. John Wiley: Hoboken, pp. 91.

Leadbeater, N.E. and M. Marco. 2002. Preparation of polymer-supported ligands and metal complexes for use in catalysis. *Chemical Reviews* 102: 3217-3273.

Lee, S.H., E.Y. Lee, D.W. Yoo, S.J. Hong, J.H. Lee, H. Kwak, Y.M. Lee, J. Kim, K.A. Cheal and J.K. Lee. 2007. Novel polymer-supported ruthenium and iron complexes that catalyze the conversion of epoxides into diols or diol mono-ethers: Clean and recyclable catalysts. *New Journal of Chemistry* 31: 1579-1582.

Ley, S.V., I.R. Baxendale, R.N. Bream, P.S. Jackson, A.G. Leach, D.A. Longbottom, M. Nesi, J.S. Scott, R.I. Storer and S.J. Taylor. 2000. Multi-step organic synthesis using solid-supported reagents and scavengers: A new paradigm in chemical library generation. *Journal of Chemical Society, Perkin Transaction 1* 23: 3815-4195.

Ley, S.V. and A. Massi. 2000. Polymer supported reagents in synthesis: Preparation of bicyclo [2.2.2] octane derivatives via Tandem Michael Addition Reactions and Subsequent Combinatorial Decoration. *Journal of Combinational Chemistry* 2: 104.

Ley, S.V., O. Schucht, A.W. Thomas and P.J. Murray. 1999. Synthesis of the alkaloids (±)-oxomaritidine and (±)-epimaritidine using an orchestrated multi-step sequence of polymer supported reagents. *Journal of the Chemical Society, Perkin Transaction 1* 10: 1251-1252.

Ley, S.V., A.W. Thomas and H. Finch. 1999. Polymer supported hypervalent iodine reagents in 'clean' organic synthesis with potential applications in combinatorial chemistry. *Journal of the Chememical Society, Perkin Transaction 1* 669-672.

Li, X., W. Chen, W. Hems, F. King and J. Xiao. 2004a. Asymmetric transfer hydrogenation of ketones with a polymer-supported chiral diamine. *Tetrahedron Letter* 45: 951-953.

Li, X., X. Wu, W. Chen, F.E. Hancock, F. King and J. Xiao. 2004b. Asymmetric transfer hydrogenation in water with a supported noyori-ikariya catalyst. *Organic Letters* 6(19): 3321-3324.

Liu, P.N., J.G. Deng, Y.Q. Tu and S.H. Wang. 2004. Highly efficient and recyclable heterogeneous asymmetric transfer hydrogenation of ketones in water. *Chemical Communications* 2070-2071.

Liu, R., T. Cheng, L. Kong, C. Chen, G. Liu and H. Li. 2013. Highly recoverable organoruthenium-functionalized mesoporous silica boosts aqueous asymmetric transfer hydrogenation reaction. *Journal of Catalysis* 307: 55-61.

Lu, P.H. 2009. Toy, organic polymer supports for synthesis and for reagent and catalyst immobilization. *Chemical Reviews* 109(2): 815-838.

Margaret Shea Chiles, Donald D. Jackson and Perry C. Reeves. 1980. Preparation and synthetic utility of phase-transfer catalysts anchored to polystyrene. *Journal of Organic Chemistry* 45(14): 2915-2918.

Martin-Matute, B., S.I. Pereira, E. Pena-Cabrera, J. Adrio, A.M.S. Silva and J.C. Carretero. 2007. Synthesis of polymer-supported Fesulphos ligands and their application in asymmetric catalysis. *Advanced Synthesis & Catalysis* 349(10): 1714-1724.

McNamara, C.A., M.J. Dixon and M. Bradley 2002. Recoverable catalysts and reagents using recyclable polystyrene-based supports. *Chemical Reviews* 102(10): 3275-3300.

Miguel, Y.R. de. 2000. Supported catalysts and their applications in synthetic organic chemistry. *Journal of Chemical Society, Perkin Transactions 1* 24: 4213-4221.

Nguyen, S.T. and R.H. Grubbs. 1995. The syntheses and activities of polystyrene-supported olefin metathesis catalysts based on $Cl_2(PR_3)_2Ru=CH?CH=CPh_2$. *Journal of Organometallic Chemistry* 497: 195-200.

Nishikubo, T., T. Iizawa, M. Shimojo, T. Kato and A. Shiina. 1990. New catalytic activity of polymer-supported quaternary onium salts. Regioselective addition reaction of oxiranes with active esters catalyzed by insoluble polystyrene-bound quaternary ammonium and phosphonium salts. *The Journal of Organic Chemistry* 55(8): 2536-2542.

Ohtani, N., Y. Inoue, A. Nomoto and S. Ohta. 1994. Polystyrene-supported ammonium fluoride as a catalyst for several base-catalyzed reactions. *Reactive Polymers* 24(1): 73-78.

Otvos, S.B., I.M. Mándity and F. Fülöp. 2012. Strategic application of residence-time control in continuous-flow reactors. *Journal of Catalysis* 295: 179-185.

Pedrosa, R., J.M. Andrés, R. Manzano and C. Pérez-López. 2013. Novel sulfonylpolystyrene-supported prolinamides as catalysts for enantioselective aldol reaction in water. *Tetrahedron Letters* 69(51): 10811-10819.

Pop, L.E., B.P. Deprez and A.L. Tartar. 1997. Versatile acylation of N-nucleophiles using a new polymer-supported 1-hydroxybenzotriazole derivative. *The Journal of Organic Chemistry* 62(8): 2594-2603.

Priego, J., O.G. Mancheno, S. Cabrera, R.G. Arrayas, T. Llamas and J.C. Carretero. 2002. 1-phosphino-2-sulfenylferrocenes: Efficient ligands in enantioselective palladium-catalyzed allylic substitutions and ring opening of 7-oxabenzonorbornadienes. *Chemical Communications* 21: 2512-2513.

Qu, C., W. Zhao, L. Zhang and Y. Cui. 2014. Preparation of immobilized L-prolinamide via enzymatic polymerization of phenolic L-prolinamide and evaluation of its catalytic performance for direct asymmetric Aldol reaction. *Chirality* 26: 209-213.

Randl, S., N. Buschmann, S.J. Connon and S. Blechert. 2001. Highly efficient and recyclable polymer-bound catalyst for olefin metathesis reactions. *Synlett* 1547-1550.

Regen, S.L. 1975. Triphase catalysis. *Journal of the American Chemical Society* 97(20): 5956-5957.

Regen, S.L. 1976. Triphase catalysis. Kinetics of cyanide displacement on 1-bromooctane. *Journal of the American Chemical Society* 98(20): 6270-6274.

Regen, S.L.J. 1977. Triphase catalysis. Applications to organic synthesis. *The Journal of Organic Chemistry* 42(5): 875-879.

Regen, S.L. 1979. Triphase catalysis (New synthetic methods). *Angewandte Chemie International Edition* 18(6): 421-429.

Rulli, G., K.A. Fredriksen, N. Duangdee, T. Bonge-Hansen, A. Berkessel and H. Gröger. 2013. Asymmetric organocatalytic Aldol reaction in water: Mechanistic insights and development of a semi-continuously-operating process. *Synthesis* 45: 2512-2519.

Ryan G. Manecke and John P. Mulhall. 1999. Medical treatment of erectile dysfunction. *Annals of Medicine*, 31(6): 388-398.

Schurer, S.C., S. Gessler, N. Buschmann and S. Blechert. 2000. Synthesis and application of a permanently immobilized olefin-metathesis catalyst. *Angewandte Chemie International Edition* 39: 3898-3901.

Shinichi Itsuno. March 2008. Polymer-supported metal lewis acids. Chapter 19. *In:* Yamamoto H. and Ishihara, K. (eds), Acid Catalysis in Modern Organic Synthesis, vol 2. pp. 1136. Wiley-VCH, Weinheim.

Stefane, B. and F. Požgan. 2014. Advances in catalyst systems for the asymmetric hydrogenation and transfer hydrogenation of ketones. *Catalysis Reviews: Science and Engineering* 56: 82-174.

Stephane Varray, René Lazaro, Jean Martinez and Frédéric Lamaty. 2003. New soluble-polymer bound ruthenium carbene catalysts: Synthesis, characterization, and application to ring-closing metathesis. *Organometallics* 22(12): 2426-2435.

Sugie, H., Y. Hashimoto, N. Haraguchi and S. Itsuno. 2014. Synthesis of polymer-immobilized TsDPEN ligand and its application in asymmetric transfer hydrogenation of cyclic sulfonimine. *Journal of Organometallic Chemistry* 751: 711-716.

Sun, Q., X. Meng, X. Liu, Y. Zhang, Q. Yang, F. Yang and X. Xiao. 2012. Mesoporous cross-linked polymer copolymerized with chiral BINAP ligand coordinated to a ruthenium species as an efficient heterogeneous catalyst for asymmetric hydrogenation. *Chemical Communications* 48: 10505-10507.

Suzuki, N., T. Inoue, T. Asada, R. Akebi, G. Kobayashi, M. Rikukawa, Y. Masuyama, M. Ogasawara, T. Takahashi and S.H. Thang. 2013. Asymmetric aldol reaction on water using an organocatalyst tethered on a thermoresponsive block copolymer. *Chemistry Letters* 42(12): 1493-1495.

Tamami, B., N. Iranpoor and M.A. Karimizarchi. 1993. Polymer-supported ceric(IV) catalyst: Catalytic ring opening of epoxides. *Polymer* 34(9): 2011-2013.

Tamami, B. and H. Mahdavi. 2001. Synthesis of azidohydrins from epoxides using quaternized amino functionalized cross-linked polyacrylamide as a new polymeric phase-transfer catalyst. *Tetrahedron Letters* 42(49): 8721-8724.

Tamami, B. and H. Mahdavi. 2002. Synthesis of halohydrins from epoxides using quaternized amino functionalized cross-linked polyacrylamide as a new solid-liquid phase transfer catalyst. *Reactive and Functional Polymers* 51: 7.

Tamami, B. and H. Mahdavi. 2002. Synthesis of thiocyanohydrins from epoxides using quaternized amino functionalized cross-linked polyacrylamide as a new solid-liquid phase-transfer catalyst. *Tetrahedron Letters* 43: 6225-6228.

Ter Halle, R., E. Schultz and M. Lemaire. 1997. Heterogeneous enantioselective catalytic reduction of ketones. *Synlett* 1257-1258.

Thvedt, T.H., Kristensen, T.E. Sundby, E., Hansen, T. and Hoff, B.H. 2011. Enantioselectivity, swelling and stability of 4-hydroxyprolinol containing acrylic polymer beads in the asymmetric reduction of ketones. *Tetrahedron: Asymmetry* 22(24): 2172-2178.

Tronchet, J.M.J., J.F. Tronchet and F. Barbalat-Rey. 1993. Synthesis of pyrazole c-glycosides by 1,3-dipolar cycloaddition of nitrilimines formed by lead tetraacetate oxidation of p-nitrophenylhydrazones of aldehydo sugars. *Heterocycles* 36(4): 833-844.

Wang, L., C.Y. Huang and C. Cai. 2010. Polymer-supported copper complex for the direct synthesis of oaryloxime ethers via cross-coupling of oximes and arylboronic acids. *Catalysis Communications* 11: 532-536.

Wang, X., J. Zhang, Y. Liu and Y. Cui. 2014. Chiral porous TADDOL-embedded organic polymers for asymmetric diethylzinc addition to aldehydes. *Bulletin of the Chemical Society of Japan* 87(3): 435-440.

Wennemers, H. 2011. Asymmetric catalysis with peptides. *Chemical Communications* 47(44): 12036-12041.

Xiao, W., R. Jin, T. Cheng, D. Xia, H. Yao, F. Gao, B. Deng and G. Liu. 2012. A bifunctionalized organic-inorganic hybrid silica: Synergistic effect enhances enantioselectivity. *Chemical Communications* 48: 11898-11900.

Xu, X., R. Wang, J. Wan, X. Ma and J. Peng. 2013. Phosphonate-containing polystyrene copolymer-supported Ru catalyst for asymmetric transfer hydrogenation in water. *RSC Advanced* 3: 6747-6751.

Yang, Y.-C. and D.E. Bergbreiter. 2013. Soluble polymer-supported organocatalysts. *Pure and Applied Chemistry* 85(3): 493-509.

Yao, Q.A. 2000. Soluble polymer-bound ruthenium carbene complex: A robust and reusable catalyst for ring-closing olefin metathesis. *Angewandte Chemie International Edition* 39: 3896-3898.

Zhang, D., C. Ren, W. Yang and J. Deng. 2012. Helical polymer as mimetic enzyme catalyzing asymmetric aldol reaction. *Macromolecular Rapid Communications* 33: 652-657.

Zhang, L., S. Luo and J.-P. Cheng. 2011. Non-covalent immobilization of asymmetric organocatalysts. *Catalysis Science and Technology* 1(4): 507-516.

Alkaloid-supported Catalysts

Poonam Kumari, Rajesh Kumar Meena and Pragati Fageria*
Department of Chemistry, University of Rajasthan, Jaipur, Rajasthan, India

1. Introduction

Plants are the most important source of traditional medicine system that has alleviated human ailment and endorsed health for thousands of years (Ishtiyak and Hussain 2017, Sadia et al. 2018). Plants are a reservoir of an array of active constituents that have considerable medicinal applications like antiviral, anticancer, analgesic, antitubercular etc. (Rupani and Chavez 2018, Uniyal et al. 2006). Among all of them, the alkaloids are significant secondary metabolites that were primarily discovered and applied as early as 4000 years ago and are well accepted for their rich medicinal potential as they have potential to harm, heal and relieve. The pain relieving alkaloids are morphine and cocaine, and alkaloids that can strain health care systems are nicotines (Amirkia and Heinrich 2014). A plethora of alkaloids which illustrate medicinal applications are paclitaxel, vincristine, the cephalosporins, the penicillins, atropine, pilocarpine, quinine, vincamine, etc. Alkaloids are known as biologically significant metabolites and necessary for life as they are highly effective, and restricted in their mode of action and application. Now it is recognized that alkaloids are ubiquitous in the biome, including plants and a variety of microorganisms, also isolated from amphibians (Dey et al. 2020).

1.1 Background

The first alkaloid isolated from opium in crude form, morphine, was reported by Serturner in 1805; it remains an important medicinal agent (Devereaux et al. 2018). A few years thereafter, the essential antimalarial agent quinine was obtained (Kaufman and Ruveda 2005). Alkaloids originated from the word

---

*Corresponding author: pragati.fageria@gmail.com

alkali. The term "alkaloids" was coined in 1819 by German pharmacist Wilhelm Meissner (Croteau et al. 2000). According to him, "alkaloids (which mean alkali-like, alk-alkali, oid-like) were defined as basic nitrogen compounds isolated from plants." In 1880, Konigs introduced that alkaloids should be defined as naturally occurring organic bases which contain a pyridine ring as shown in Figure 1 (Chatwal 2019).

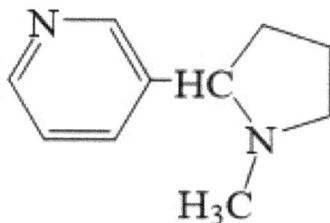

Figure 1. Alkaloid containing pyridine ring

A broad definition of alkaloid is a class of naturally occurring compounds with a basic character: containing one nitrogen atom in a heterocyclic ring structure, which is frequently found in plant kingdom (Dey et al. 2020). When alkaloids react with acid, nitrogen atoms behave as a base and form the salts like the inorganic alkalis in an acid-base reaction. Alkaloids in pure form are usually colourless, odourless crystalline solids, but sometimes they can be yellowish liquids. Quite often, they also have a bitter taste. 3000 alkaloids are known in more than 4000 plant species (Debnath et al. 2018).

1.2   Classification of alkaloids

The conventional definition of alkaloids gives emphasis to their bitter taste, basic nature, origin from plant, and their physiological actions (Eguchi et al. 2019). The current definition states that the basic nitrogenous plant products are mostly optically active and possess nitrogen heterocycle as their structural units with a pronounced physiological action. Since the amino acid skeleton is mainly retained in the alkaloid structure, alkaloids originating from the same amino acid show similar structural features (Kaur and Arora 2015). The presence of a basic nitrogen atom at any position in the molecule, which does not contain nitrogen in an amide or peptide bond, is the most important constraint for categorization as an alkaloid. In this connection, alkaloids can be classified according to their chemical structure; for example, alkaloids that contain an indole ring system are known as indole alkaloids. The major classes of alkaloids are the tropanes, pyrrolidines, purines, indoles, quinolines, isoquinolines, imidazole, steroids, and terpenoids (Figure 2) (Roy 2017). Based on their biological system, alkaloids are also classified based on the biological system in which they are found, like opium alkaloids occur in the opium poppy (*Papaver somniferum*) (Devereaux et al. 2018, Debnath et al. 2018).

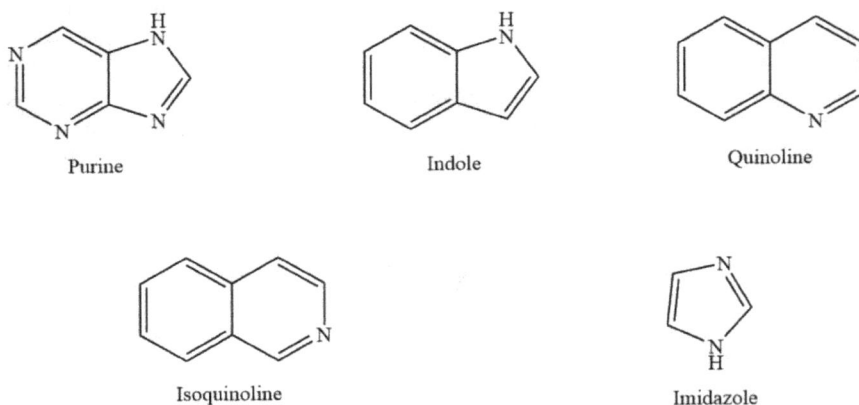

Purine   Indole   Quinoline

Isoquinoline   Imidazole

Figure 2. The major class of alkaloids

Often alkaloids are classified based on their biosynthetic origins in the three central classes: (1) true, (2) protoalkaloids and (3) pseudoalkaloids (Eguchi et al. 2019). The "true alkaloids" contain the nitrogen content in the heterocycle ring and originate from the amino acids like nicotine, cocaine, quinine, morphine, and atropine (Figure 3). True alkaloids are biologically active and extremely reactive in nature (Dewick 2002, Funayama and Cordell 2015). They form water-soluble salts, and are generally crystalline in nature, which conjugate with acid and form

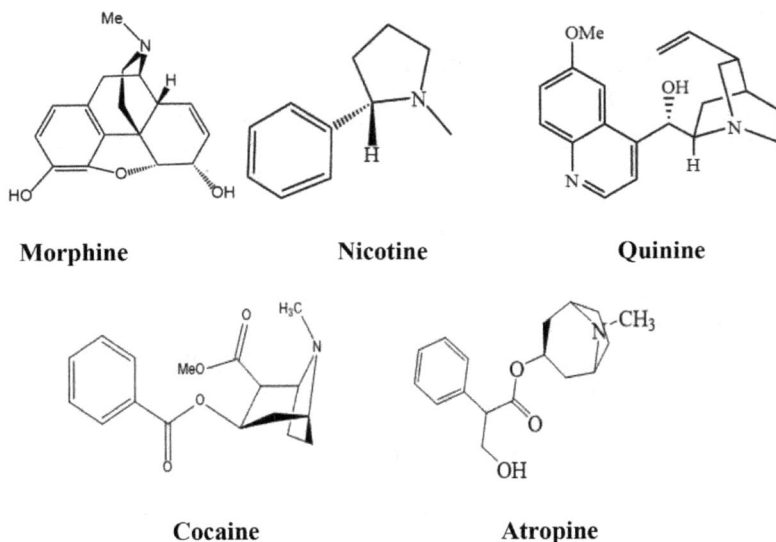

Morphine   **Nicotine**   **Quinine**

Cocaine   **Atropine**

Figure 3. Few examples of true alkaloids

a salt. Majority of true alkaloids are bitter in taste and solid, excluding nicotine, which is a brown liquid. This group also contains a few alkaloids that, besides nitrogen heterocycle, contain peptide fragments like ergotamine or terpene (e.g. evonine). The piperidine alkaloids coniceine and coniine can be considered as the true alkaloids, even though they are not originated from the amino acids. These alkaloids are found in plants: (a) in free-state, (b) as N-oxide, or (c) as salts. Various amino acids like L-phenylalanine/L-tyrosine, L-ornithine, L-histidine, L-lysine are the key sources of true alkaloids (Chatwal 2019, Eguchi et al. 2019, Dewick 2002, Funayama and Cordell 2015).

The protoalkaloids are other nitrogen-containing alkaloids, derived from amino acids, peptide and cyclopeptide alkaloids or polyamine alkaloids but are not part of the heterocyclic ring system. For example, L-Tryptophan and L-tyrosine are the main precursors of this type of alkaloids. This class of alkaloids is applied in a variety of health disorders, including mental illness, pain, and neuralgia. (Funayama and Cordell 2014, 2015).

The last group of alkaloid-like substances is compounds that include steroid, purine and terpene-like structures as well as capsaicin, ephedrine and caffeine. (Cordell 1981, Aniszewski 2015). Another class of alkaloids which is known as Pseudoalkaloids, not directly originated from amino acids; instead, they are connected with amino acid pathways where they are derived from by amination or transamination reaction, from forerunners or post-cursors of amino acid (Cappillino and Sattely 2014).

1.3 Pharmacological applications of alkaloids

Alkaloids were foundational in the development of organic chemistry and were among the first pharmaceuticals developed. They remain in a privileged position in present medicine, used extensively to treat pain, cancer, dementia and innumerable other ailments (Figure 4) (Aniszewski 2015).

Plant alkaloids are one of the principal groups of natural products. Despite the recent advances in the field of synthetic drug discovery in recent decades, plant alkaloids have stayed a feasible source of bioactive compounds with significant medicinal potential (Bribi 2018). More than 12,000 alkaloids are isolated from a variety of plant species. Many of these compounds possess rich therapeutic aspects. For example, the substantiated plant alkaloids include the narcotic analgesics, morphine and codeine, apomorphine (morphine derivative) used in Parkinson's disease, the muscle relaxant papaverine, and have various antimicrobial activities (Kartsev 2004). The bark of homonym trees cultivated in equatorial climatic zones has more than 30 alkaloids, generally known as Cinchona alkaloids. Approximately 50% of alkaloids are represented by quinine, quinidine, cinchonidine and cinchonine (Hiemstra and Wynberg 1981). Quinine is the most well-known alkaloid and was used as the anti-malarial drug of choice for over 400 years until chloroquine was discovered, while quinidine is employed as an anti-arrhythmic agent. In chemistry, quinine, quinidine, cinchonidine and cinchonine

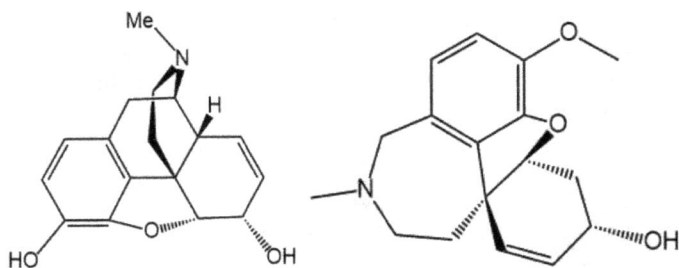

Morphine: Analgesic   Galanthamine (dementia treatment)

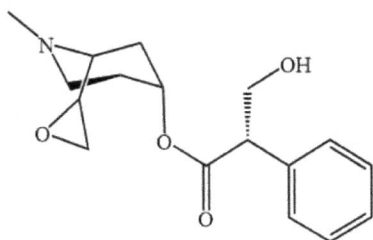

Scopolamine (anti-nausea)

Figure 4. Examples of pharmaceutical alkaloids with their therapeutic uses

are used as economical chiral source. Many naturally occurring and synthesized alkaloids illustrated the enormous applications in the field of medicine, organic reactions and are well recognized as supported catalysts. Here in this chapter, we will explore the catalysts supported by alkaloids in organic reactions.

2.   Alkaloids in organic reactions

Pasteur in 1853 firmly recognized the role of cinchona alkaloids in organic chemistry which ushered in an era of racemate resolutions by the crystallization of diastereomeric salts (Flack 2009). Organic catalysts supported by alkaloids are utilized as catalysts in a variety of organic synthesis reaction. Due to their insolubility, alkaloids are frequently used as a catalyst (Lichman 2021). Cinchona alkaloids and their derivatives, once known for the popular antimalarial drug, emerged as versatile organocatalysts in reactions as they activate the nucleophile by enamine and carbanion formation, and electrophile is activated via hydrogen bond. This can be also attributed to the plenty of Cinchona alkaloids in nature, their commercial availability at reasonable prices, nontoxicity, stability and effortless handling in laboratory, and their convenient modification by simple reactions (Kartsev 2004, Bribi 2018).

2.1 Cinchona alkaloids and their derivatives: Versatile catalysts in asymmetric synthesis

Remarkable progress in the area of asymmetric organocatalysis has been achieved in the last decades. Cinchona alkaloids are influential organocatalysts in the realm of asymmetric organocatalysis during the last 2 decades (Pellisier 2007). Owing to high enantioselectivity and the presence of tunable functional groups, cinchona alkaloids are able to catalyze a broad range of reactions (Tanriver et al. 2016). The compounds of Cinchona are quinine (QN), quinidine (QD), cinchonidine (CD) and cinchonine (CN), shown in Figure 5 (Kacprzak and Gawronski 2001).

As we can notice from the above Figure 5, the Cinchona skeleton consists of two rigid rings: an aliphatic quinuclidine and an aromatic quinoline ring joined together by two carbon-carbon single bonds. These compounds are diastereomers and having five stereogenic centers and a quinuclidine ring, bearing a tertiary and chiral nitrogen atom. The chiral quinuclidinyl nitrogen is the most central feature of these compounds as it is responsible for the direct transfer of chirality during catalytic reactions (Marcelli and Hiemstra 2010). Furthermore, the quinuclidinyl nitrogen atom is 103 times more basic in comparison to quinoline nitrogen and is thus responsible for the basic character of the catalyst. Moreover, OH group present at C-9 position acts as Bronsted acid and several studies have shown that Cinchona compounds work as bifunctional catalysts as shown in Figure 6.

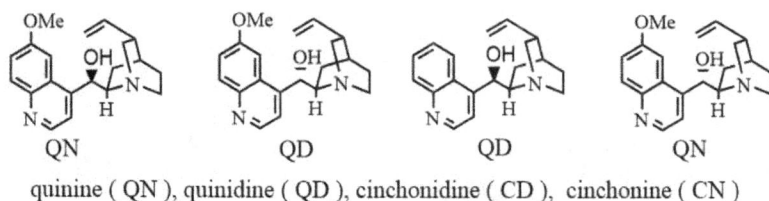

quinine (QN), quinidine (QD), cinchonidine (CD), cinchonine (CN)

Figure 5. Compounds of Cinchona alkaloids

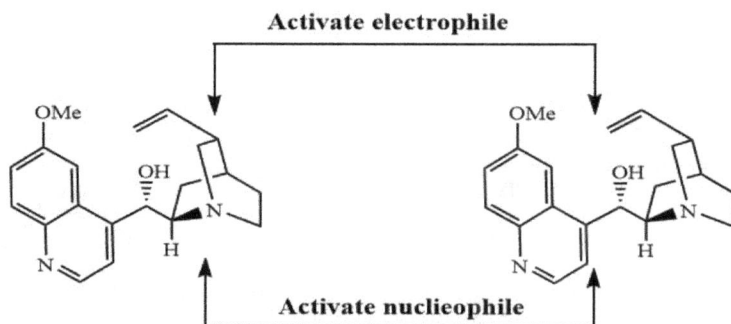

Figure 6. Bifunctional organocatalysts (both nucleophile and electrophile activation by basic quinine and quinidine)

The acid and base coexist in these molecules, and thus, it is possible to activate both the nucleophile and the electrophile simultaneously to use as bifunctional organocatalysts (Yeboah et al. 2011, Kacprzak and Gawronski 2001, Marcelli and Hiemstra 2010).

The Cinchona alkaloids occur in pairs, which differ in configurations at C8, C9, and N1 positions. For example, Quinine vs quinidine and cinchonidine vs cinchonine have opposite absolute configuration, which means that very often these pairs of diastereomers act as enantiomers at C-9 position (Stegbauer 2012, Duan 2014). The eight major Cinchona alkaloids (Figure 7) are diastereomers, usually referred to as pseudoenantiomers, because they propose enantiomeric products when used as catalysts. The H-bonding groups, such as the hydroxyl group, urea, and thiourea at the C-9 position, activate the electrophile by hydrogen bonding. The relative orientation of the two rings, quinoline and quinuclidine, is responsible to create a "chiral pocket" around the reactive site, which further forces a specific approach of the substrates and also formation of enantioselective product. When employed in diverse types of enantioselective syntheses by asymmetric catalysis, the yield, diastereoselectivity and enantioselectivity of products are determined by structural features of Cinchona alkaloids and their derivatives (Kacprzak and Gawro 2001, Salvadori et al. 1999).

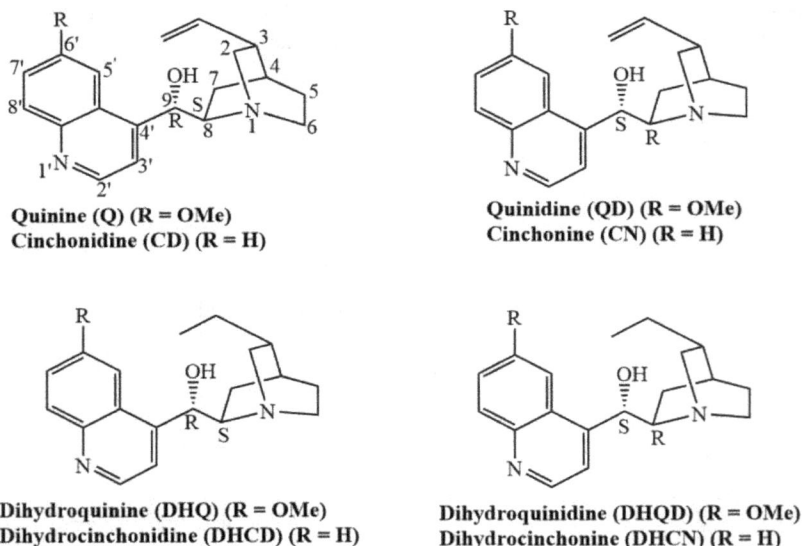

Quinine (Q) (R = OMe)
Cinchonidine (CD) (R = H)

Quinidine (QD) (R = OMe)
Cinchonine (CN) (R = H)

Dihydroquinine (DHQ) (R = OMe)
Dihydrocinchonidine (DHCD) (R = H)

Dihydroquinidine (DHQD) (R = OMe)
Dihydrocinchonine (DHCN) (R = H)

Figure 7. Structures of eight major Cinchona alkaloids

2.1.1 Alkaloid-supported catalysts on cinchonidine and cinchonine

The platinum catalyst modified by cinchonine (CN) or cinchonidine (CD) is the most widely studied system among the heterogeneous enantioselective catalysts, and has been applied to industrial processes on moderately large scales (Orito et al. 1979 a & b, Borszeky et al. 1997). Following the hydrogenation of ethyl benzoylformate, the product was obtained in quantitative yield with optical purities up to 80% (see Scheme 1). One notable property of the Orito reaction is that two chiral modifiers, CD and CN, are diastereomers of each other (Orito et al. 1979a & b, Borszeky et al. 1997).

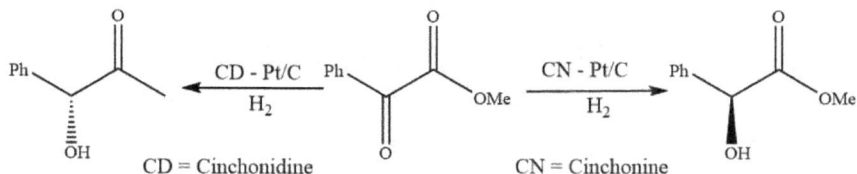

Scheme 1. The Orito reaction; enantioselective hydrogenation of ketone over CD- or CN-modified Pd

2.1.2 Aliphatic α,β-unsubstituted carboxylic Acids

Huck et al. (1996) began to publish results for the enantioselective hydrogenation of aliphatic α,β-unsubstituted carboxylic acids shortly after the initial report by Hall et al. (Hall et al. 1996, Huck et al. 2000). The product ee-value was moderately high at 52% by using the CD-modified Pd/Al_2O_3 in a nonpolar solvent such as hexane under high-pressure hydrogen. One notable property of the aliphatic substrates, represented by (E)-2-methyl-2-butenoic acid (tiglic acid), was formation of the S-product (see Scheme 2).

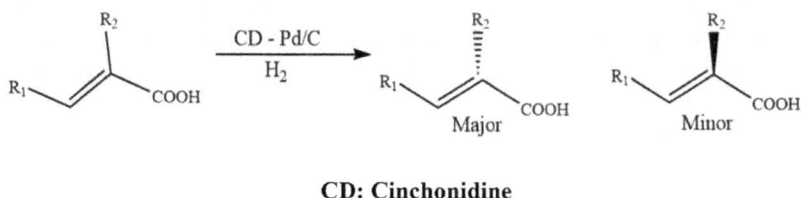

CD: Cinchonidine

Scheme 2. Stereo direction during the hydrogenation of α,β-dialkylpropenoic

2.1.3 Hydrogenation of pyrone derivatives

The hydrogenation of 2-pyrone derivatives is a special case of the chiral modified palladium catalysis, for multiple reasons. The enantioselective hydrogenation of 4-hydroxy-6-methyl-2-pyrone over the CD-modified palladium catalyst was first reported by Huck and colleagues (Scheme 3) (Hall et al. 1996, Huck et al. 2000). The hydrogenation was performed with Pd/TiO_2 in a protic solvent under atmospheric hydrogen pressure (the same as for the PCA family). The highest ee-value in the initial report was up to 85%, but this was obtained at only a 2% conversion. Such a low conversion was not due to suppression of the overreaction; rather, the main reason was the very slow hydrogenation of the substrate. Hydrogenation with the CD-modified Pd/TiO$_2$ catalyst was 25-fold slower than the corresponding reaction with the unmodified Pd/TiO$_2$.

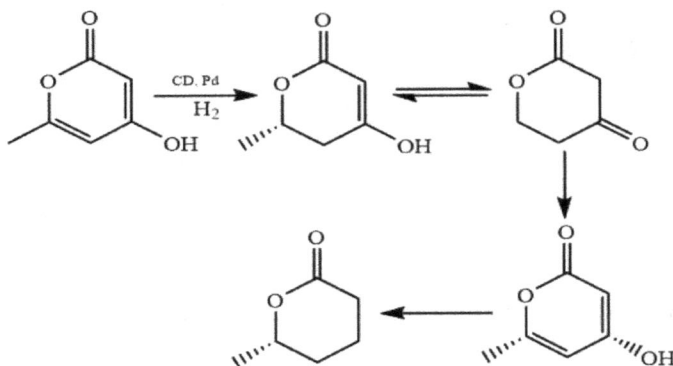

Scheme 3. Enantioselective hydrogenation of 4-hydroxy-6-methyl-2-pyrone over the CD-modified Pd/TiO2 and succeeding diastereoselective hydrogenation

Georg Breding studied the decarboxylation reaction in the presence of chiral alkaloids, such as nicotine or quinidine. The first asymmetric C–C bond forming reaction is accredited also to Georg Breding (Breding 1912). This milestone achievement was related to Rosenthaler's work, who was able to prepare mandelonitrile by the addition of HCN to benzaldehyde in the presence of an isolated enzyme, emulsin (Rosentahler 1908). Breding was also able to perform this reaction in the presence of alkaloids as catalysts, such as the pseudo enantiomeric quinine and quinidine (Breding and Fiske 1912) (See Scheme 4). The determination of enantioselectivity was hampered by a lack of methods to

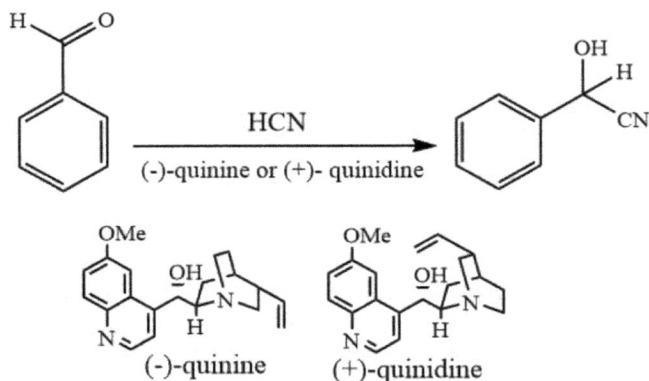

Scheme 4. Preparation of Mandelonitrile by the addition of HCN to benzaldehyde using quinine and quinidine

achieve not only efficient purification but also consistent analyses. Therefore, the presence of a chiral impurity, which often occurred from the catalyst, can spoil the determination of the correct, optical rotation-based ee-values (Prelog 1954).

Breding and Fiske reinvestigated the synthesis of asymmetric cyanohydrin in the way of the more proficient reaction (Breding and Fiske 1912). Pracejus demonstrated that methyl phenyl ketene could be converted to (-)-α-phenyl methyl propionate in 74% ee by using O-acetyl quinine as catalyst (Pracejus 1960) (Scheme 5).

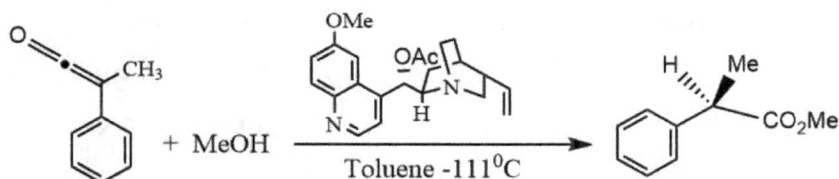

Scheme 5. Pracejus' enantioselective ester synthesis from phenyl methyl ketene

3. Platinum cinchona alkaloid catalyst for enantioselective α-ketoester hydrogenation

However, the many variables associated with the physical and chemical properties of supported catalysts make mechanistic elucidation difficult. Catalyst modification is a strategy that has been widely applied in heterogeneous hydrogenations (Erathodiyil et al. 2011). However, this strategy has been successful only in a limited number of reactions due to the high substrate specificity of such catalysts, so that only a particular combination of a metal, a modifier and a substrate type would give rise to good enantioselectivity (Blaser et al. 1997). Metal nanostructures are of particular interest in this case because of their high activity under mild conditions which is associated with their large surface area, and because of their selectivity for catalytic transformations. Small variations in the metal, the modifier and the substrate type can lead to significant changes in enantiodiscrimination.

Enantioselective hydrogenation is one of the most important industrial asymmetric processes to produce chiral molecules with excellent selectivity (Scheme 6) (Nandanan et al. 2011).

Platinum nanoparticle catalysts supported on silica, alumina, titania and dendritic architectures are mainly used in the hydrogenation of activated α-ketoesters (Blaser et al. 1997). Orito and coworkers first reported the enantioselettive hydrogenation of α-ketoesters catalysed by a cinchona alkaloid modified supported platinum catalyst in 1978 Scheme 7 (Orito et al. 1979).

Since then, the extensive research by Wells (Webb et al. 1992), Blaser (Blaser et al. 1991), Bradley (Bradley et al. 1987) and their co-workers has established the optimal reaction conditions and suggested a reaction mechanism. Platinum

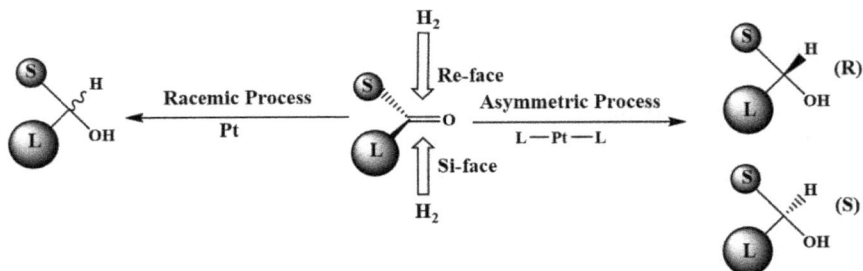

Scheme 6. General reaction of hydrogenation of α-ketoesters

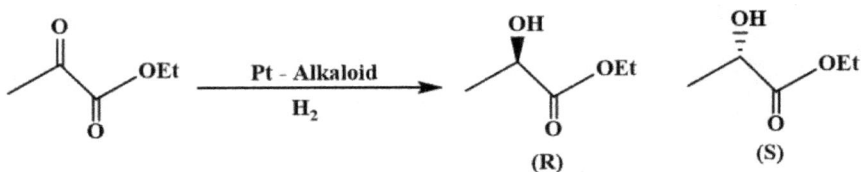

Scheme 7. Hydrogenation of ethyl pyruvate using Orito's catalysts

catalysts modified with cinchona alkaloids for the hydrogenation of activated ketones have demonstrated ligand acceleration as a heterogeneous catalyst system (Erathodiyil et al. 2011). Modification of the platinum surface by naturally occurring cinchona alkaloids not only affects the product enantioselectivity, cinchonidine gives the (R)-product, whilst cinchonine, its pseudo enantiomer, furnishes the (S)-product (Schemes 8), but it also enhances the reaction rate by up to one hundred times over the racemic reaction (Collier et al. 1995, Arx et al.

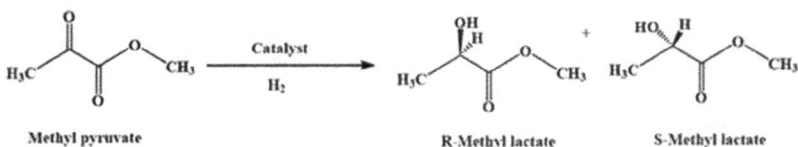

Methyl pyruvate   R-Methyl lactate   S-Methyl lactate

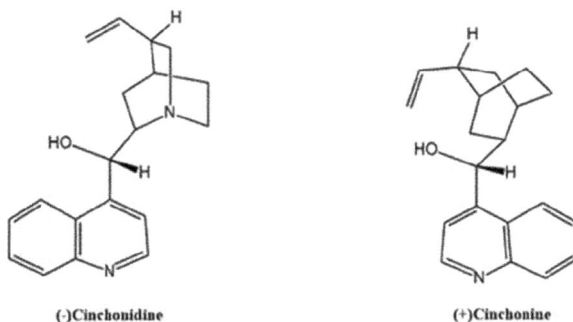

(-)Cinchonidine   (+)Cinchonine

Schemes 8. Here catalyst is conventional unmodified supported platinum catalyst

2002, Vayner et al. 2004). The reaction is well behaved giving quite respectable optical yields, as high as 94% under optimised conditions (Erathodiyil et al. 2011, Toukoniitty et al. 2004).

In the initial studies, Astruc and co-workers (Astruc et al. 2005) focused on the racemic hydrogenation of ethyl pyruvate. Quantitative conversion to ethyl lactate was achieved under a hydrogen pressure of 100 psi in 6 h with 1 mol% of nanowire catalyst (Scheme 9). Hydrogenation of other activated ketones also proceeded well, producing the corresponding alcohols in quantitative yields. The proposed catalytic cycle consisted of a fast adsorption of ketone and hydrogen on the Pt surface, stepwise addition of the two adsorbed hydrogen atoms to the C–O bond with the half hydrogenated intermediate, followed by the fast desorption of the alcohol.

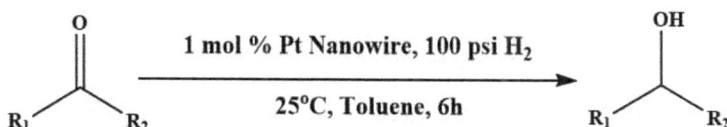

Scheme 9. Hydrogenation of ketoesters

3.1 Hydrogenation of 2,4-diketo acid derivatives

Martin et al. (2000) reported the application of a cinchona-modified Pt catalysts use in a synthesis of the enantio- and chemoselective hydrogenation of several 2,4-diketo acid derivatives to the corresponding 2-hydroxy compounds. In presence of catalyst the reaction can be carried out with chemoselectivities of >99% and enantioselectivities up to 86% (R) and 68% (S), respectively (Martin et al. 2000) (see Scheme 10).

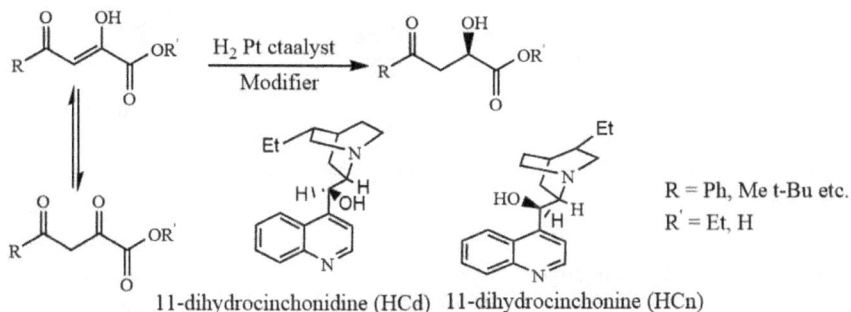

11-dihydrocinchonidine (HCd) 11-dihydrocinchonine (HCn)

Scheme 10. Hydrogenation of 2,4-diketo acid derivatives, structure of substrates and modifiers

3.2 Asymmetric ring-opening of prochiral acid anhydrides

Jun et al. (1987) reported that the asymmetric ring-opening of prochiral acid anhydrides with methanol has been achieved by a catalytic quantity of cinchona alkaloids. The product, the optically active half-ester and reaction rate of the ring-opening selectivity dependent on the nature of the reaction medium, polarity of solvent, and substrate concentration (Scheme 11). They obtained enantiomeric excess of the reaction up to 70% and the reaction proceeds via general-base catalysis by the quinuclidine moiety of the base (Jun et al. 1987).

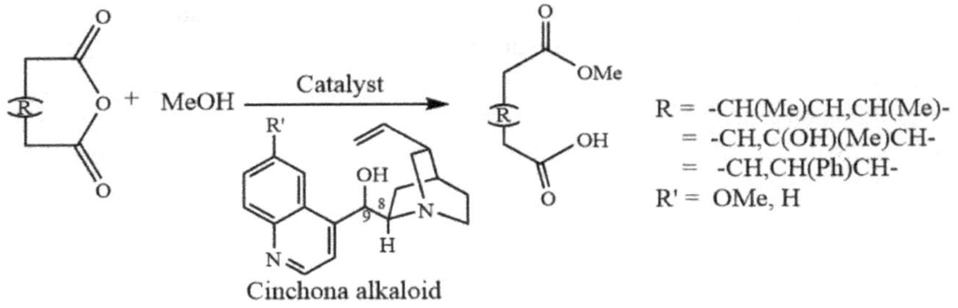

R = -CH(Me)CH,CH(Me)-
= -CH,C(OH)(Me)CH-
= -CH,CH(Ph)CH-
R' = OMe, H

Cinchona alkaloid

Scheme 11. Cyclic acid anhydrides afforded ring-opened half-ester

Origin of enantioselectivity of the cinchona alkaloid modified Pt nanoparticles has been extensively studied by Ma and co-workers. The surface chemistry and interactions between the ligands, Pt nanowires and substrates were critical in these reactions and have to be examined in the future (Ma et al. 2005, 2006, 2007, Konigsmann et al. 2007). Using a similar approach for the hydrogenation of ethylbenzoyl formate, the corresponding ethyl mandelates were obtained in quantitative conversions and enantioselectivities of 54% to 57% (see Scheme 12).

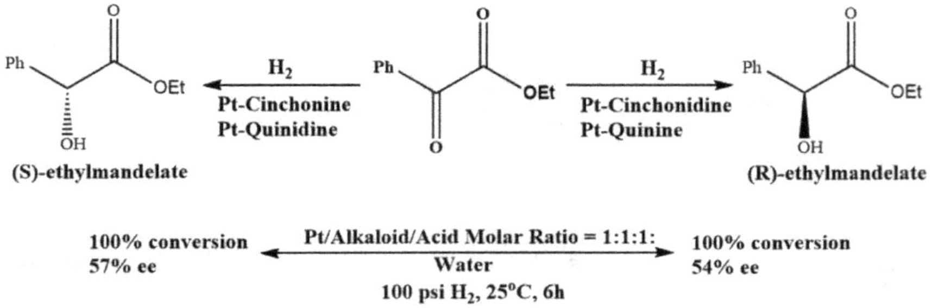

(S)-ethylmandelate   (R)-ethylmandelate

100% conversion   Pt/Alkaloid/Acid Molar Ratio = 1:1:1:   100% conversion
57% ee   Water   54% ee
                              100 psi H$_2$, 25°C, 6h

Scheme 12. Asymmetric hydrogenation of ethylbenzoyl formate over alkaloid-modified Pt nanowires

4. Conjugate addition reactions via cinchona alkaloid catalysts

Recently, cinchona alkaloids have been identified as effective catalysts for the enantioselective conjugate addition of dimethyl malonate, ethyl acetoacetate to nitroalkenes, and Michael addition between nitromethane, chalcones, thiols and various α,β-unsaturated ketones with high yields and enantiomeric excess (Li et al. 2004, Ding et al. 2011, Vakulya et al. 2005).

McDaid and colleagues reported the catalytic asymmetric 1,4-addition of thiols to cyclic enones with modified cinchona alkaloid (C-1), which has been illustrated in Scheme 13 (Rana et al. 2010, McDaid et al. 2002). The Michael products can be isolated with high yield and enantioselectivity for a variety of derivatives with selectivities ranging from 93 to >99% ee.

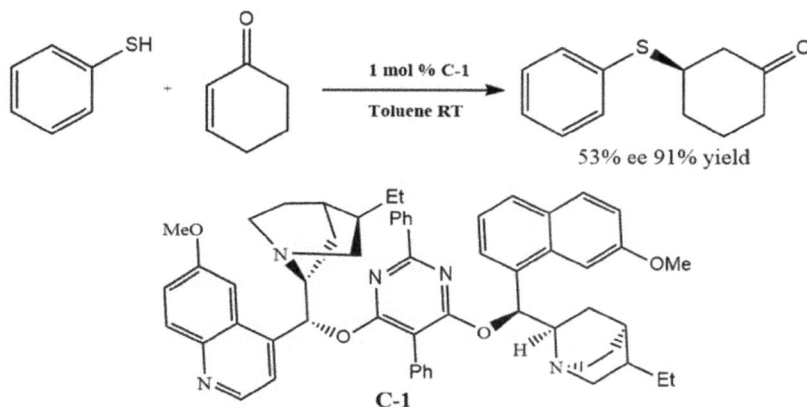

53% ee 91% yield

Scheme 13. Enantioselective Michael addition of thiophenols to enones

Later, highly enantio- and diastereoselective tandem Michael-aldol reactions, proficiently catalyzed by a 1 mol% cinchona alkaloid thiourea (C-2), were reported by Zu and others (Zu et al. 2007). This new one-pot method proceeds via synergistic noncovalent hydrogen-bonding activation of both the Michael donor and acceptor. The methodology succeeded in the construction of versatile chiral thiochromanes with the formation of three stereogenic centers from simple achiral compounds as shown in Scheme 14.

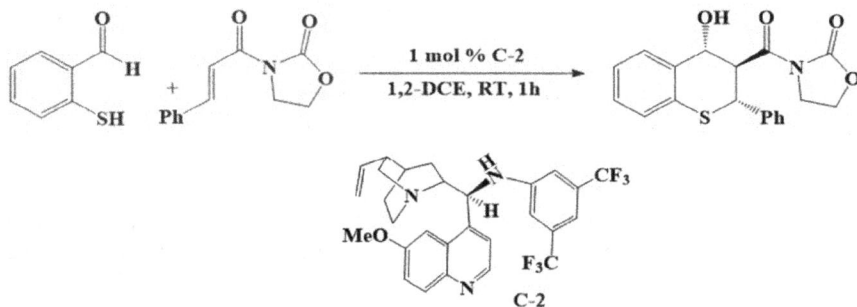

Scheme 14. Reaction of 2-Mercaptobenzaldehyde with α,β-unsaturated oxazolidinone

These results prompted us to examine conjugate additions involving trisubstituted carbon nucleophiles, especially those that could be conveniently generated *in situ* from readily available racemic carbonyl compounds. Similarly, the conjugate addition has been reported with catalyst C-3 for a direct, stereo controlled production of adjacent carbon- or heteroatom-substituted quaternary and tertiary stereocenters from readily available starting β-ketoester. Catalyst C-3 provides excellent enantioselectivities (94 and 95% ee) for both diastereomers of the 1,4-adducts (see Scheme 15) (Li et al. 2005).

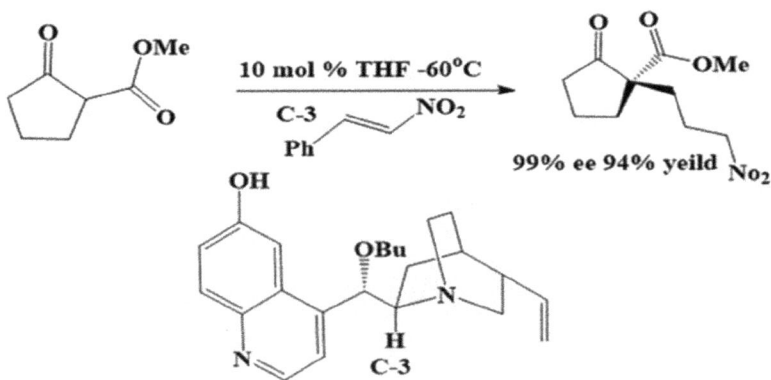

Scheme 15. Asymmetric Michael addition reaction

Chiral oxacyclic structures such as tetrahydrofuran rings are commonly found in many bioactive compounds. Cinchona-alkaloid-thiourea C-4 catalyzes the cycloetherification of ε-hydroxy-α,β-unsaturated ketones with exceptional enantioselectivity, even with low catalyst loadings at room temperature. The plausible path of the reaction suggest that activation intermediate might go through the transition state (TS-1) illustrated in Scheme 16 (Connon 2008). Connon and co-workers exhibited a novel asymmetric synthesis method for THF via the catalytic cycloetherification of ε-hydroxy-α,β-unsaturated ketones supported by cinchona-alkaloid-thiourea-based bifunctional organocatalysts. This catalytic process represents a highly practical cycloetherification method that provides excellent enantioselectivities (Asano and Seijiro 2011).

The catalyst C-4 can also catalyze the domino aza-Michael-Michael reactions of anilines with nitroolefin enoates to afford chiral 4-aminobenzopyrans bearing two consecutive stereogenic centers and one quaternary stereocenter (Scheme 17). The products can be isolated with high yield and enantioselectivity (Wang et al. 2011). They optimized the reaction with the best results and found that in the presence of 2-propanol as solvent and at room temperature in the presence of 10 mol% of catalyst, C-4 gives 96% yield with 96% ee.

Motoba et al. synthesized trifluoromethyl-substituted 2-isoxazolines by using a domino Michael-cyclization-dehydration reaction of hydroxylamine (NH$_2$OH) with a range of (E)-trifluoromethylated enone derivatives in the presence of N-3,5-bis(trifluoromethyl benzyl) quinidinium bromide catalyst C-5 as a chiral phase transfer catalyst (Scheme 18) (Motoba et al. 2010).

Molleti et al. reported an efficient enantioselective conjugate addition of malononitrile to a range of β-substituted 2-enoylpyridines catalyzed by

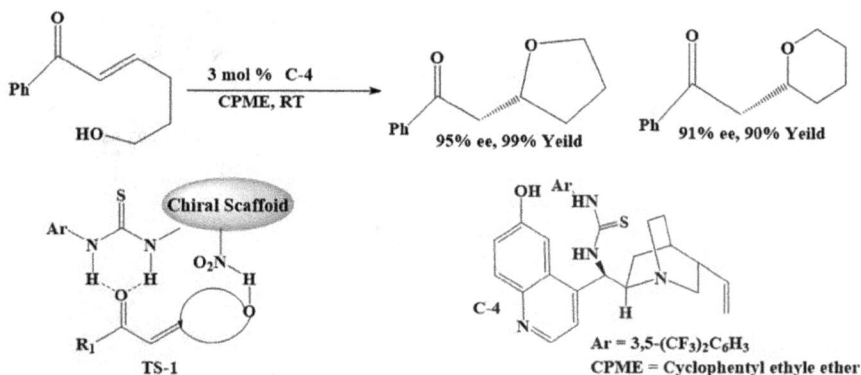

Scheme 16. Cycloetherification via intramolecular oxy-Michael addition reaction mediated by bifunctional organocatalyst

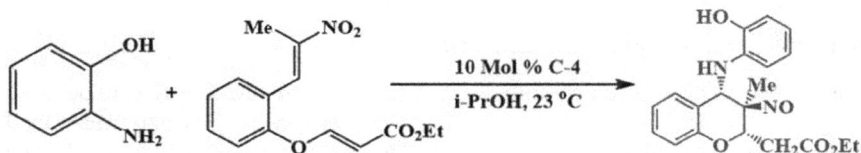

Scheme 17. Asymmetric Aza-Michael-Michael addition reaction of aniline with nitroolefin enoate

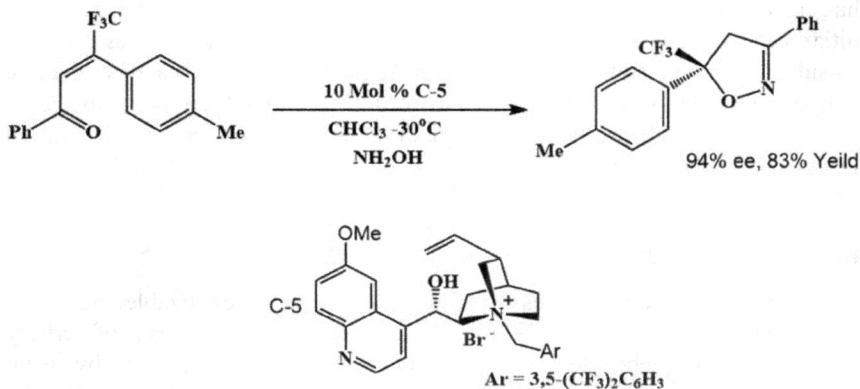

Scheme 18. Synthesis of trifluoromethyl substituted 2-isoxazolines

cinchona alkaloid-based bifunctional urea catalysts C-6. Both enantiomers of the products could be achieved with the same level of enantioselectivity by using pseudoenantiomeric catalysts in up to 97% ee and in excellent yields. One of the enantioenriched products has been transformed to a highly functionalized piperidone derivative as shown in Scheme 19 where R_1 is Ph (Molleti et al. 2012).

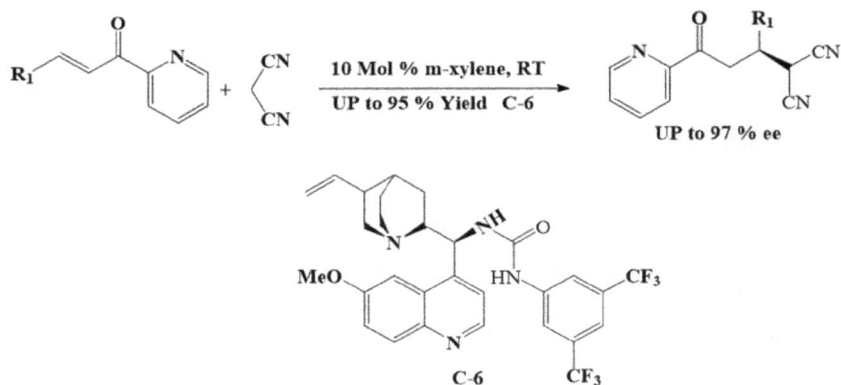

Scheme 19. Enantioselective conjugate addition of malononitrile to β-substituted 2-enoylpyridines (R1: Ph)

4.1 Hydroxyalkylation reaction

The readily available cinchonidine (C-7) and cinchonine (C-8) can be used for the catalysis of the hydroxyalkylation of heteroaromatics and was found to be a brilliant catalyst for the hydroxyalkylation reaction. For example, the hydroxyalkylation of indoles with ethyl-3,3,3-trifluoropyruvate occurs to afford corresponding 3-substituted products. The reaction conditions (temperature, solvent, cinchona/trifluoropyruvate/indole ratios) were optimized and after that, reactivity and selectivity of cinchona derivatives were analyzed. The initial reaction rates depicted that the support of cinchona alkaloids as catalysts considerably enhanced the rate of reaction. While applying C-7 catalyst, the rate of hydroxyalkylation is higher than two orders of magnitude as compared to without catalyst and results in higher yield as well as enantioselectivity. In the presence of C-8 catalyst, the reaction proceeded with high yields and ee values (98% and 90%) as shown in Scheme 20 (Torok et al. 2005).

4.2 Aldol reaction

Perera et al. reported the cross-aldol reaction between enolizable aldehydes and α-ketophosphonates for the highly enantioselective synthesis of tertiary β-formyl-α-hydroxyphosphonates. The process can be accomplished by using 9-amino-9-deoxy-*epi*-quinine (C-9) as a catalyst, as shown in Scheme 21. The reaction works especially well with acetaldehyde, which is a tough substrate

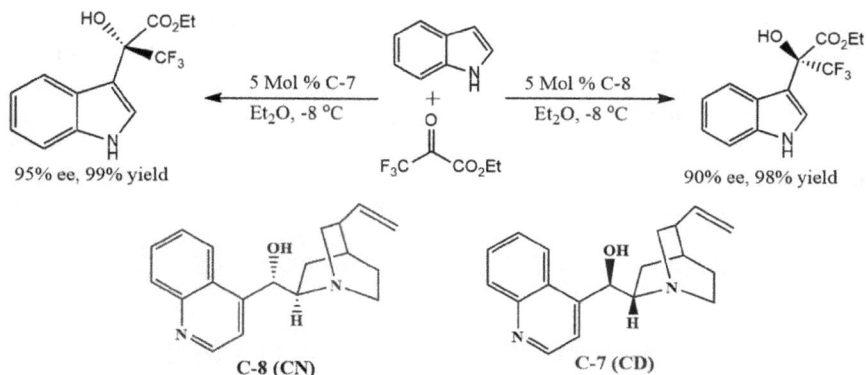

Scheme 20. Enantioselective hydroxyalkylation of Indoles with ethyl 3,3,3-trifluropyruvate catalyzed by C-7 and C-8

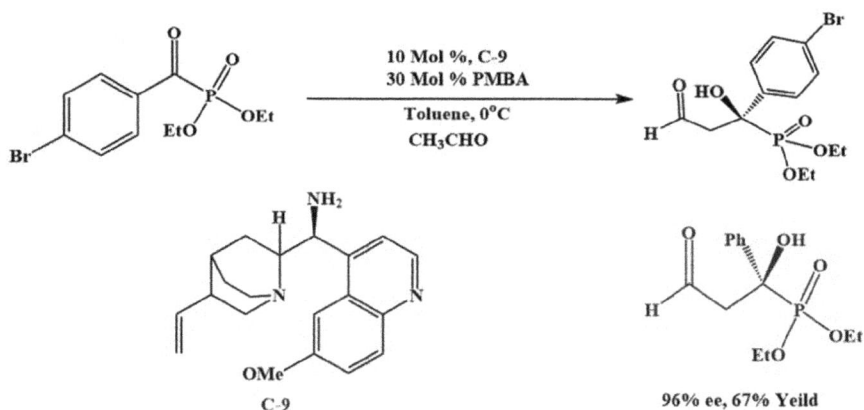

Scheme 21. Synthesis of β-formyl-α-hydroxyphosphonates using C-9 catalyst

for organocatalyzed cross-aldol reaction (Perera et al. 2011). Remarkable enantioselectivity with 96% ee and 67% yield was achieved for both electron-withdrawing and electron-donating substituents and the end products have anticancer activities.

4.3 Henry reaction

The Henry reaction or nitroaldol reaction comprises a significant class of C–C bond forming reactions that afford easy access to important synthetic intermediates from readily accessible nitroalkanes and carbonyl compounds.

The expansion of highly enantioselective and general catalytic nitroaldol reactions with ketones is a demanding yet enviable task in organic synthetic chemistry. Hongming and co-workers reported, for the first time, an asymmetric nitroaldol reaction with α-ketoesters catalyzed by a new C6'-OH cinchona alkaloid (C-10) catalyst. C-10 is an outstanding catalyst to synthesize highly enantioenriched products from the reaction of α-ketoesters with nitromethane as shown in Scheme 22. The products of Henry reaction could be elaborated to aziridines, β-lactams and α-alkylcysteines. This reaction is operationally simple and affords high enantioselectivity as well as good to excellent yield for a broad range of α-ketoesters (Hongming et al. 2006 a & b, McCooey et al. 2007). This reaction is operationally simple and affords high enantioselectivity as well as good to excellent yield for a broad range of α-ketoesters.

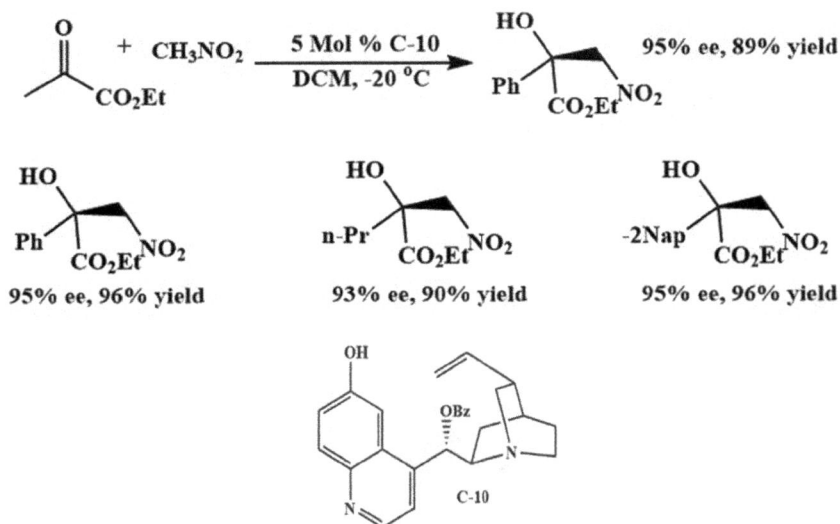

95% ee, 89% yield

95% ee, 96% yield   93% ee, 90% yield   95% ee, 96% yield

C-10

Scheme 22. Cinchona-catalyzed in Henry reaction

4.4 Asymmetric Friedel-Crafts reaction

The high efficiency of bifunctional cinchona alkaloid catalysts in the endorsement of mechanistically unrelated C–C bond formations led researchers to assume that they might function as efficient catalysts for enantioselective Friedel-Crafts reactions of indoles with carbonyl compounds (Wang et al. 2006, Magraner et al. 2016). To validate this theory, Hongming and co-workers examined the cinchona alkaloid-catalyzed reaction of indole with an alkynyl α-ketoester as shown in Scheme 23 (Hongming et al. 2006a). Hongming and co-workers illustrated an efficient asymmetric Friedel-Crafts reaction that, unprecedently, is applicable to a wide range of both indoles and carbonyls. The use of a readily accessible catalyst in combination with a high enantioselectivity that is insensitive to reaction concentration, temperature, air, and moisture should allow this reaction to provide useful enantioselective access to new chiral indole derivatives. Here in

Scheme 23, we can see that in Friedel-Crafts addition of Indoles to α-ketoesters, a complete reaction is proficient with C-11 catalyst to produce the corresponding Friedel-Crafts adduct in 96% yield and 88% ee (Hongming et al. 2006b).

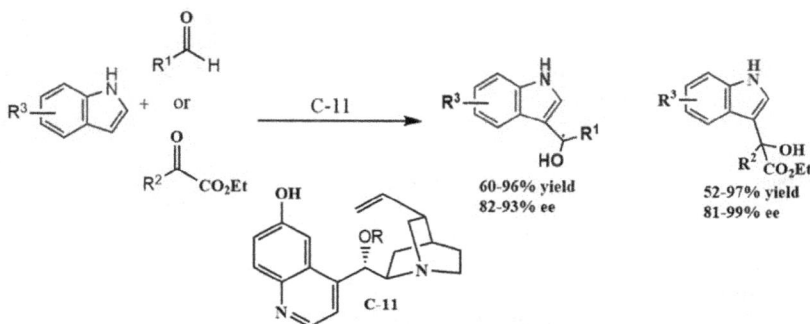

60-96% yield
82-93% ee

52-97% yield
81-99% ee

Scheme 23. Friedel-Crafts addition of Indoles to α-ketoesters with Cinchona based catalyst C-11

5. Conclusion

Our extensive studies of alkaloid-supported catalysis have led to the conclusion that alkaloid-supported catalysts can catalyze a wide variety of new, highly enantioselective, general, and clean reactions. Here in the chapter, we have studied various alkaloid and their derivative-supported asymmetric synthetic reactions. Bifunctional nature of Cinchona derivatives extends the application of alkaloids in vast catalytic reactions. Platinum based Cinchona derivatives are also studied in detail, and are used in the organic reactions as a catalyst, which provide significant enantioselectivity. In addition, we have explored a variety of modified cinchona alkaloids discovered from the Cinchona species used in conjugate addition reactions, Michael addition, Aldol reaction, Friedal-Crafts, and Henry reaction. Cinchona alkaloid chemistry continues to contribute to medicinal chemistry, supramolecular chemistry, cross coupling reactions and asymmetric synthesis. The whole study opens up the scope of more alkaloid-supported reaction and enormous use of alkaloids in the field of organocatalysis.

References

Amirkia, V. and M. Heinrich. 2014. Alkaloids as drug leads – A predictive structural and biodiversity-based analysis. *Phytochemistry Letters* 10: xlviii-liii. https://doi.org/10.1016/j.phytol.2014.06.015

Anarat, C.G. and E.S. Sattely. 2014. The chemical logic of plant natural product biosynthesis. *Current Opinion in Plant Biology* 19: 51-58. https://doi.org/10.1016/j.pbi.2014.03.007

Aniszewski, T. 2015. Alkaloids: Chemistry, Biology, Ecology, and Applications. Elsevier. ISBN: 9780444594624

Arx, M.V., T. Mallat and A. Baiker. 2002. Asymmetric hydrogenation of activated ketones on platinum: Relevant and spectator species. *Topics in Catalysis* 19(1): 75-87. https://doi.org/10.1023/A:1013885300523

Asano, K. and S. Matsubara. 2011. Asymmetric catalytic cycloetherification mediated by bifunctional organocatalysts. *Journal of the American Chemical Society* 133(42): 16711-16713. https://doi.org/10.1021/ja207322d

Astruc, D., F. Lu and J.R. Aranzaes. 2005. Nanoparticles as recyclable catalysts: The frontier between homogeneous and heterogeneous catalysis. *Angewandte Chemie International Edition* 44(48): 7852-7872. https://doi.org/10.1002/anie.200500766

Bela, T., M. Abid, G. London, J. Esquibel, M. Torok, S.C. Mhadgut, P. Yan and G.K. Surya Prakash. 2005. Highly enantioselective organocatalytic hydroxyalkylation of indoles with ethyl trifluoropyruvate. *Angewandte Chemie* 117(20): 3146-3149. https://doi.org/10.1002/ange.200462877

Blaser, H.U. 1991. Enantioselective synthesis using chiral heterogeneous catalysts. *Tetrahedron: Asymmetry* 2(9): 843-866. https://doi.org/10.1016/S0957-4166(00)82195-3

Blaser H.U., H.P. Jalett, M. Muller and M. Studer. 1997. Enantioselective hydrogenation of c-ketoesters using cinchona modified platinum catalysts and related systems: A review. *Catalysis Today* 37(4): 441-463. https://doi.org/10.1016/s0920-5861(97)00026-6

Bond, G., K.E. Simons, A. Lbbotson, P.B. Wells and D.A. Whan. 1992. Platinum-catalysed enantioselective hydrogenation: Effects of low coverage of modifier. *Catalysis Today* 12(4): 421-425. https://doi.org/10.1016/0920-5861(92): 80058-u

Borszeky, K., T. Mallat and A. Baiker. 1997. Enantioselective hydrogenation of α,β-unsaturated acids. Substrate-modifier interaction over cinchonidine modified PdAl$_2$O$_3$. *Tetrahedron: Asymmetry* 8(22): 3745-3753. https://doi.org/10.1016/S0957-4166(97)00526-0

Bradley, J.S., E. Hill, M.E. Leonowicz and H. Witzke. 1987. Clusters, colloids and catalysis. *Journal of Molecular Catalyst* 41(1-2): 59-74. https://doi.org/10.1016/0304-5102(87)80019-6

Breding, G. and P.S. Fiske. 1912. Beiträge zur chemischen Physiologie und Pathologie. *Biochemistry Z.* 46: 7.

Bribi, N. 2018. Pharmacological activity of alkaloids: A review. *Asian Journal of Botany* 1(1): 1-6. https://doi:10.63019/ajb.v1i2.467

Chatwal, G.R. 2019. Organic Chemistry of Natural Products. Volume I. pp. 658. Himalaya Publishing House Pvt. Ltd. ISBN: 978-93-5024-664-1

Collier, P.J., Goulding T. Iggo and J.A.R. Whyman. 1995. Studies of the platinum cinchona alkaloid catalyst for enantioselective α-ketoester hydrogenation. *Chiral Reactions in Heterogeneous Catalysis* 105-110. Springer, Boston, MA. https://doi.org/10.1007/978-1-4615-1909-6_11

Connon, S.J. 2008. Asymmetric catalysis with bifunctional cinchona alkaloid based urea and thiourea organocatalysts. *Chemical Communications* 22: 2499-2510. https://doi.org/10.1039/B719249E

Cordell, G.A., 1981. Introduction to alkaloids: A biological approach. *Journal of Pharmaceutical Sciences* 72(3): 328. Wiley Interscience, New York. https://doi.org/10.1002/jps.2600720334

Croteau, R., T.M. Kutchan and N.G. Lewis. 2000. Natural products (secondary metabolites). *Biochemistry and Molecular Biology of Plants* 24: 1250-1319.

Debnath, B., W.S. Singh, M. Das, S. Goswami, M.K. Singh, D. Maiti and K. Manna. 2018. Role of plant alkaloids on human health: A review of biological activities. *Materials Today Chemistry* 9: 56-72. https://doi.org/10.1016/j.mtchem.2018.05.001

Devereaux, A.L., S.L. Mercer and C.W. Cunningham. 2018. Dark classics in chemical neuroscience: Morphine. *ACS Chemical Neuroscience* 9(10): 2395-2407. https://doi.org/10.1021/acschemneuro.8b00150

Dewick, P.M. 2002. Medicinal Natural Products: A Biosynthetic Approach. John Wiley and Sons. ISBN: 978-0-470-74168-9

Dey, P., A. Kundu, A. Kumar, M. Gupta, B.M. Lee, T. Bhakta, S. Dash and H.S. Kim. 2020. Analysis of alkaloids (indole alkaloids, isoquinoline alkaloids, tropane alkaloids). *Recent Advances in Natural Products Analysis* 505-567. https://doi.org/10.1016/B978-0-12-816455-6.00015-9

Ding, M., F. Zhou, Y.L. Liu, C.H. Wang, X.L. Zhao and J. Zhou. 2011. Cinchona alkaloid-based phosphoramide catalyzed highly enantioselective Michael addition of unprotected 3-substituted oxindoles to nitroolefins. *Chemical Science* 2(10): 2035-2039. https://doi.org/10.1039/C1SC00390A

Duan, J.D. and P.F. Li. 2014. Asymmetric catalysis mediated by primary amines derived from Cinchona alkaloids: Recent advances. *Catalysis Science & Technology* 4(2): 311-320. https://doi.org/10.1039/C3CY00739A

Eguchi, R., N. Ono, A.H. Morita, T. Katsuragi, S. Nakamura, M. Huang, M.A.-U. Amin and S. Kanaya. 2019. Classification of alkaloids according to the starting substances of their biosynthetic pathways using graph convolutional neural networks. *BMC Bioinformatics* 20(1): 1-13. https://doi.org/10.1186/s12859-019-2963-6

Erathodiyil, N., H. Gu, H. Shao, J.J. and J.Y. Ying. 2011. Enantioselective hydrogenation of α-ketoesters over alkaloid-modified platinum nanowires. *Green Chemistry* 13(11): 3070-3074. https://doi.org/10.1039/C1GC15606C

Funayama, S. and G.A. Cordell. 2014. Alkaloids: A treasury of poisons and medicines. Elsevier. ISBN: 9780124173149.

Hall, T.J., P. Johnston, W.A.H. Vermeer, S.R. Watson and P.B. Wells. 1996. Enantioselective hydrogenation catalysed by palladium. *Studies in Surface Science and Catalysis* 101: 221-230. https://doi.org/10.1016/S0167-2991(96)80232-1

Hiemstra, H. and H. Wynberg. 1981. Addition of aromatic thiols to conjugated cycloalkenones, catalyzed by chiral beta-hydroxy amines, a mechanistic study of homogeneous catalytic asymmetric synthesis. *American Chemical Society* 103(2): 417-430. https://doi.org/10.1021/ja00392a029

Hiratake, J., M. Inagaki, Y. Yamamoto and J. Oda. 1987. Enantiotopic-group differentiation, catalytic asymmetric ring-opening of prochiral cyclic acid anhydrides with methanol, using cinchona alkaloids. *Journal of the Chemical Society Perkin Transactions, 1* 1053-1058. https://doi.org/10.1039/P19870001053

Hongming, L., B. Wang and L. Deng. 2006a. Enantioselective nitroaldol reaction of α-ketoesters catalyzed by cinchona alkaloids. *Journal of the American Chemical Society* 128(3): 732-733. https://doi.org/10.1021/ja057237l

Hongming, L., Y.-Q. Wang and L. Deng. 2006b. Enantioselective Friedel-Crafts reaction of indoles with carbonyl compounds catalyzed by bifunctional cinchona alkaloids. *Organic Letters* 8(18): 4063-4065. https://doi.org/10.1021/ol061552a

Hoxha, F., L. Konigsmann, A. Vargas, D. Ferri, T. Mallat and A. Baiker. 2007. Role of guiding groups in cinchona-modified platinum for controlling the sense of enantiodifferentiation in the hydrogenation of ketones. *Journal of the American Chemical Society* 129(34): 10582-10590. https://doi.org/10.1021/ja073446p

Huck, W.R., T. Mallat and A. Baiker. 2000. Potential and limitations of palladium–cinchona catalyst for the enantioselective hydrogenation of a hydroxymethylpyrone. *Journal of Catalysis* 193(1): 1-4. https://doi.org/10.1006/jcat.2000.2890

Ishtiyak, P. and S.A. Hussain. 2017. Traditional use of medicinal plants among tribal communities of Bangus Valley, Kashmir Himalaya, India. *Studies on Ethno-Medicine* 11(4): 318-331. https://doi.org/10.1080/09735070.2017.1335123

Kacprzak, K. and J. Gawronski. 2001. Cinchona alkaloids and their derivatives: Versatile catalysts and ligands in asymmetric synthesis. *Synthesis Stuttgart* 2001(7): 961-998. https://doi.org/10.1055/s-2001-14560

Kartsev, V.G. 2004. Natural compounds in drug discovery. Biological activity and new trends in the chemistry of isoquinoline alkaloids. *Medicinal Chemistry Research* 13(7): 325-336. https://doi.org/10.1007/s00044-004-0038-2

Kaufman, T.S. and E.A. Rúveda. 2005. The quest for quinine: Those who won the battles and those who won the war. *Angewandte Chemie International Edition* 44(6): 854-885. https://doi.org/10.1002/anie.200400663

Kaur, R. and S. Arora. 2015. Alkaloids-important therapeutic secondary metabolites of plant origin. *Journal of Critical Reviews* 2(3): 1-8. https://tarjomefa.com/wp-content/uploads/2018/05/9114-English-TarjomeFa.pdf

Kazutaka, M., H. Kawai, T. Furukawa, A. Kusuda, E. Tokunaga, S. Nakamura, M. Shiro and N. Shibata. 2010. Enantioselective synthesis of trifluoromethyl-substituted 2-isoxazolines: Asymmetric hydroxylamine/enone cascade reaction. *Angewandte Chemie* 122(33): 5898-5902. https://doi.org/10.1002/ange.201002065

Li, H., Y. Wang, L. Tang and L. Deng. 2004. Highly enantioselective conjugate addition of malonate and β-ketoester to nitroalkenes: Asymmetric C–C bond formation with new bifunctional organic catalysts based on cinchona alkaloids. *Journal of the American Chemical Society* 126(32): 9906-9907. https://doi.org/10.1021/ja047281l

Li, H., Y. Wang, L. Tang, F. Wu, X. Liu, C. Guo, B.M. Foxman and L. Deng. 2005. Stereocontrolled creation of adjacent quaternary and tertiary stereocenters by a catalytic conjugate addition. *Angewandte Chemie International Edition* 117(1): 107-110. https://doi.org/10.1002/anie.200461923

Lichman, B.R. 2021. The scaffold-forming steps of plant alkaloid biosynthesis. *Natural Product Reports* 38(1): 103-129. https://doi.org/10.1039/d0np00031k

Ma, Z. and F. Zaera. 2005. Role of the solvent in the adsorption-desorption equilibrium of cinchona alkaloids between solution and a platinum surface: Correlations among solvent polarity, cinchona solubility, and catalytic performance. *The Journal of Physical Chemistry B* 109(1): 406-414. https://doi.org/10.1021/jp046017b

Ma, Z. and F. Zaera. 2006. Competitive chemisorption between pairs of cinchona alkaloids and related compounds from solution onto platinum surfaces. *Journal of the American Chemical Society* 128(51): 16414-16415. https://doi.org/10.1021/ja0659323

Ma, Z., I. Lee and F. Zaera. 2007. Factors controlling adsorption equilibria from solution onto solid surfaces: The uptake of cinchona alkaloids on platinum surfaces. *Journal of the American Chemical Society* 129(51): 16083-16090. https://doi.org/10.1021/ja076011a

Marcelli, T. and H. Hiemstra. 2010. Cinchona alkaloids in asymmetric organocatalysis. *Synthesis Stuttgart* 2010(8): 1229-1279. https://doi.org/ 10.1055/s-0029-1218699

McCooey, H. Seamus and J.C. Stephen. 2007. Readily accessible 9-epi-amino cinchona alkaloid derivatives promote efficient, highly enantioselective additions of aldehydes and ketones to nitroolefins. *Organic Letters* 9(4): 599-602. https://doi.org/10.1021/ol0628006

McDaid, P., Y. Chen and L. Deng. 2002. A highly enantioselective and general conjugate addition of thiols to cyclic enones with an organic catalyst. *Angewandte Chemie International Edition* 41(2): 338-340. https://doi.org/10.1002/1521-3773

Nagaraju, M., N.K. Rana and V.K. Singh. 2012. Highly enantioselective conjugate addition of malononitrile to 2-enoylpyridines with bifunctional organocatalyst. *Organic Letters* 14(17): 4322-4325. https://doi.org/10.1021/ol3015607

Orito, Y., S. Imai, S. Niwa and G.H. Nguyen. 1979a. Asymmetric hydrogenation of methyl benzoylformate using platinum-carbon catalysts modified with cinchonidine. *Journal of Synthetic Organic Chemistry of Japan* 37(2): 173-174. https://doi.org/10.5059/yukigoseikyokaishi.37.173

Orito, Y., S. Imai and S. Niwa. 1979b. Asymmetric hydrogenation of methyl pyruvate using Pt-C catalyst modified with cinchonidine. *Nippon Kagaku Kaishi* 1979(8): 1118-1120. https://doi.org/10.1246/nikkashi.1979.1118

Pellisier, H. 2007. Asymmetric organocatalysis. *Tetrahedron* 38(63): 9267-9331. https://doi.org/10.1016/j.tet.2007.06.024

Pracejus, H. 1960. Organische Katalysatoren, LXI. Asymmetrische Synthesen mit Ketenen, I. Alkaloid-katalysierte asymmetrische Synthesen von α-Phenyl-propionsäureestern. *Justus Liebigs Annalen der Chemie* 634(1): 9-22. https://doi.org/10.1002/jlac.19606340103

Prelog, V. and M. Wilhelm. 1954. Untersuchungen über asymmetrische Synthesen VI. Der Reaktionsmechanismus undder sterische Verlauf der asymmetrischen Cyanhydrin-Synthese. *Helvetica Chimica Acta.* 37(6): 1634-1660. https://doi.org/10.1002/hlca.19540370608

Rana, N.K., S. Selvakumar and V.K. Singh. 2010. Highly enantioselective organocatalytic sulfa-Michael addition to α, β-unsaturated ketones. *The Journal of Organic Chemistry* 75(6): 2089-2091. https://doi.org/10.1021/jo902634a

Rosentahler, L. 1908. Durch enzyme bewirkte asymmetrische synthesen. *Biochemistry Z.* 14: 238-253.

Roy, A. 2017. A review on the alkaloids an important therapeutic compound from plants. *International Journal of Plant Biotechnology* 3(2): 1-9. https://www.researchgate.net/publication/320098967

Rupani, R. and A. Chavez. 2018. Medicinal plants with traditional use: Ethnobotany in the Indian subcontinent. *Clinics in Dermatology* 36(3): 306-309. https://doi.org/10.1016/j.clindermatol.2018.03.005

Sadia, S., A. Tariq, S. Shaheen, K. Malik, F. Khan, M. Ahmad, H. Qureshi and B.G. Nayyar. 2018. Ethnopharmacological profile of anti-arthritic plants of Asia – A systematic review. *Journal of Herbal Medicine* 13(1): 8-25. https://doi.org/10.1016/j.hermed.2018.08.003

Salvadori, P., D. Pini and A. Petri. 1999. MCM-41 anchored cinchona alkaloid for catalytic asymmetric dihydroxylation of olefins: A clean protocol for chiral diols using molecular oxygen. *Synlett* 1999(8): 1181-1190. https://doi.org/10.1055/s-1999-2791

Sandun, P., V.K. Naganaboina, L. Wang, B. Zhang, Q. Guo, L. Rout and C.G. Zhao. 2011. Organocatalytic highly enantioselective synthesis of β-formyl-α-hydroxyphosphonates. *Advanced Synthesis & Catalysis* 353(10): 1729-1734. https://doi.org/10.1002/adsc.201000835

Stegbauer, L., F. Sladojevich and D.J. Dixon. 2012. Bifunctional organo/metal cooperative catalysis with cinchona alkaloid scaffolds. *Chemical Science* 3(4): 942-958. https://doi.org/10.1039/C1SC00416F

Studer, M., S. Burkhardt, A.F. Indolese and H.U. Blaser. 2000. Enantio- and chemoselective reduction of 2,4-diketo acid derivatives with cinchona modified Pt-catalyst-synthesis of (R)-2-hydroxy-4-phenylbutyric acid ethyl ester. *Chemical Communications* 14: 1327-1328. https://doi.org/10.1039/b002538k

Toukoniitty, E., P.M. Arvela, N. Kumar, T. Salmi and D.Yu. Murzin. 2004. Continuous enantioselective hydrogenation of ethylbenzoylformate over Pt/Al₂O₃ catalyst: Bed dilution effects and cinchonidine adsorption study. *Catalysis Letters* 95: 179-183. https://doi.org/10.1023/B:CATL.0000027292.23132.5c

Uniyal, S.K., K.N. Singh, P. Jamwal and B. Lal. 2006. Traditional use of medicinal plants among the tribal communities of Chhota Bhangal, Western Himalaya. *Journal of Ethnobiology and Ethnomedicine* 2(14): 1-8. https://doi:10.1186/1746-4269-2-14

Vakulya, B., S. Varga, A. Csampai and T. Soos. 2005. Highly enantioselective conjugate addition of nitromethane to chalcones using bifunctional cinchona organocatalysts. *Organic Letters* 7(10): 1967-1969. https://doi.org/10.1021/ol050431s

Vayner, G., K.N. Houk and Y.K. Sun. 2004. Origins of enantioselectivity in reductions of ketones on cinchona alkaloid modified platinum. *Journal of American Society* 126(1): 199-203. https://doi.org/10.1021/ja035147f

Wang, X.F., J. An, X.X. Zhang, F. Tan, J.R. Chen and W.J. Xiao. 2011. Catalytic asymmetric Aza-Michael – Michael addition cascade: Enantioselective synthesis of polysubstituted 4-aminobenzopyrans. *Organic Letters* 13(4): 808-811. https://doi.org/10.1021/ol1031188

Yeboah, E.M., S.O. Yeboah and G.S. Singh. 2011. Recent applications of cinchona alkaloids and their derivatives as catalysts in metal-free asymmetric synthesis. *Tetrahedron* 67(10): 1725-1762. http://dx.doi.org/10.1016/j.tet.2010.12.050

Zu, L., J. Wang, H. Li, H. Xie, W. Jiang and W. Wang. 2007. Cascade Michael-aldol reactions promoted by hydrogen bonding mediated catalysis. *Journal of the American Chemical Society* 129(5): 1036-1037. https://doi.org/10.1021/ja067781+

Polystyrene-supported Catalysts

Deepali Khokhar[1*], Manjinder Kour[2], Ruby Phul[3], Alpesh Kumar Sharma[4]
and Sapana Jadoun[5*]

[1] School of Basic & Applied Sciences, Department of Chemistry, Lingaya's
Vidyapeeth, Faridabad - 121002, Haryana, India
[2] Microbiology and Cell Biology Department, Montana State University,
Bozeman, MT, 59715, USA
[3] Department of Chemistry, Bilkent University, Ankara 06800, Turkey
[4] Department of Chemistry, University of Chicago, USA 5735 S Ellis Ave,
Chicago, IL 60637
[5] Departamento de Química, Facultad de Ciencias, Universidad de Tarapacá,
Avda. General Velásquez, 1775, Arica, Chile

1. Introduction

The advancement of eco-friendly synthetic procedures and technologies are
necessary for improving the present state of living of the mankind. There is
a need to reuse the catalysts for the optimization of resources (Drabina et al.
2017, Islam et al. 2016, Qi et al. 2010). Binding of homogeneous catalysts
with the mobile solids helps in procuring the renewable catalysts. Amongst the
remunerative accessible organic polymers like conducting conjugated polymers
and copolymers (Khokhar et al. 2021a, 2021b, 2022), cross-linked polymers of
styrene pertain to the notable competent carriers. Polymer-supported catalyst
facilitates several benefits such as recyclability, easy to separate, economical and
their utilization in particular continuous flow processes (Ahadi et al. 2019, Hong
et al. 2020, Islam et al. 2012, Iwai et al. 2013, Mane et al. 2018, Mikhaylov
et al. 2016, 2018, Mohammadi and Movassagh 2016, Movassagh et al. 2018,
Mpungose et al. 2018, Nejati et al. 2018, Roy and Uozumi 2018, Sedghi et al.

*Corresponding authors: deepalikhokhar003@gmail.com; sjadoun022@gmail.com

2019), complex organic synthesis (Alza et al. 2011, Clot-Almenara et al. 2016, Howard et al. 2019, Islam et al. 2016, Kasaplar et al. 2012, Kazemi et al. 2018, Kitanosono et al. 2018, Nasrollahzadeh et al. 2019, Qi et al. 2010, Samuels et al. 2016, Whiteoak et al. 2014, Xu et al. 2018, 2010), precursors to natural materials (Albukhari et al. 2019, Murugesan et al. 2020), and pharmaceuticals (Albukhari et al. 2019, Burange et al. 2022, Li et al. 2019). The significance, ambit, and prospect of cross-coupling reactions for twain purposes that is in research and industries have seen remarkable interest lately and polymer-supported catalysts (Nasrollahzadeh et al. 2019) have given them tremendous boom, especially the heterogeneous polymer-supported catalysts (Howard et al. 2019, Ju et al. 2019, Mohammadi and Movassagh 2018), offering notable advantages over homogeneous catalysts whose major limitation is the possibility of metal cessation with the end product, which hinders its recovery and reusability as a catalyst (Mikhaylov et al. 2018). To overcome these difficulties, a more promising solution is the availability of corresponding heterogeneous solid supports (Bartáček et al. 2019). The immobilized support has many advantages such as the ease of approach, stability, and catalyst recycling potential (Jia et al. 2019) with continual processing and easy segregation (Mikhaylov et al. 2016). Amongst the economically accessible organic polymers, cross-linked polymers of styrene pertain to the highly relevant supported catalyst (Alza et al. 2011, Bartáček et al. 2020, Clot-Almenara et al. 2016, Islam et al. 2016, Kasaplar et al. 2012, Mikhaylov et al. 2016, 2018, Qi et al. 2010, Whiteoak et al. 2014, Xu et al. 2010) catalyst with various transition metals' ions like zinc (Biswas et al. 2018, 2019, Ghosh et al. 2018, Kim et al. 2018), silver (Jiang et al. 2021, Peng et al. 2016), Pd (Ahadi et al. 2019, Ju et al. 2019, Mohammadi and Movassagh 2016, 2018, Movassagh et al. 2018, Sedghi et al. 2019), Ru (Izgi et al. 2020, Şen et al. 2018, 2019, Wang et al. 2020), Cu (Bahsis et al. 2019, Dehbanipour et al. 2018, Islam et al. 2016, Kodicherla and Mandapati 2014, Perumgani et al. 2016, Sun et al. 2013, Yan et al. 2019), and many more helps in catalyzing the reaction by promoting bond formation. This book chapter presents multifarious applications of polystyrene-supported catalysts in cross-coupling and organic synthesis reactions defining their studied recycling potential, stability, selectivity, and catalytic activity.

2. Cross-coupling reactions

A heterogeneous catalytic system of palladium(II) acyclic diaminocarbene complex on polystyrene support was prepared by Mikhaylov and co-workers (Mikhaylov et al. 2016) via nucleophilic addition of the amine group of polystyrene to the (Pd(II)-ADC) complex. The synthesized catalytic system exhibited high and stable catalytic activity in sonogashira as well as Suzuki-Miyura cross-coupling. PS-supported catalyst is retrievable and recycled many times besides notable efficacy loss. The enhanced reaction rate was discerned in the Sonogashira reaction with reduced level of palladium leaching and decreased catalytic amount (Mikhaylov et al. 2016).

Based on theophylline, Mohammadi and Movassagh (Mohammadi and Movassagh 2016) reported a sustainable catalyst, N-heterocyclic carbine-Pd(II) complex (NHC-Pd(II)) immobilized onto polystyrene resin. This catalysis system is added in the Suzuki-Miyaura process of Ar-N_2^+ BF_4^- salts with Ar-$(BOH)_2$ acid to biaryl derivative, gave adequate yield with no additives, showed enhanced life cycles, and was easily filtered from the reaction blend using the filtration process, Figure 1.

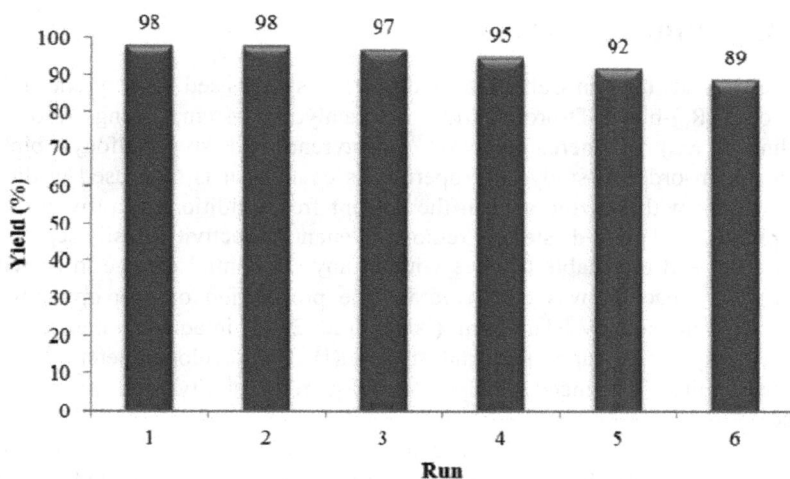

Figure 1. Recycling activity of the PS-NHC-Pd(II) catalyst in the reaction of 4-bromobenzenediazonium tetrafluoroborate with phenylboronic acid. With permission from Elsevier (Mohammadi and Movassagh 2016)

Iwai et al. (Iwai et al. 2013) synthesized polystyrene-phosphane (PS-Ph_3P) hybrid, in which styrenes, divinylbenzene, and a tris(p-vinylphenyl)phosphane, are covalently attached via radical copolymerization. This hybrid catalyst system increased reactivity of reaction forming metal complex with minimum synthetic work. Its usage in different heterogeneous applications such as favouring Pd-catalysed cross-coupling reactions and Ir-catalysed and Rh-catalysed borylation process makes it versatile and attractive for developing more such types of catalysis.

A novel heterogeneous and reusable polystyrene aided [PdCl-(SeCSe)], pincer type, complex catalyst was prepared by Mohammadi and co-workers (Mohammadi and Movassagh 2018) and the catalytic activity was evaluated. It is further used in Sonogashira reaction of 1,2-disubstituted alkynes with 1,3-enynes and resulted in high yield percentage of acetylenes and greatly enhanced the rate of reaction with high E-Z ratios, as well as readily isolated from reaction blend using filtration and reusable for six cycles successively without any substantial change in activity.

The catalysis activity of Pd(II)-N-heterocyclic carbene complex was investigated by Movassagh et al. (Movassagh et al. 2018) in the Suzuki-Miyaura reaction of arylboronic acid with aroyl chlorides, in the presence K_2CO_3 and acetone-water, to give aryl ketones. The good yield of the end product was received in the reaction and found that heterogeneous stimulant can be reusable up to 4 cycles with no major variation in its catalytic property and can easily be recovered from reaction.

3.　Organic synthesis

Caballero et al. (Bañón-Caballero et al. 2010) synthesised polystyrene aided N-sulfonyl-(R_a)-binam-D-prolinamide, a catalytic system, using thiol-ene coupling (TEC) of mercaptomethyl polystyrene and styrylsulfonyl binam derivative. In order to study its properties as catalyst, it is then used in direct aldol process with benzoic acid in the solvent free conditions to provide high yield product and good diastereo-, regio- and enantioselectivity. Easily separable by filtration and recyclable 6 times without any substantial change in activity, oxidative A^3 reaction was employed for the preparation of propargylamines with benzyl alcohols by Islam et al. (Islam et al. 2016) in aqueous media using prepared polystyrene-supported catalyst of Cu(I). The developed heterogeneous stimulated offers enhanced catalytic activity, recyclability, and appreciable efficacy.

Biswas and co-workers (Biswas et al. 2018) explored the catalytic efficiency of chloromethylated polystyrene-supported [PS-Zn(II)L] complex in the formation process of organic cyclic carbonates, via CO_2 fixation under 1 atm pressure and at room temperature, in solvent free conditions, from epoxides as well as organic carbamates which was synthesized from amines using green agent, dimethyl carbonate (DMC). High catalytic efficiency and recyclability up to 5 times was observed (Figures 2 and 3).

Lai et al. (Lai et al. 2018) prepared novel PS supported cis-4-hydroxydiphenylprolinol and examined its catalytic behaviour by reacting N-protected hydroxylamines and α,β-unsaturated aldehydes together and noticed that the synthesized catalyst is highly enantioselective, favouring enantioselective domino reaction with 83% yield, and reusable up to 10 cycles repeatedly without any significant loss in its activity.

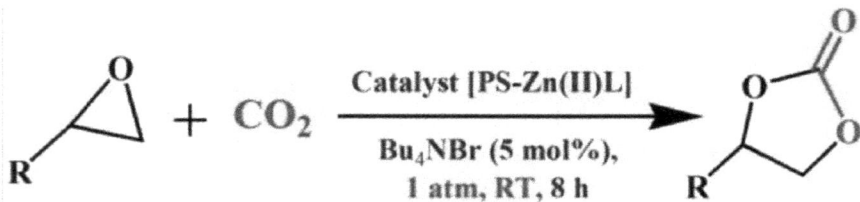

Figure 2. Catalytic formation of cyclic carbonates from epoxides and CO_2 by [PS-Zn(II)L]. With permission from Elsevier (Biswas et al. 2018)

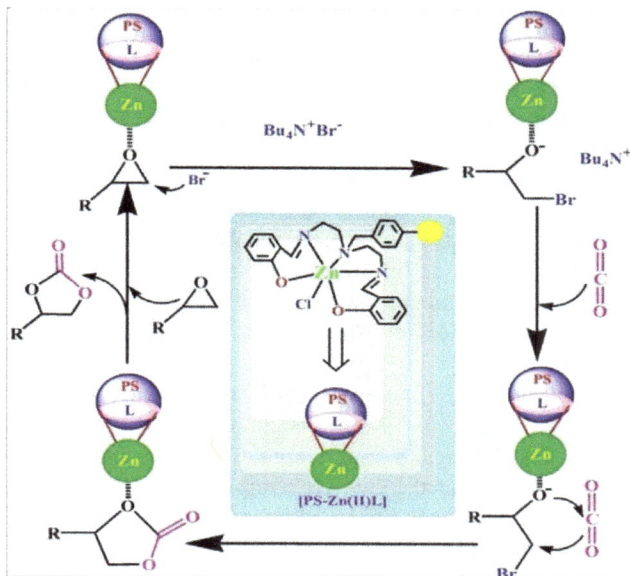

Figure 3. Probable mechanism of catalytic formation of cyclic carbonate by [PS-Zn(II)L]. With permission from Elsevier (Biswas et al. 2018)

Alza et al. (Alza et al. 2011) prepared CuAAC catalyzed diarylprolinol silyl ether, with the help of polystyrene as an insoluble supporting catalyst, and underwent Michael addition reactions via both enamine and iminium ion intermediates. In enamine ion catalysis, linear aldehydes are favoured over branched aldehydes. The mediation of dialkyl malonates and nitromethane through iminium ions leads to unsaturated aldehydes. The setting of CuAAC catalyzed ether on polystyrene provides for high catalytic activity and easy separation by simple method of filtration and expanding the reusable cycles.

A copper-catalyzed azide-alkyne cycloaddition (CuAAC) of a chiral squaramide with the support of polystyrene (PS) was successfully reported by Kasaplar and co-workers (Kasaplar et al. 2012). The characteristics of the prepared organocatalyst include high catalytic behaviour, and easily filterable from its reaction mixture. In addition to this, extended recyclability (can be reused 10 times) is also observed. In the asymmetric Michael addition reaction of 1,3-dicarbonyl compounds and β-nitrostyrenes, when added with the organic catalyst provides good yield and enantioselectivity.

Whiteoak and co-workers (Whiteoak et al. 2014) prepared pyrogallol (i.e. 1,2,3-trihydroxybenzene) based and PS supported catalyst which is active at much lower temperatures (45°C) than earlier organocatalyst systems (>100°C) for the CO_2 transformation and oxiranes into cyclic carbonates. Its reusability and regeneration of lost catalytic activity by treating with methyl iodide make it more promising and sustainable.

Clot-Almenara et al. (Clot-Almenara et al. 2016) studied the catalytic properties of polystyrene aided TRIP catalyst and found high activity and highly enantioselective behavior of linear aldehydes. The immobilized organocatalyst onto PS resin, when put under rigorous test through 28 h flow experiment, obtained a high yield of the product without any activity loss. Another characteristic of high recyclability makes it an attractive choice.

PS-benzotriazole-gold (I) complex, prepared by immobilizing gold (I) catalyst on supported polystyrene by Cao and co-workers (Cao and Yu 2011), exhibited potential catalytic activity for the first time. They tested recyclability for three different reaction models, which includes Nazarov reaction of enynyl acetate, 1,6-enyne cyclization and the tandem 3,3-rearrangement and third is furan rearrangement.

Samuels and co-workers (Samuels et al. 2016) developed polystyrene resin-supported diphosphine catalysis by using facile method of processing solid phase synthesis with compatible ligand architecture. The yield and activity efficiency was tested by making use of the catalyst complex in Rh-catalysed asymmetric hydrogenation and observed good to excellent yield quantity and raised catalytic activity with selectivity in intermediate range in the process.

Zuo et al. (Zuo et al. 2014) prepared novel chloromethyl polystyrene based catalyst by replacing chlorine atoms partly by sulfonic groups, containing acid as well as cellulose sites. High yield and high catalytic activity were observed after using catalyst compound in the conversion of microcrystalline cellulose. High catalytic behaviour is due to the presence of large amount of chlorine (–Cl) groups and sulfonic (–SO$_3$H) groups over the catalyst moiety.

Xu et al. (Xu et al. 2010) prepared exceptional PS-supported ionic liquid catalyst using esterification reaction chains showing augmented recyclability, thermal stability, and catalytic activity, Figures 4 and 5.

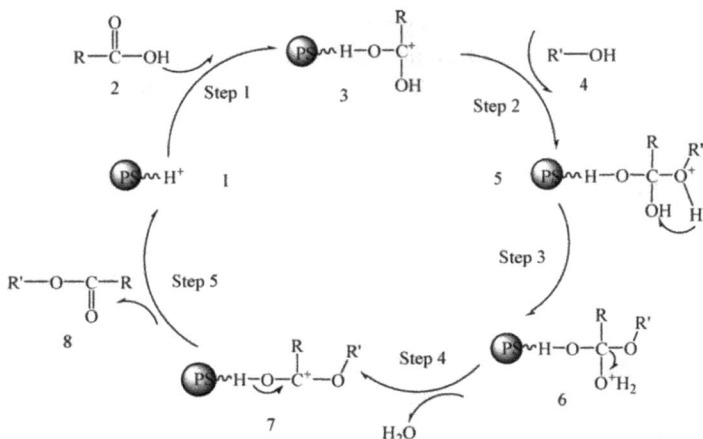

Figure 4. Mechanism of esterification catalyzed over PS-CH$_2$-[SO$_3$H-pIM][HSO$_4$] catalyst. With permission from Elsevier (Xu et al. 2010)

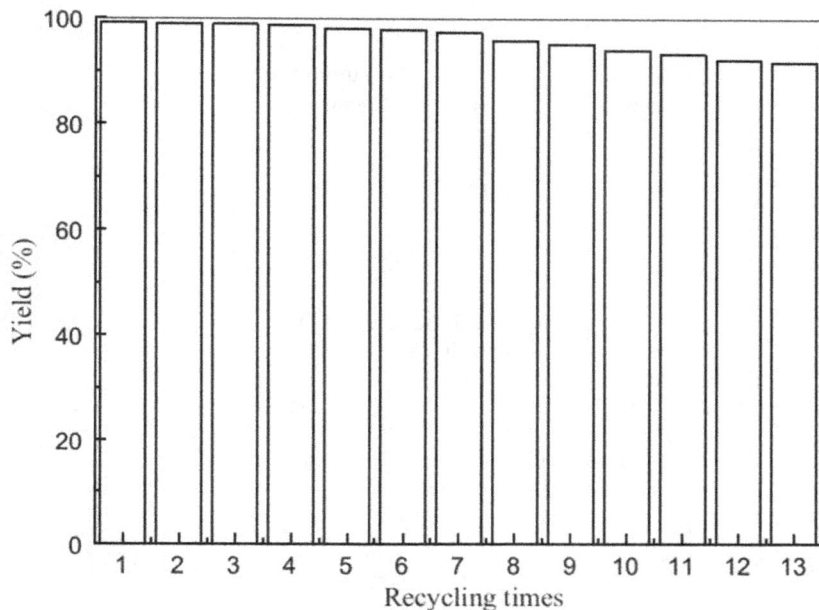

Figure 5. Recycling of PS-CH$_2$-[SO$_3$H-pIM][HSO$_4$] catalyst for the synthesis of n-butyl acetate. With permission from Elsevier (Xu et al. 2010)

An easy, potent and ecological process was explored by Qi et al. (Qi et al. 2010) for carboxylation of several aziridines and epoxides providing a high transformation rate, selectivity, recyclability, and stability, by employing PS supported amino acids acting as rigid activator support.

4. Conclusion

This particular subject of polymer-supported catalysts covers a quick advancing sphere that encloses fundamentals from almost all arenas of chemistry. It adumbrates a term that covers an entire class of synthesis that can be assisted with reprocessable catalysts and reusable reagents. Yet all academics in this sphere are aiming for exemplary retrievable catalysts having eminent performance attributed to reaction rates and their regio-, and stereoselectivity apart from leaching, reduction, etc., which can be efficiently restored. These immobilized polystyrene supports accord many more appealing and exhilarating features such as materialistic and environmental dynamic impulses that are functioning in coaction. Altogether, this is appealing and exceptional progress in integral chemistry and describes alluring prospects for futuristic investigation for developing ideal polystyrene immobilized support for green chemistry.

Acknowledgement

Author Sapana Jadoun is grateful for the support of National Research and Development Agency of Chile (ANID) for the project FONDECYT Postdoctoral 3200850.

References

Ahadi, Arefeh, Sadegh Rostamnia, Paria Panahi, Lee D. Wilson, Qingshan Kong, Zengjian An and Mohammadreza Shokouhimehr. 2019. Palladium comprising dicationic bipyridinium supported periodic mesoporous organosilica (PMO): Pd@ Bipy–PMO as an efficient hybrid catalyst for Suzuki–Miyaura cross-coupling reaction in water. *Catalysts* 9(2): 140.

Albukhari, Soha M., Muhammad Ismail, Kalsoom Akhtar and Ekram Y. Danish. 2019. Catalytic reduction of nitrophenols and dyes using silver nanoparticles@ cellulose polymer paper for the resolution of waste water treatment challenges. *Colloids and Surfaces A: Physicochemical and Engineering Aspects* 577: 548–561.

Alza, Esther, Sonia Sayalero, Pinar Kasaplar, Diana Almasi and Miquel A. Perics. 2011. Polystyrene-supported diarylprolinol ethers as highly efficient organocatalysts for Michael-type reactions. *Chemistry – A European Journal* 17(41): 11585–11595.

Bahsis, Lahoucine, Hicham Ben El Ayouchia, Hafid Anane, Alejandro Pascual-Álvarez, Giovanni De Munno, Miguel Julve, Salah-Eddine Stiriba. 2019. A reusable polymer-supported copper (I) catalyst for triazole click reaction on water: An experimental and computational study. *Applied Organometallic Chemistry* 33(4): e4669.

Bañón-Caballero, Abraham, Gabriela Guillena and Carmen Nájera. 2010. Solvent-free direct enantioselective aldol reaction using polystyrene-supported N-sulfonyl-(Ra)-binam-D-prolinamide as a catalyst. *Green Chemistry* 12(9): 1599–1606.

Bartáček, Jan, Pavel Drabina, Jiří Váňa and Miloš Sedlák. 2019. Recoverable polystyrene-supported catalysts for sharpless allylic alcohols epoxidations. *Reactive and Functional Polymers* 137: 123–132.

Bartáček, Jan, Jiří Váňa, Pavel Drabina, Jan Svoboda, Martin Kocúrik and Miloš Sedlák. 2020. Recoverable polystyrene-supported palladium catalyst for construction of all-carbon quaternary stereocenters via asymmetric 1,4-addition of arylboronic acids to cyclic enones. *Reactive and Functional Polymers* 153: 104615.

Biswas, Imdadul Haque, Surajit Biswas, Md Sarikul Islam, Sk Riyajuddin, Priyanka Sarkar, Kaushik Ghosh, Sk Manirul Islam. 2019. Catalytic synthesis of benzimidazoles and organic carbamates using a polymer supported zinc catalyst through CO_2 fixation. *New Journal of Chemistry* 43(36): 14643–14652.

Biswas, Surajit, Resmin Khatun, Manideepa Sengupta and Sk Manirul Islam. 2018. Polystyrene supported zinc complex as an efficient catalyst for cyclic carbonate formation via CO_2 fixation under atmospheric pressure and organic carbamates production. *Molecular Catalysis* 452: 129–137.

Burange, Anand S., Sameh M. Osman and Rafael Luque. 2022. Understanding flow chemistry for the production of active pharmaceutical ingredients. *iScience* 103892.

Cao, Wenjie and Biao Yu. 2011. A recyclable polystyrene-supported gold (I) catalyst. *Advanced Synthesis & Catalysis* 353(11-12): 1903–1907.

Clot-Almenara, Lidia, Carles Rodríguez-Escrich, Laura Osorio-Planes and Miquel A. Pericàs. 2016. Polystyrene-supported TRIP: A highly recyclable catalyst for batch and flow enantioselective allylation of aldehydes. *ACS Catalysis* 6(11): 7647–7651.

Dehbanipour, Zahra, Majid Moghadam, Shahram Tangestaninejad, Valiollah Mirkhani, Iraj Mohammadpoor-Baltork. 2018. Chloromethylated polystyrene supported copper (II) bis–thiazole complex: Preparation, characterization and its application as a heterogeneous catalyst for chemoselective and homoselective synthesis of aryl azides. *Applied Organometallic Chemistry* 32(9): e4436.

Drabina, Pavel, Jan Svoboda and Miloš Sedlák. 2017. Recent advances in C–C and C–N bond forming reactions catalysed by polystyrene-supported copper complexes. *Molecules* 22(6): 865.

Ghosh, Swarbhanu, Paramita Mondala, Debasis Das, Kazi Tuhinac and Sk. Manirul Islam. 2018. Use of PS-Zn-anthra complex as an efficient heterogeneous recyclable catalyst for carbon dioxide fixation reaction at atmospheric pressure and synthesis of dicoumarols under greener pathway. *Journal of Organometallic Chemistry* 866: 1–12.

Hong, Kootak, Mohaddeseh Sajjadi, Jun Min Suh, Kaiqiang Zhang, Mahmoud Nasrollahzadeh, Ho Won Jang, Rajender S. Varma and Mohammadreza Shokouhimehr. 2020. Palladium nanoparticles on assorted nanostructured supports: Applications for Suzuki, Heck, and Sonogashira cross-coupling reactions. *ACS Applied Nano Materials* 3(3): 2070–2103.

Howard, Ioli C., Ceri Hammond and Antoine Buchard. 2019. Polymer-supported metal catalysts for the heterogeneous polymerisation of lactones. *Polymer Chemistry* 10(43): 5894–5904.

Islam, Md, Anupam Singha Roy and Sk Islam. 2016. Functionalized polystyrene supported copper (I) complex as an effective and reusable catalyst for propargylamines synthesis in aqueous medium. *Catalysis Letters* 146(6): 1128–1138.

Islam, Sk Manirul, Sanchita Mondal, Paramita Mondal, Anupam Singha Roy, K. Tuhina, Noor Salam and Manir Mobarak. 2012. A reusable polymer supported copper catalyst for the C–N and C–O bond cross-coupling reaction of aryl halides as well as arylboronic acids. *Journal of Organometallic Chemistry* 696(26): 4264–4274.

Iwai, Tomohiro, Tomoya Harada, Kenji Hara and Masaya Sawamura. 2013. Threefold cross-linked polystyrene–triphenylphosphane hybrids: Mono-P-ligating behavior and catalytic applications for aryl chloride cross-coupling and C (Sp3)-H borylation. *Angewandte Chemie International Edition* 52(47): 12322–12326.

Izgi, Mehmet Sait, Ömer Şahin, Erhan Onat and Cafer Saka. 2020. Epoxy-activated acrylic particulate polymer-supported Co–Fe–Ru–B catalyst to produce H2 from hydrolysis of NH3BH3. *International Journal of Hydrogen Energy* 45(43): 22638–22648.

Jia, Xiaofei, Zuyu Liang, Jianbin Chen, Jinhe Lv, Kai Zhang, Mingjie Gao, Lingbo Zong, and Congxia Xie. 2019. Porous organic polymer supported rhodium as a reusable heterogeneous catalyst for hydroformylation of olefins. *Organic Letters* 21(7): 2147–2150.

Jiang, Shuai, Lin Wang, Yandong Duan, Jing An, Qingzhi Luo, Yumei Zhang, Yongfu Tang, Jianyu Huang, Bingkai Zhang, Jing Liu and Desong Wang. 2021. A novel strategy to construct supported silver nanocomposite as an ultra-high efficient catalyst. *Applied Catalysis B: Environmental* 283: 119592.

Ju, Pengyao, Shujie Wu, Qing Su, Xiaodong Li, Ziqian Liu, Guanghua Li and Qiaolin Wu. 2019. Salen–porphyrin-based conjugated microporous polymer supported Pd nanoparticles: Highly efficient heterogeneous catalysts for aqueous C–C coupling reactions. *Journal of Materials Chemistry A* 7(6): 2660–2666.

Kasaplar, Pinar, Paola Riente, Caroline Hartmann and Miquel A. Pericas. 2012. A polystyrene-supported, highly recyclable squaramide organocatalyst for the enantioselective Michael addition of 1,3-dicarbonyl compounds to B-nitrostyrenes. *Advanced Synthesis & Catalysis* 354(16): 2905–2910.

Kazemi, Mosstafa, Massoud Ghobadi and Ali Mirzaie. 2018. Cobalt ferrite nanoparticles (CoFe$_2$O$_4$ MNPs) as catalyst and support: Magnetically recoverable nanocatalysts in organic synthesis. *Nanotechnology Reviews* 7(1): 43–68.

Khokhar, Deepali, Sapana Jadoun, Rizwan Arif, Shagufta Jabin and Vaibhav Budhiraja. 2021a. Copolymerization of O-phenylenediamine and 3-amino-5-methylthio-1H-1,2,4-triazole for tuned optoelectronic properties and its antioxidant studies. *Journal of Molecular Structure* 1228: 129738.

Khokhar, Deepali, Sapana Jadoun, Rizwan Arif and Shagufta Jabin. 2021b. Functionalization of conducting polymers and their applications in optoelectronics. *Polymer-Plastics Technology and Materials* 60(5): 465–487.

Khokhar, Deepali, Sapana Jadouna, Rizwan Arif, Shagufta Jabin and Dhirendra Singh Rathore. 2022. Facile synthesis of the chemically oxidative grafted copolymer of 2,6-diaminopyridine (DAP) and thiophene (Th) for optoelectronic and antioxidant studies. *Journal of Molecular Structure* 1248: 131453.

Kim, Dongwoo. Hoon Ji, Moon Young Hur, Wonjoo Lee, Tea Soon Kim and Deug-Hee Cho. 2018. Polymer-supported Zn-containing imidazolium salt ionic liquids as sustainable catalysts for the cycloaddition of CO$_2$: A kinetic study and response surface methodology. *ACS Sustainable Chemistry & Engineering* 6(11): 14743–14750.

Kitanosono, Taku, Koichiro Masuda, Pengyu Xu and Shu-Kobayashi. 2018. Catalytic organic reactions in water toward sustainable society. *Chemical Reviews* 118(2): 679–746.

Kodicherla, Balaswamy and Mohan Rao Mandapati. 2014. A reusable polystyrene-supported copper (II) Catalytic system for N-arylation of Indoles and Sonogashira coupling reactions in water. *Applied Catalysis A: General* 483: 110–115.

Lai, Junshan, Sonia Sayalero, Alessandro Ferrali, Laura Osorio-Planes, Fernando Bravo, Carles Rodríguez-Escrich and Miquel A. Pericàs. 2018. Immobilization of Cis-4-hydroxydiphenylprolinol silyl ethers onto polystyrene. Application in the catalytic enantioselective synthesis of 5-hydroxyisoxazolidines in batch and flow. *Advanced Synthesis & Catalysis* 360(15): 2914–2924.

Li, Xiaoyang, Yufei Cao, Kai Luo, Yunze Sun, Jiarong Xiong, Licheng Wang, Zheng Liu, Jun Li, Jingyuan Ma, Jun Ge, Hai Xiao and Richard N. Zare. 2019. Highly active enzyme–metal nanohybrids synthesized in protein–polymer conjugates. *Nature Catalysis* 2(8): 718–725.

Mane, Sachin, Yu-Xia Li, Xiao-Qin Liu, Ming Bo Yue and Lin-Bing Sun. 2018. Development of adsorbents for selective carbon capture: Role of homo- and cross-coupling in conjugated microporous polymers and their carbonized derivatives. *ACS Sustainable Chemistry & Engineering* 6(12): 17419–17426.

Mikhaylov, Vladimir N., Viktor N. Sorokoumov, Kirill A. Korvinson, Alexander S. Novikov and Irina A. Balova. 2016. Synthesis and simple immobilization of palladium (II) acyclic diaminocarbene complexes on polystyrene support as efficient catalysts for Sonogashira and Suzuki–Miyaura cross-coupling. *Organometallics* 35(11): 1684–1697.

Mikhaylov, Vladimir N., Viktor N. Sorokoumov, Denis Martin Liakhov, Alexander G. Tskhovrebov and Irina A. Balova. 2018. Polystyrene-supported acyclic diaminocarbene palladium complexes in Sonogashira cross-coupling: Stability vs. catalytic activity. *Catalysts* 8(4): 141.

Mohammadi, Elmira and Barahman Movassagh. 2016. Polystyrene-resin supported N-heterocyclic carbene-Pd (II) complex based on plant-derived theophylline: A reusable and effective catalyst for the Suzuki-Miyaura cross-coupling reaction of arenediazonium tetrafluoroborate salts with arylboronic acids. *Journal of Organometallic Chemistry* 822: 62–66.

Mohammadi and Movassagh 2018. A polystyrene supported [PdCl–(SeCSe)] complex: A novel, reusable and robust heterogeneous catalyst for the Sonogashira synthesis of 1,2-disubstituted alkynes and 1,3-enynes. *New Journal of Chemistry* 42(14): 11471–11479.

Movassagh, Barahman, Fatemeh Hajizadeh and Elmira Mohammadi. 2018. Polystyrene-supported Pd (II)–N-heterocyclic carbene complex as a heterogeneous and recyclable precatalyst for cross-coupling of acyl chlorides with arylboronic acids. *Applied Organometallic Chemistry* 32(1): e3982.

Mpungose, Philani P., Zanele P. Vundla, Glenn E.M. Maguire and Holger B. Friedrich. 2018. The current status of heterogeneous palladium catalysed Heck and Suzuki cross-coupling reactions. *Molecules* 23(7): 1676.

Murugesan, Kathiravan, Thirusangumurugan Senthamarai, Vishwas G. Chandrashekhar, Kishore Natte, Paul CJ Kamer, Matthias Beller and Rajenahally V. Jagadeesh. 2020. Catalytic reductive aminations using molecular hydrogen for synthesis of different kinds of amines. *Chemical Society Reviews* 49(17): 6273–6328.

Nasrollahzadeh, Mahmoud, Mohaddeseh Sajjadi, Mohammadreza Shokouhimehr and Rajender S. Varma. 2019. Recent developments in palladium (nano) catalysts supported on polymers for selective and sustainable oxidation processes. *Coordination Chemistry Reviews* 397: 54–75.

Nejati, Kamellia, Sheida Ahmadi, Mohammad Nikpassand, Parvaneh Delir Kheirollahi Nezhad and Esmail Vessally. 2018. Diaryl ethers synthesis: Nano-catalysts in carbon-oxygen cross-coupling reactions. *RSC Advances* 8(34): 19125–19143.

Peng, Fang, Qi Wang, Rongjia Shi, Zeyi Wang, Xin You, Yuhong Liu, Fenghe Wang, Jay Gao and Chun Mao. 2016. Fabrication of sesame sticks-like silver nanoparticles/polystyrene hybridnanotubes and their catalytic effects. *Scientific Reports* 6(1): 1–9.

Perumgani, Pullaiah C., Srinivas Keesara, Saiprathima Parvathaneni and Mohan Rao Mandapati. 2016. Polystyrene supported N-phenylpiperazine–Cu (Ii) complex: An efficient and reusable catalyst for KA 2-coupling reactions under solvent-free conditions. *New Journal of Chemistry* 40(6): 5113–5120.

Qi, Chaorong, Jinwu Ye, Wei Zeng and Huanfeng Jiang. 2010. Polystyrene-supported amino acids as efficient catalyst for chemical fixation of carbon dioxide. *Advanced Synthesis & Catalysis* 352(11-12): 1925–1933.

Roy, David and Yasuhiro Uozumi. 2018. Recent advances in palladium-catalyzed cross-coupling reactions at Ppm to Ppb molar catalyst loadings. *Advanced Synthesis & Catalysis* 360(4): 602–625.

Samuels, M.C., F.J.L. Heutz, A. Grabulosa and P.C.J. Kamer. 2016. Solid-phase synthesis and catalytic screening of polystyrene supported diphosphines. *Topics in Catalysis* 59(19): 1793–1799.

Sedghi, Roya, Bahareh Heidari, Hatef Shahmohamadi, Pourya Zarshenas and Rajender S. Varma. 2019. Pd Nanocatalyst adorned on magnetic chitosan@ N-heterocyclic carbene: Eco-compatible Suzuki cross-coupling reaction. *Molecules* 24(17): 3048.

Şen, Betül, Buse Demirkan, Aysun Savk, Remziye Kartop, Mehmet Salih Nas, Mehmet
   Hakkı Alma, Sedat Sürdem and Fatih Şen. 2018. High-performance graphite-
   supported ruthenium nanocatalyst for hydrogen evolution reaction. *Journal of
   Molecular Liquids* 268: 807–812.
Şen, Betül, Ayşenur Aygün, Aysun Şavk, Sibel Duman, Mehmet Harbi Calimli, Ela Bulut
   and Fatih Şen. 2019. Polymer-graphene hybrid stabilized ruthenium nanocatalysts
   for the dimethylamine-borane dehydrogenation at ambient conditions. *Journal of
   Molecular Liquids* 279: 578–583.
Sun, Qi, Zhonfei Lv, Yuyang Du, Qinming Wu, Liang Wang, Longfeng Zhu, Xiangju
   Meng, Wanzhi Chen and Feng-Shou Xiao. 2013. Recyclable porous polymer-
   supported copper catalysts for Glaser and Huisgen 1,3-diolar cycloaddition reactions.
   Chemistry – An Asian Journal 8(11): 2822–2827.
Wang, Rong, Li Qin, Xin Wang, Bihua Chen, Yun Zhao and Guohua Gao. 2020.
   Polymer supported N-heterocyclic carbene ruthenium complex catalyzed transfer
   hydrogenation of ketones. *Catalysis Communications* 138: 105924.
Whiteoak, Christopher J., Andrea H. Henseler, Carles Ayats, Arjan W. Kleij and Miquel
   A. Pericàs. 2014. Conversion of Oxiranes and CO_2 to organic cyclic carbonates using
   a recyclable, bifunctional polystyrene-supported organocatalyst. *Green Chemistry*
   16(3): 1552–1559.
Xu, Caili, Min Hu, Qi Wang, Guangyin Fan, Yi Wang, Yun Zhang, Daojiang Gao and Jian
   Bi. 2018. Hyper-cross-linked polymer supported rhodium: An effective catalyst for
   hydrogen evolution from ammonia borane. *Dalton Transactions* 47(8): 2561–2567.
Xu, Zhenjin, Hui Wan, Jinmei Miao, Mingjuan Han, Cao Yang and Guofeng Guan.
   2010. Reusable and efficient polystyrene-supported acidic ionic liquid catalyst for
   esterifications. *Journal of Molecular Catalysis A: Chemical* 332(1–2): 152–157.
Yan, Shuo, Shiguang Pan, Takao Osako and Yasuhiro Uozumi. 2019. Solvent-free A3 and
   KA2 coupling reactions with Mol Ppm level loadings of a polymer-supported copper
   (II)–bipyridine complex for green synthesis of propargylamines. *ACS Sustainable
   Chemistry & Engineering* 7(10): 9097–9102.
Zuo, Yong, Ying Zhang and Yao Fu. 2014. Catalytic conversion of cellulose into levulinic
   acid by a sulfonated chloromethyl polystyrene solid acid catalyst. *ChemCatChem*
   6(3): 753–757.

Polyacrylic Acid and Its Derivatives-supported Catalyst

Anurakshee Verma[1*]**, Rizwan Arif**[1]**, Sapana Jadoun**[2] **and Jamal A. Siddiqui**[3]

[1] Department of Chemistry, School of Basic and Applied Sciences,
Lingaya's Vidyapeeth, Faridabad - 121002, Haryana, India
[2] Departamento de Química, Facultad de Ciencias, Universidad de Tarapacá, Avda.
General Velásquez, 1775, Arica, Chile
[3] Interdisciplinary Institute of Nanotechnology, Aligarh Muslim University, Aligarh, India

1. Introduction

1.1 Polyacrylic acid

Every year, millions of polyacrylic acid and its derivative are synthesized with other hydrophilic monomers. This type of polymers are widely utilized in applications like cleaning detergents, adhesives, sewage treatment, cosmetics, and medicine delivery. They are the chemical industry's polymers (Buchholz 2012). Polyacrylic acid has a low degradability due to its all-carbon backbone, although its widespread use in both practical and fundamental research. There is a demand for bio-degradable and bio-compatible polymers in the pharmaceutical and bio-medical industries for applications ranging from drug delivery to tissue engineering (Swift 1998, Gross 2002). Different type of techniques have been investigated to address these issues, including oligomer chain extension, vinyl polymerization, and so on (Paik et al. 1996, Gancet 1999, Chiellini 2003). These methods, however, are still acrylate-based, with the goal of adding fragile spots into the all type of carbon polymer backbone. In spite of these efforts, marketable success has been partial, and poly (acrylic acid) remains the industry's preferred polymer. Many scientists have discovered different techniques from earlier approaches in that it incorporates a degradable carbonate linkage into the polymer backbone, resulting in a structural simulator of poly(acrylic acid),

*Corresponding author: anurakshee@gmail.com

namely, poly(acrylic acid) (glyceric acid carbonate). This method is atom-efficient and delivers a degradation effect at per repeating unit, and yields CO_2 and glyceric acid as nontoxic, safe, and renewable degradation products.

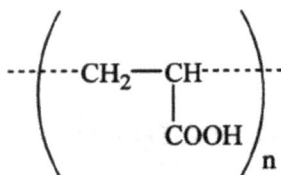

$$\left(-\!-\!-CH_2-\underset{\underset{COOH}{|}}{CH}-\!-\!-\!-\right)_n$$

Figure 1. Structure of polyacrylic acid

Polyacrylic acid and its derivatives are simple polymers made up of acrylic acid, that are unchanging in strong acids and bases. PAA is a frail anionic poly-electrolyte that can form composites with non-ionic polymers or with oppositely charged polymers. PAA has a gas transition temperature of 106 degrees Fahrenheit (Zhou 2008). On the other hand, it is a fluoro polymer-copolymer based on sulfonated tetra fluoro ethylene that is likewise chemically resistant (Mauritz 2004).

1.2 Supported catalyst

Most common transition metal cation complexes have insoluble supports and the advantages of homogeneous and heterogeneous catalysis can be united. Polymeric supports are particularly useful due to the different type of functional groups that may be produced for binding the catalytic group and the comfort with which the reaction products can be separated from the insoluble support. In several processes, natural polymers like cellulose, and synthetic compounds such as polyacrylic acid have been utilized as catalytic supports. Noble catalysts are mostly employed in industry as nanoparticles sustained on silica, zeolite, alumina or porous carbon surfaces (Tsonis 1984). Dispersed catalysts can be made using a variety of industrial techniques (Campbell 1988). They've been made in the centers of diblock copolymer micelles (Spatz 1995, Antonietti et al. 1995, Watson 1999), cores of triblock nanospheres (Underhill 2000), cores of dendrimers (Balogh 1998, Zhao 1998, Vassilev 1999), hydrophilic area of diblock copolymer films (Ciebien 1998), conditions of homopolymers (Vorontsov 1998) or surfaces of polymer microspheres in more fresh academic endeavours (Pathal 2000). They were made more easily by disperse reducing metal salts in solvent with polymer (Hirai 1998, Mayer 1996, Ahmadi et al. 1996).

2. Polyacrylic acid and its derivatives-supported catalyst

In sustainable chemistry, green catalysis for synthetic procedures and reactions is both thrilling and challenging. The development of numerous approaches to improve catalyst efficiency has sparked interest (Chang 2013). Immobilizing catalyst is a popular method for reusing catalyst and simplifying the purification process (Polshettiwar 2011). Catalyst immobilization on various support materials boosts catalyst production while also preventing dangerous catalysts since escaping from the environment (Biag 2013). Multiplicities of catalyst-supported materials have recently been created for immobilizing catalysts on beads, surfaces, and membranes, including polymer supports (Biag 2013, Bradshaw 2018, Kasaplar 2013, Wu 2013, Strappaveccia 2013, Yang 2013, Hickey 2015), ionic liquid supports (Khedkar 2013, Shen 2012), inorganic supports (Zhao 2011, Arya 2012, Genelot 2012, Wang 2013, Machado 2015, Mittal 2015), metal organic frame supports (Genna 2013), and nanocomposite supports (Pourjavadi 2015, Wang 2013, Zhu 2012). Magnetic iron oxide nanocomposites (Lu 2009) have been the most extensively utilized support material because they provide a large surface area with extra binding sites, reducing the drawbacks of heterogeneous catalysis (Zhang 2016).

2.1 Polyacrylic acid and its derivatives-supported platinum catalyst

Platinum (Pt) is one of the most often utilized materials as a catalyst for polymer support owing to its brilliant catalytic ability, solidity, and great conductivity. Unfortunately, because of the high cost of noble metal, it is difficult to scale up manufacturing of the devices. Many scientists and researchers, such as Neo and Ouyang in 2013, Li in 2016, Yang and Tang in 2016, are focusing on catalytic materials, that have flexible, quasi-solid-state, and solid electrolytes to overcome the problems. Shuyang et al. (2008) synthesized Pt nanoparticles, which were electroless dumped on grafted multiwall carbon nanotubes made of poly (acrylic acid). The thickness and homogeneity of Pt nanoparticles on MWNTs was found to be proportional to the density of PAA grafting. Pt on MWNTs showed the highest ratio of geranial/nerol, 5.2, which is 3 times higher than that of Pt/active carbon (Guo 2008). Wu (2012) used a doping, dedoping–redoping technique for doping PANI with PAA and poly(styrenesulfonic acid) (PSS). These groups may be more conducive to Pt electro deposition than PANI–CO_2H, PAA's group. The activity and stability of the PANI–PSS–Pt electrode toward methanol oxidation are superior to those of the PANI/PAA/Pt electrode, according to cyclic voltammetry and chronoamperometric response studies. Kuo (2011, 2012) explained the PANI–PAA–Pt composite exhibit considerable improvement of electro activity for methanol oxidation. Although CPs are routinely utilized as catalyst supports, the precise role of the CP and the comparative influence of the dopant in a CP matrix on the development of the composite's catalytic activity is yet unknown.

2.2 Polyacrylic acid and its derivatives-supported rhodium catalyst

Rhodium is used as catalyst for the hydrogenation of phenol and its nanoparticles supported by polyacrylic acid (PAA). An identical catalytic system was previously employed in an aqueous-alcoholic solution and ionic liquids (Maksimov 2013). The rhodium nanoparticles were made in a tetra-alkyl-ammonium ionic liquid by reducing an aqueous solution of rhodium tri-chloride with sodium borohydride in the presence of polymer, then adding -CD.

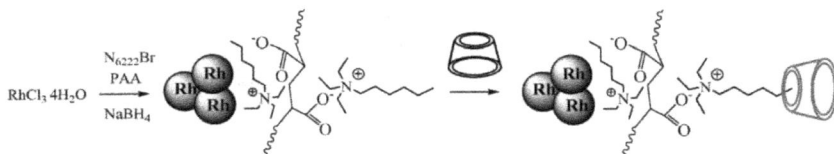

Figure 2. Rhodium nanoparticles

Because rhodium metal nanomaterials are notably agents in supported states for a range of catalytic applications, with CO hydrogenation (Kim 2012), hydrogen generation (Celok 2012, Rogatis 2008), and chemo-selective hydrogenations (Jacinto 2008, Maksimov 2013, Nakamula 2011, Huang 2012). It's of scientific interest to produce nano-sized particles on a fusion material, particularly if it results in particles or dispersions with exclusive features. Polymers were utilized as stabilizers in prior colloidal Rh particle studies. Valentini (1985) studied the catalytic efficiency of a polymer-bound catalyst may be greatly increased by optimizing the preparative approach and changing the characteristics of the support via a crosslinking procedure linking the polymer's reactive clusters. In the case of polyacrylic acid, a great scattering of metal parts within the reactant has no promising impact on the catalytic movement of the committed metal species. Edersson (1998) synthesized the coupling of 2-diphenyl phosphino-2-diphenyl phosphino-methyl pyrrolidine (PPM) to polyacrylic acid outcomes in a simple water-soluble equivalent of the parental ligand. The macromolecular ligand has a high water solubility. The reaction of the macromolecular ligand with bis-norbornadiene rhodium triflate Rh produces a polymer-bound cationic rhodium. Phosphine compound catalyst may be used for enantioselective hydrogenation in water or under biphasic conditions. Maksimov (2013) detected the hydrogenation of phenols in ionic liquids, and original catalyst scheme based on rhodium nanoparticles stabilized by polyacrylic acid has been proposed. The introduction of ionic liquid cations hooked on the surface layer self-possessed with polyacrylic acid has been exposed to expressively reduce particle size and diminish nanoparticle aggregation using TEM and XPS methods. Reyed (2014) prepared poly vinylpyrrolidone (PVP) (Jaini 2012, Ashida 2007), poly vinyl alcohol (PVA) (Jaini 2012), and poly acrylic acid (PAA) (Maksimov 2013) polymers that are often employed as protective agents. This type of polymer has effectiveness and characteristics property as protective agents in mixed-media colloids. As a result, it's crucial to look at how hybrid materials work as rhodium nanoparticle support protectors. The hydrogenation of nitrobenzene was examined by SIHN produced from PAA/Al$_2$O$_3$ and various concentrations of 3-tri-methoxysilyl propyl methacrylate (TMPM) as a connecter agent. Kuklin (2015) observed the hydrogenation of phenol in aqueous solution and ionic liquid; the inspiration of cyclodextrins on the doings and selectivity of a catalytic scheme based on rhodium nanoparticles stabilized by polyacrylic acid was examined. The reaction intermediate and the composition of the cyclodextrin (CD) were revealed to have an important impact on the pace of reaction and the circulation of reaction products.

2.3 Polyacrylic acid and its derivatives-supported palladium catalyst

The use of supported heterogeneous catalysts has a brief history. Catalysts made of palladium are a good example of this sort of material. Pd–C catalysts, for example, are frequently used in alkene hydrogenation (Augustine 1995). They have the benefits of wide availability, low handling and preparation costs, and activity that is often on par with or better than the more "modern" hydrogenation catalysts that are rarely reported. While hydrogenation is important, other palladium-catalyzed processes, such as allylic substitutions and cross-coupling chemistry, are currently of greater interest. A number of palladium catalysts have been developed for various purposes (Farina 2004). At very mild conditions, the author discovered a few years ago that classical Pd–C catalysts have reactivity similar to the well-known homogeneous Pd catalyst, tetrakis (triphenylphosphine) palladium (Bergbreiter 1992, 1983). The allylic substitution of allyl acetates by nucleophiles such as secondary amines might be catalyzed by a supported Pd species produced from an organometallic derivative of cross-linked polystyrene (Bergbreiter 1983). In these circumstances, catalysis was most effective in the presence of a soluble phosphine, with reactions occurring at room temperature on occasion. The importance of innovative forms of catalysts, such as colloidal metal catalysts, for various catalytic processes, including cross-coupling reactions, has recently increased (Reetz 2004). Pd(0) colloidal catalysts, for example, exhibit a high reactivity, which is likely synthesized in situ during cross-coupling chemistry from Pd(II) palladacycle molecules (Consorty 2005).

The technique of electroless copper deposition onto dielectric substrates utilizing pd+2/poly(acrylic acid) thin films is provided by Jackson (1990), who noticed the catalyst production mechanism. Dip-coating dielectric substrates in aqueous poly (acrylic acid) forms a 50-300 nm thick polymer layer on the surface, which is then dipped in water, where the process is catalyzed by PdSO$_4$. Palladium absorption in the poly(acrylic acid) film is revealed to be owing to the ion exchange H+/Pd+2. In most situations, the polymer ligand used in polymer-metal complex catalysts is made up of only one type of polymer, according to Jiang (1994). Polymethacrylic acid-palladium (PMAA-Pd) (Guo 1984) and chitosan-palladium (CS-Pd) (Wang 1992) complexes have been produced and

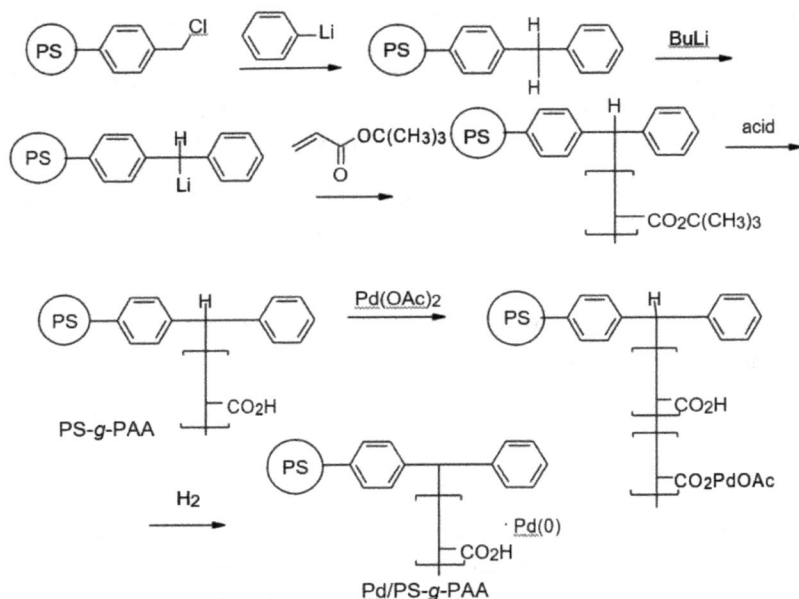

Figure 3. Synthesis of Pd–PAA-*g*-PS catalysts

used as heterogeneous hydrogenation catalysts. As is well known, enzymes are natural polymeric catalysts with extremely high activity and selectivity at mild conditions. A bipolymer palladium complex was used as a homogenous catalyst (Lio 2001). The Pd nanoparticles were manufactured inside a new form of porous polymeric support, which is reported by Liu (2001). The support is made up of diblock microspheres with block-segregated inner domains. The diblock microspheres were created using an oil-in-water emulsion process that has previously been used to make homopolymer preform spheres (Nakachi 2000) PtBA-b-PCEMA (poly(t-butyl acrylate)-block-poly (2-cinnamoyloxyethyl). Erman (2004) discovered electrospun copolymers of acrylonitrile and acrylic acid (PAN-AA) mats; catalytic palladium (Pd) nanoparticles were produced by reducing PdCl2 with hydrazine. PAN-AA and PdCl2 solutions in dimethyl formamide were used to make electrospun fibre mats (DMF). Electrospun fibres with Pd particles have 4.5 times the catalytic activity of current fibres. Gröschel et al. (2005) describes catalytically active membranes based on poly(acrylic acid) networks that incorporate palladium nanoparticles as a suitable catalyst for a gas-phase hydrogenation process. In the presence of a block copolymer, palladium particles were produced in organic solutions using reducing agents such as $NaBH_4$ or $LiAlH_4$ from $Pd(OAc)_2$. After combining the metal dispersion with a polymer dispersion having a predefined proportion of polymer, catalytically active membranes were created by cross-linking it with a difunctional epoxide (acrylic acid). The effects of membrane porosity, catalyst loading, and reactant mixture flow velocity on membrane catalytic behaviour are investigated. Chen (2006) observed that Pd(0) crystallites can be supported by grafts of poly(acrylic acid) on polyethylene powder (PE-g-PAA) or polystyrene (PS-g-PAA) in some procedures, acting as a homogeneous Pd(0) catalyst. These Pd–PE-g-PAA catalysts were active in allylic substitution reactions in the presence of a phosphine ligand. In these reactions, the active catalysts are leached from the support. The allylic substitution procedure requires external triphenylphosphine and substrate for the chemistry and Pd leaching (Bergbreiter 2006).

2.4 Polyacrylic acid and its derivatives-supported TEMPO catalyst

The stable nitroxyl radical TEMPO (2,2,6,6-tetramethylpiperidine-1-oxyl) is extensively used in polymer research as a spin-label for assessing polymer chain mobility and in controlled radical polymerizations (Nicolas 2013, Lappan 2015). TEMPO-grafted water-soluble polymers are redox active and could be engaged in energy storage. TEMPO is often used as an oxidation mediator for polysaccharides because it favourably oxidizes primary alcohols to the related aldehydes and carboxylic acids. Brag (2013) focused on the use of TEMPO grafted on water-soluble polymers to oxidize cellulose. In one variation, polyvinylamine with grafted TEMPO (PVAm-T) adsorbs on cellulose and promotes oxidation, resulting in surface aldehydes, which then form covalent bonds with amine groups on PVAm-T chains (Pelton 2011). A single procedure is used for oxidation and grafting at output. In a different experiment, author made poly(acrylic acid-g-TEMPO) (PAA-T) and showed that it may be a beneficial cellulose oxidant (Pelton 2014).

The breakdown of layer-by-layer (LbL) thin films formed of 2,2,6,6-tetramethylpiperidine-1-oxyl free radical-appended poly(acrylic acid) (TEMPO-PAA) and poly(ethylenimine) (PEI) was examined using a quartz crystal microbalance (QCM) and cyclic voltammetry by Anzai (2014). The shift in the resonance frequency of the QCM increased after electrolysis, indicating that the film was decomposed. Pelton (2016) tested the water solubility properties over the pH range as part of our ongoing study with PAA-T. Under acidic circumstances (pH 24.5), author was started to see adjustable stage separation of PAA-T polymers, dependent on the polymer content and degree of TEMPO substitution. Author offered a systematic set of experiments in which PAA-T stage performances in water were described, the theory of the powerful factor for phase separation is supported by an after sign. Huang (2017) seems that addition of catalysts to magnetic polyvalent provisions to help with catalyst reusing and recapture can be viable option. Author offered a simple and effective approach for immobilizing TEMPO onto polymer-modified magnetic nanoparticles. The resultant poly (acrylic acid) magnetic nanocomposite is an excellent material for immobilizing the organocatalyst H-TEMPO by an esterification reaction with the polymer chain's pendant carboxyl group. Recycling tests revealed that the

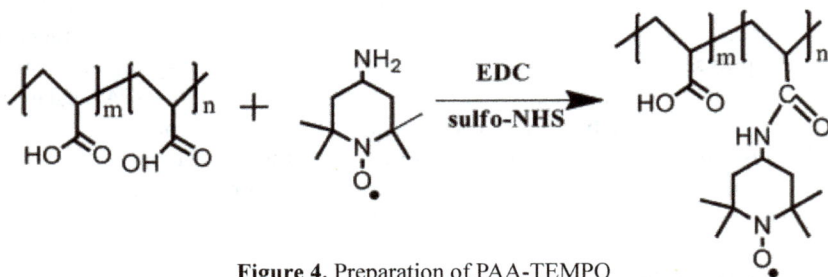

Figure 4. Preparation of PAA-TEMPO

nanocomposite catalyst's stability and reusability met expectations. Liu (2018) used solution polymerization to make water-soluble polyacrylic acid; P(AA-co-TA) was employed in place of free TEMPO as a recoverable catalyst. The use of water-soluble supported TEMPO in the selective oxidation of cellulose was initially described by Pelton (2011), Liu (2013), Shi (2014), and Fu (2017). A water-dissoluble nitroxide polymer of PVAm-TEMPO was formed by loading 4-COOH TEMPO onto water-dissoluble polyvinyl amine (PVAm). PVAm TEMPO impulsively adsorbed on cellulose and oxidized the C6 hydroxyl to aldehyde groups, which cooperated to generate covalent connections with primary amines on PVAm, according to the findings. Temporarily, they immobilized TEMPO on PAA via acylation between the amino groups of 4-NH_2-TEMPO and the carboxyl groups of PAA, resulting in a water-dissoluble nitroxide polymer of PAA-TEMPO. To make the PAA TEMPO performance as a catalytic oxidation catalyst, it was important to adsorb a layer of PVAm on the surface of cellulose, unlike PVAm-TEMPO.

Li (2018) created extremely biocompatible DO30 nanoassemblies based on oxidized starch self-assembly at biological concentrations for improved anti-cancer behaviour. DO30 nanoassemblies were revealed to be non-toxic in numerous cell lines. When compared to DOX alone or DOX-loaded PAA-PEG-RGD nano-assemblies, DOX-loaded DO30-PEG-RGD nano-assemblies demonstrated special buildup and enhanced inhibitory propagation of HepG2 cells via v3

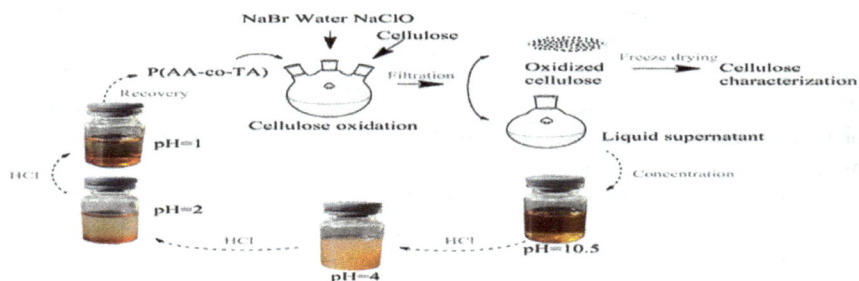

Figure 5. Polyacrylic acid TEMPO-mediated oxidation of cellulose

integrin-mediated gratitude. The demonstration of the RGD modified DO30 nano-assembly is a great opportunity to develop a new method for constructing safe and effective anti-cancer nanocarriers.

2.5 Polyacrylic acid and its derivatives-supported other catalyst

Molybdenum complexes formed via dispersion of MoCl5 in a polyacrylic acid matrix. The particulars of taster preparation, heat treatment and adsorbate experience all mark the exact bonding of the cation to the support. After activation at 373 K, molybdenum is prepared to the polymer matrix through mono and bi-dentate oxygen ligands, exposing the progressive replacement of chlorine ligands in activated material by oxygen ligands from the polymer. The Mo-Cl bond is resistant to dioxygen and hydrolysis in the PAA medium, implying strong matrix bonding and the presence of coordinately saturated octahedral Mo (V) complexes (Dyrek 1993).

A simple method to a degradable form of poly (acrylic acid) – poly (glyceric acid carbonate) by the copolymerization of benzyl glycidate and CO_2 using the bi-functional catalyst followed by hydrogenolysis. The polymerization method provides a strong selectivity for carbonate linkages (>99%) as well as a polymer/ cyclic carbonate selectivity. Poly (glyceric acid carbonate) is generated when the resultant polymer is deprotected, and it degrades in aqueous solution. The PGAC hydrogel rapidly dissolves when a degradable site is inserted into each repeating unit (Zhang 2015). Only a few well-defined structures, such as poly(malic acid) s, make up the existing pool of acidic polymers, which is both small and diverse (Guerin 1985). Along with its biocompatibility and degradability, the authors believe that poly (glyceric acid carbonate) will be useful to chemists and material scientists working on a variety of chemical, biological, and pharmacological difficulties. The nanocomposite material was characterized using a variety of spectroscopic approaches. Powder X-ray diffraction patterns indicated the formation of silver nanoparticles, while a transmission electron microscopy image revealed that Ag nanoparticles are formed and equally diffused in mesoporous polyacrylic acid (Mandi 2016). With glycerol as a hydrogen basis, the Ag-MCP-1 nanocomposite may be working as an effective heterogeneous catalyst in the reductive coupling of nitrobenzenes and alcohols.

Poly (acrylic acid) may be utilized as a stabilizer for nanoparticle production to increase the dispersion of nanoparticles within the hydrogel matrix and partially avoid the development of aggregates. PAA's carboxyl functional groups have the capability to cooperate with an extensive variety of inorganic materials. Our own research has shown that reinforced silver nanoparticles have higher catalytic movement in a variety of processes. Silver possesses the greatest electrical conductivity, as well as improved catalytic activity, antimicrobial characteristics, and biocompatibility. As a result of the right selection of organic mesoporous PAA and inorganic Ag particles phases, Brodt (2014) have produced a unique material that shows both organic and inorganic features.

The electrodes' stability was shown to be much greater than that of electrodes made from the same catalyst using traditional procedures (Brodt et al. 2014). Brodt et al. suggested that PAA occupies a crucial role in increasing the electrode's stability based on their findings. The electrospinning and solution characteristics of mixed Nafion and PAA solutions were investigated by Chen et al. (2008). The least quantity of PAA required for bead-free electrospinning while retaining sufficient conductivity has been observed to be about 25% by weight (Chen et al. 2008).

The use of PAA-Nafion composites as a catalyst support for PEM fuel cell electrodes was investigated. A modified UV light driven photochemical reaction of the Pt precursor with a modest quantity of Nafion as surfactant was employed to manufacture the catalyst. The freshly produced Pt nanoparticles were treated with a PAA-Nafion solution. Carbon was only added afterwards to increase the electrical conductivity. The stability of the catalyst was tested in a fuel cell test bench using potential cycling in aqueous solution and accelerated degradation experiments. A formula for electrospinning was included in the manufactured catalyst suspension.

3. Application of polyacrylic acid-supported catalyst

Polyacrylic acid-supported catalyst prospective uses in chemical and biological devices, layer-by-layer (LbL) thin films have gotten a lot of interest in the field of polymer science and technology. Alternate deposition of polymeric materials by electrostatic bonding, hydrogen bonding, or biological affinity can be used to make LbL thin films, linear polymers, dendrimers, proteins, and polysaccharides (Decher 1997), which are examples of polymeric materials for LbL deposition.

In a pH 8.4 aqueous buffer solution, cross-linked hydrogels made from poly(glyceric acid carbonate) and poly(ethylene glycol) diaziridine degrade significantly more than hydrogels made from poly(acrylic acid) and poly(ethylene glycol) diaziridine. For example: ion-exchange resins, emulsifiers, chelating agents, scale inhibitors, superabsorbents, and dispersants. Many of these applications use linear or cross-linked PAA designs, the PAA backbone's flexibility allows for multivalent interactions with a variety of surfaces or ions, which may be influenced by branching. Tunable synthetic approaches for the insertion of controlled branching into well-defined PAA designs are necessary to understand the structure–property correlations for these significant commercial systems. Authors have produced a library of PAA polymers with systematic architectural alterations to solve this difficulty (Inoue 2005).

Polymer brushes have emerged as a useful and promising tool for immobilizing various specimens and moieties on macroscopic surfaces, such as macromolecules, proteins, and nanoparticles. When compared to alternative methods for chemical grafting of various species, the incorporation of polymer brushes into specimens on silicon substrates, such as electron-beam lithography, block copolymer micelle nanolithography (BCML), in situ nucleation and growth, layer by layer deposition, or self-assembled monolayers (SAMs), has a number of advantages (Verma 2018, Jadoun 2018, Verma 2017, Riaz 2015).

4. Conclusion and future prospects

Polyacrylic acid and its derivatives are produced in various ways with the help of catalyst. Platinum (Pt) is one of the most commonly used catalysts for polymer support because of its remarkable catalytic activity, stability, and conductivity. For aqueous-alcoholic solution and ionic liquids, rhodium has the same catalytic mechanism. TEMPO (2,2,6,6-tetramethylpiperidine-1-oxyl) is a spin-label used in polymer research to measure polymer chain mobility and in controlled radical polymerizations. Pd–C catalytic support is often used in the hydrogenation of alkenes. A wide spectrum of polyacrylic derivatives is tolerated by the whole catalytic system. The aforementioned catalytic method has numerous advantages, including a simple set-up procedure and the ability to reuse the catalyst for multiple reaction cycles. These qualities make the method inexpensive and environmentally benign, and it has a lot of potential for polymer value addition synthesis.

Acknowledgement

Author Sapana Jadoun is grateful for the support of National Research and Development Agency of Chile (ANID) for the project FONDECYT Postdoctoral 3200850.

References

Ahmadi, T.S., Z.L. Wang, T.C. Green, A. Henglein and M.A. El-Sayed. 1996. Shape-controlled synthesis of colloidal platinum nanoparticles. *Science Journal* 272: 1924-1925.

Antonietti, M., M. Wenz, L. Bronstein and M.V. Seregina. 1995. Synthesis and characterization of noble metal colloids in block copolymer micelles. *Advanced Material* 7: 1000.

Arya, K., U.C. Rajesh and D.S. Rawat. 2012. Proline confined FAU zeolite: Heterogeneous hybrid catalyst for the synthesis of spiroheterocycles via a mannich type reaction. *Green Chemistry* 14(12): 3344–3351.

Ashida, T., K. Miura, T. Nomoto, S. Yagi, H. Sumida, G. Kutluk, K. Soda, H. Namatame and M. Taniguchi. 2007. Synthesis and characterization of Rh (PVP) nanoparticles studied by XPS and NEXAFS. *Surface Science* 601: 3898-3901.

Augustine, R.L. 1995. Heterogeneous Catalysis for the Synthetic Chemist. Dekker, New York.

Baig, R.B.N. and R.S. Varma. 2013. Magnetically retrievable catalysts for organic synthesis. *Chemical Communications* 49(8): 752–770.

Baig, R.B.N. and R.S. Verma. 2013. Copper on chitosan: A recyclable heterogeneous catalyst for azide-alkyne cycloaddition reactions in water. *Green Chemistry* 15(7): 1839–1843.

Balogh, L. and D.A. Tomalia. 1998. Poly(amidoamine) dendrimer-templated nanocomposites. 1. Synthesis of zerovalent copper nanoclusters. *Journal of American Chemical Society* 120: 7355.

Bergbreiter, D.E. and B. Chen. 1983. Liquid/liquid biphasic recovery/reuse of soluble polymer-supported catalysts. *Journal of the Chemical Society, Chemical Communications* 1238.

Bergbreiter, D.E., B. Chen and T.J. Lynch. 1983. Palladium/polystyrene catalysts. *The Journal of Organic Chemistry* 48: 4179.

Bergbreiter, D.E., B. Chen and D. Weatherford. 1992. New strategies in using macromolecular catalysts in organic synthesis. *Journal of Molecular Catalysis* 74: 409.

Bergbreiter, D.E., A. Kippenberger and Z. Zhong. 2006. Catalysis with palladium colloids supported in poly (acrylic acid)-grafted polyethylene and polystyrene. *Canadian Journal of Chemistry* 84: 1343–1350.

Bradshaw, M., J. Zou, L. Byrne, K. Swaminathan Iyer, S.G. Stewart and C.L. Raston. 2011. Pd(ii) conjugated chitosan nanofibre mats for application in Heck cross-coupling reactions. *Chemical Communication* 47(45): 12292–12294.

Bragd, P.L., H. Van Bekkum and A.C. Besemer. 2004. Tempo-mediated oxidation of polysaccharides: Survey of methods and applications. *Topics in Catalysis* 27: 49–66.

Brodt, M., T. Han, N. Dale, E. Niangar, R. Wycisk and P.J. Pintauro. 2014. Electrospun nanofiber electrodes for hydrogen/air proton exchange membrane fuel cells. *Journal of The Electrochemical Society* 20(162): 84–91.

Buchholz, F.L. 2012. Polyacrylamides and Poly (Acrylic Acids). Wiley-VCH Verlag GmbH & Co. KGaA: Weinheim, Germany. 28.

Campbell, I.M. 1988. Catalysis at Surfaces. Chapman and Hall: London.

Çelik, D., S. Karahan, M. Zahmakıran and S. Özkara. 2012. Hydrogen generation from the hydrolysis of hydrazine-borane catalyzed by rhodium (0) nanoparticles supported on hydroxyapatite. *International Journal of Hydrogen Energy* 37: 5143-5151.

Chen, H., J.D. Snyder and Y.A. Elabd. 2008. Electrospinning and solution properties of nafion and poly(acrylic acid). *Macromolecules* 41: 128–135.

Chen, J., A.D. Del Genio, B.E. Carlson and M.G. Bosilovich. 2008. The spatiotemporal structure of twentieth-century climate variations in observations and reanalyses. Part I: Long-term trend. *Journal of Climate* 21: 2611–2633.

Chiellini, E., A. Corti, S. D'Antone and R. Solaro. 2003. Biodegradation of poly (vinyl alcohol) based materials. *Progress in Polymer Science* 28: 963.

Chng, L.L., N. Erathodiyil and J.Y. Ying. 2013. Nanostructured catalysts for organic transformations. *Accounts of Chemical Research* 46(8): 1825–1837.

Ciebien, J.F., R.E. Cohen and A. Duran. 1998. Catalytic properties of palladium nanoclusters synthesized within diblock copolymer films: Hydrogenation of ethylene and propylene. *Supramolecular Science* 5: 31.

Consorti, C.S., F.R. Flores and J. Dupont. 2005. Kinetics and mechanistic aspects of the Heck reaction promoted by a CN-palladacycle. *Journal of American Chemical Society* 127: 12054.

De Rogatis, L., T. Montini, M.F. Casula and P. Fornasiero. 2008. Design of nanomaterials based on complex Al-Zr-Ce-oxides for bio ethanol transformations. *Journal of Alloys and Compounds* 451: 516-520.

Decher, G. 1997. Fuzzy nanoassemblies: Toward layered polymeric multicomposites. *Science*, 277: 1232-1237.

Dyrek, K., K. Kruczala and Z. Sojka. 1993. Catalysis on Polymer Supports. ESR of Mo(V) dispersed in poly(acrylic acid) matrices. *The Journal of Physical Chemistry* 97: 9196-9200

Farina, V. 2004. High-turnover palladium catalysts in cross-coupling and Heck chemistry: A critical overview. *Advanced Synthesis and Catalysis* 346: 1553.

Fu, Q., Z.R. Gray, A. Vander and R.H. Pelton. 2016. Phase behavior of aqueous poly(acrylic acid-g-TEMPO). *Macromolecule* 49(13): 4935-4939.

Gancet, C., R. Pirri, J.M. Dalens, B. Boutevin, B. Guyot, C. Loubat, J. Le Petit, A.M. Farnet and S. Tagger. 1999. Methodology in studying improvement of polyacrylates biodegradability. *Macromolecular Symposia* 144: 211

Genelot, M., N. Villandier, A. Bendjeriou, P. Jaithong, L. Djakovitch and V. Dufaud. 2012. Palladium complexes grafted onto mesoporous silica catalysed the double carbonylation of aryl iodides with amines to give α-ketoamides. *Catalysis Science & Technology* 2(9): 1886-1893.

Genna, D.T., A.G. Wong-Foy, A.J. Matzger and M.S. Sanford. 2013. Heterogenization of homogeneous catalysts in metal-organic frameworks via cation exchange. *Journal of American Chemical Society* 135(29): 10586-10589.

Gröschel, Lothar, Rami Haidar, Andreas Beyer, Helmut Cölfen, Benjamin Frank, and Reinhard Schomäcker. 2005. Hydrogenation of propyne in palladium-containing polyacrylic acid membranes and its characterization. *Industrial & Engineering Chemistry Research* 44(24): 9064-9070.

Gross, R.A. and B. Kalra. 2002. Biodegradable polymers for the environment. *Science* 297: 807.

Guerin, P., M. Vert, C. Braud and R.W. Lenz. 1985. Optically active malic acid. *Polymer Bulletin* 14: 187.

Guo, G., F. Qin, D. Yang, C. Wang, H. Xu and S. Yang. 2008. Synthesis of platinum nanoparticles supported on poly(acrylic acid) grafted MWNTs and their hydrogenation of citral. *Chemistry of Materials* 20(6): 2291-2297.

Guo, X.Y., H.J. Zong, Y.J. Li and Y.Y. Jiang. 1984. Catalytic hydrogenation behaviours of palladium complexes of chitosan-polyacrylic acid and chitosan-polymethacrylic acid. *Macromolecular Rapid Communications* 5: 507.

Hickey, D.P., R.D. Milton, D. Chen, M.S. Sigman and S.D. Minteer. 2015. TEMPO-modified linear poly(ethylenimine) for immobilization-enhanced electrocatalytic oxidation of alcohols. *ACS Catalysis* 5(9): 5519-5524.

Hirai, H., N. Yakura, N. Seta and S. Hodoshima. 1998. Characterization of palladium nanoparticles protected with polymer as hydrogenation catalyst. *Reactive & Functional Polymers* 37: 121.

Huang, G., X. Liu, Y. Bei and H. Ma. 2017. Facile TEMPO immobilization onto poly(acrylic acid)-modified magnetic nanoparticles: Preparation and property. *International Journal of Polymer Science* 2017: https://doi.org/10.1155/2017/9621635

Huang, L., P. Luo, W. Pei, X. Liu, Y. Wang, J. Wang, W. Xing and J. Huang. 2012. Selective hydrogenation of nitroarenes and olefins over rhodium nanoparticles on hydroxyapatite. *Advanced Synthesis & Catalysis* 354: 2689-2694.

Inoue, H. and J. Anzai. 2005. Stimuli-sensitive thin films prepared by a layer-by-layer deposition of 2-iminobiotin-labeled poly(ethyleneimine) and avidin. *Langmuir* 21(18): 8354–8359.

Jacinto, M.J., P.K. Kiyohara, S.H. Masunaga, R.F. Jardim and L.M. Rossi. 2008. Recoverable rhodium nanoparticles: Synthesis, characterization and catalytic performance in hydrogenation reactions. *Applied Catalysis*, A 338: 52-57.

Jackson, R.L. 1990. Pd+2/poly (acrylic acid) thin films as catalysts for electroless copper deposition: Mechanism of catalyst formation. *Journal of the Electrochemical Society* 137.

Jadaun, S., A. Verma, S.M. Asharaf and U. Riaz. 2017. A short review on the synthesis, characterization, and application 5 studies of poly(1-naphthylamine) a seldom explored 6 polyaniline derivative. *Colloid and Polymer Science* 275(9): 1443-1445.

Jadaun, S., A. Verma and U. Riaz. 2018. Luminol modified polycarbazole and poly (o-anisidine): Theoretical insights compared with experimental data. *Spectrochimica Acta Part A: Molecular and Biomolecular Spectroscopy* 204: 64-72.

Jaine, J.E. and M.R. Mucalo. 2012. Synthesis and characterization of polymer-protected rhodium and palladium sols in mixed media. *Journal of Colloid and Interface Science* 375: 12-22.

Jiang, Y., T. Li, M. Lu, D. Li, F. Ren, H. Zhao and Y. Li. 2018. TEMPO-oxidized starch nanoassemblies of negligible toxicity compared with polyacylic acids for high performance anti-cancer therapy. *International Journal of Pharmaceutics* 21: 8354–8359.

Jin, J.J., G.C. Chen, M.Y. Huang and Y.Y. Jiang. 1994. Catalytic hydrogenation behaviours of palladium complexes of chitosan-polyacrylic acid and chitosan-polymethacrylic acid. *Reactive Polymers* 23: 95-100.

Kasaplar, P., C. Rodríguez-Escrich and M.A. Pericas. 2013. Continuous flow, highly enantioselective Michael additions catalyzed by a PS-supported squaramide. *Organic Letters* 15(14): 3498–3501.

Khedkar, M.V., T. Sasaki and B.M. Bhanage. 2013. Immobilized palladium metal-containing ionic liquid-catalyzed alkoxycarbonylation, phenoxycarbonylation, and aminocarbonylation reactions. *ACS Catalysis* 3(3): 287–293.

Kim, S., K. Qadir, S. Jin, A. Satyanarayana Reddy, B. Seo, B.S. Mun, S.H. Joo and J.Y. Park. 2012. Trend of catalytic activity of CO oxidation on Rh and Ru nanoparticles: Role of surface oxide. *Catalysis Today* 185: 131-137.

Kuklin, S., A. Maximov, A. Zolotukhina and E. Karakhanov. 2016. New approach for highly selective hydrogenation of phenol to cyclohexanone: Combination of rhodium nanoparticles and cyclodextrins. *Catalaysis Communications* 73: 63-68.

Kuo, C.W., C.C. Yang and T.Y. Wu. 2011. Facile synthesis of composite electrodes containing platinum particles distributed in nanowires of polyaniline-poly(acrylic acid) for methanol oxidation. *International Journal of Electrochemical Science* 6.

Kuo, C.W., B.K. Chen, Y.H. Tseng, T.H. Hsieh, K. Shan, T.Y. Wu and H.R. Chen. 2012. A comparative study of poly(acrylic acid) and poly(styrenesulfonic acid) doped into polyaniline as platinum catalyst support for methanol electro-oxidation. *Journal of Taiwan Institute of Chemical Engineering* 43: 798–805.

Lappan, U., B. Wiesner and U. Scheler. 2015. Rotational dynamics of spin-labeled polyacid chain segments in polyelectrolyte complexes studied by Cw Epr spectroscopy. *Macromolecules* 48: 3577–3581.

Li, Y., Y. Wu, Q. Xu, Y. Gao, G. Cao, Z. Meng and C. Yang. 2013. Facile and controllable synthesis of polystyrene/palladium nanoparticle@polypyrrole nanocomposite particles. *Polymer Chemistry* 4(17): 4655–4662.

Liu, S., T. Sun, D. Yang, M. Cao and H. Liang. 2018. Polyacrylic acid supported TEMPO for selective catalytic oxidation of cellulose: Recovered by its pH sensitivity. *Cellulose* 25: 5687–5696.

Lu, J. and P.H. Toy. 2009. Organic polymer supports for synthesis and for reagent and catalyst immobilization. *Chemical Reviews* 109(2): 815–838.

Lu, Z., G. Liu, H. Phillips, J.M. Hill, J. Chang and R.A. Kydd. 2001. Palladium nanoparticle catalyst prepared in poly(acrylic acid)-lined channels of diblock copolymer microspheres. *Nano Letters* 1: 683-687.

Machado, A., M.H. Casimiro, L.M. Ferreira, J.E. Castanheiro, A.M. Ramos, I.M. Fonseca and J. Vital. 2015. New method for the immobilization of nitroxyl radical on mesoporous silica. *Microporous and Mesoporous Materials* 203: 63–72.

Maksimov, A.L., S.N. Kuklin, Y.S. Kardasheva and E.A. Karakhanov. 2013. Hydrogenation of phenol in ionic liquid on rhodium nanoparticles. *PET Chemistry* 53: 157-163.

Malmstrom, T. and C. Andersson. 1999. Enantioselective hydrogenation in water catalysed by rhodium phosphine complexes bound to polyacrylic acid. *Journal of Molecular Catalysis A: Chemical* 139: 259–270.

Mandi, U., A. Singha Roy, Sudipta K. Kundu, S. Roy, A. Bhaumik and Sk. Manirul Islam. 2016. Mesoporous polyacrylic acid supported silver nanoparticles as an efficient catalyst for reductive coupling of nitrobenzenes and alcohols using glycerol as hydrogen source. *Journal of Colloid and Interface Science* 472: 202–209.

Mauritz, K.A. and R.B. Moore. 2004. State of understanding of Nafion. *Chemical Reviews* 104: 4535–4585.

Mayer, A.B.R. and J.E. Mark. 1996. Polymer-protected palladium nanoparticles and their use in catalysis. *Macromolecular Research* 33: 451.

Mittal, N., G.M. Nisola, J.G. Seo, S.-P. Lee and W.-J. Chung. 2015. Organic radical functionalized SBA-15 as a heterogeneous catalyst for facile oxidation of 5-hydroxymethylfurfural to 2,5-diformylfuran. *Journal of Molecular Catalysis A: Chemical* 404-405: 106–114.

Mustafa, M., A. Demir, A. Mehmet, Gulgun Z. Yusuf and M.B. Erman. 2004. Palladium nanoparticles by electrospinning from poly(acrylonitrile-co-acrylic acid)-PdCl2 solutions. *Relations between Preparation Conditions, Particle Size, and Catalytic Activity Macromolecules* 37: 1787-1792.

Nakache, E., N. Poulain, F. Candau, A.M. Orecchioni and J.M. Irache. 2000. pp. 577-635. *In:* Nalwa, H.S. (Ed.), Handbook of Nanostructured Materials and Nanotechnology. Academic Press: San Diego.

Nakamula, I., Y. Yamanoi, T. Imaoka, K. Yamamoto and H. Nishihara. 2011. A uniform bimetallic rhodium/iron nanoparticle catalyst for the hydrogenation of olefins and nitroarenes. *Angewandte Chemie International Edition* 50: 5830-5833.

Nicolas, J., Y. Guillaneuf, C. Lefay, D. Bertin, D. Gigmes and B. Charleux. 2013. Nitroxide-mediated polymerization. *Progress in Polymer Science* 38: 63–235.

Paik, Y.H., E.S. Simon and G. Swift. 1996. Production of polysuccinimide by thermal polymerization of maleamic acid. *Hydropol Polymers* 248: 79.

Pathak, S., M.T. Greci, R.C. Kwong, K. Mercado, G.K.S. Prakash, G.A. Olah and M.M.E. Thompson. 2000. Synthesis and applications of palladium-coated poly(vinyl pyridine) nano spheres. *Chemistry of Material* 12: 1985.

Pelton, R., P.R. Ren, J. Liu and D. Mijolovic. 2011. Polyvinylamine graft-tempo adsorbs onto, oxidizes and covalently bonds to wet cellulose. *Biomacromolecules* 12(4): 942–948.

Polshettiwar, V., R. Luque, A. Fihri, H.B. Zhu, M. Bouhrara and J.M. Basset. 2011. Magnetically recoverable nanocatalysts. *Chemical Reviews* 111(5): 3036–3075.

Pourjavadi, A., S.H. Hosseini, M. Doulabi, S.M. Fakoorpoor and F. Seidi. 2012. Multi-layer functionalized poly(ionic liquid) coated magnetic nanoparticles: Highly recoverable and magnetically separable brønsted acid catalyst. *ACS Catalysis* 2(6): 1259–1266.

Reetz, M.T. and J.G. de Vries. 2004. Sustainable Mizoroki–Heck reaction in water: Remarkably high activity of Pd(OAc)2 immobilized on reversed phase silica gel with the aid of an ionic liquid *Chemical Communications* (Cambridge), 1559.

Riaz, U., S.M. Ashraf and A. Verma. 2016. Influence of conducting polymer as filler and matrix on the spectral morphological and fluorescent properties of sonochemically intercalated poly(o-phenylenediamine)/montmorillonite nanocomposites recent patents on nanotechnology. *Recent Patenets on Nanotechnology* 10(1): 66–76.

Riaz, U., S.M. Ashraf and A. Verma. 2015. Recent advances in the development of conducting polymer intercalated clay nanocomposites. *Current Organic Chemistry* 19(17): 1275-1291.

Schel, L.G., R. Haidar, A. Beyer and H. Collfen. 2005. Hydrogenation of propyne in palladium-containing polyacrylic acid membranes and its characterization. *Industrial & Engineering Chemistry Research* 44: 9064-9070.

Shen, Z.L., H.L. Cheong, Y.C. Lai, W.Y. Loo and T.P. Loh. 2012. Application of recyclable ionic liquid-supported imidazolidinone catalyst in enantioselective Diels-Alder reactions. *Green Chemistry* 14(9): 2626–2630.

Shi, S., R. Pelton, Q. Fu and S. Yang. 2014. Comparing polymer-supported TEMPO mediators for cellulose oxidation and subsequent polyvinylamine grafting. *Industrial & Engineering Chemistry Research* 53: 4748–4754.

Spatz, J.P., A. Roescher, S. Sheiko, G. Krausch and M. Moller. 1995. Synthesis and characterization of noble metal colloids in block copolymer micelles. *Advanced Materials* 7: 1000.

Strappaveccia, G., D. Lanari and D. Gelman. 2013. Efficient synthesis of cyanohydrin trimethylsilyl ethers via 1,2-chemoselective cyanosilylation of carbonyls. *Green Chemistry* 15(1): 199-204.

Swift, G. 1994. Water-soluble polymers. *Polymer Degradation & Stability* 45: 215.

Swift, G. 1998. Requirements for biodegradable water-soluble polymers. *Polymer Degradation & Stability* 59(19).

Takahashi, S., Y. Aikawa, T. Kudo, T. Ono, Y. Kashiwagi and J. Anzai. 2014. Electrochemical decomposition of layer-by-layer thin films composed of TEMPO-modified poly(acrylic acid) and poly(ethyleneimine). *Colloid & Polymer Science* 292: 771–776.

Tsonis, C.P.J. 1984. Catalysis on polymeric matrices. *Journal of Chemical Education* 61: 479.

Underhill, R.S. and G.J. Liu. 2000. Preparation and performance of Pd particles encapsulated in block copolymer nanospheres as a hydrogenation catalyst. *Chemistry of Materials* 12: 3633.

Valentini, G., G. Sbrana, G. Braca and P.D. Prato. 1985. Postcrossi-inking of rhodium catalysts bound to polyacrylic acid: A useful technique to enhance and preserve catalytic activity. *Journal of Catalysis* 96: 41-50.

Vassilev, K., J. Kreider, P.D. Miller and W.T. Ford. 1999. Poly (propylene imine) dendrimer complexes of Cu (II), Zn (II), and Co (III) as catalysts of hydrolysis of bis-(p-nitrophenyl) phosphate react. *Reactive & Functional Polymers* 41: 205.

Verma, A. and U. Riaz. 2017. Mechanochemically synthesized poly(o-toluidine) intercalated montmorillonite nanocomposites as anti-tuberculosis drug carriers. *International Journal of Polymeric Materials & Polymeric Biomaterials* 67(4): 221-228.

Verma, A. and U. Riaz. 2018. Synthesis, characterization and in-vitro drug release studies of sonolytically intercalated poly(o-anisidine)/montmorillonite nano composites. *Macromolecular Reseacrh* 27(2): 140-152.

Verma, A. and U. Riaz. 2018. Sonolytically intercalated poly(anisidine-co-toluidine)/bentonite nanocomposites: pH responsive drug release characteristics. *Journal of Drug Delivery Science and Technology* 48: 49-58.

Verma, A. and U. Riaz. 2018. Spectral, thermal and morphological characteristics of ultrasonically synthesized poly(anisidine-co-phenylenediamine)/bentonite nanocomposites: A potential anti-diabetic drug carrier. *Journal of Molecular Liquids* 261: 1-13.

Vorontsov, P.S., G.N. Gerasimov, E.N. Golubeva, E.I. Grigorev, S.A. Zavyalov, L.M. Zav'yalova and L.I.R. Trakhtenberg. 1998. Gas-sensitive and catalytic properties of ensembles of interacting palladium nanoparticles. *The Journal of Physical Chemistry* 72: 1742.

Wang, S., Q. Zhao and H. Wei. 2013. Aggregation-free gold nanoparticles in ordered mesoporous carbons: Toward highly active and stable heterogeneous catalysts. *Journal of American Chemical Society* 135(32) 11849–11860.

Wang, S., Z. Zhang, B. Liu and J. Li. 2013. Silica coated magnetic Fe_3O_4 nanoparticles supported phosphor tungstic acid: A novel environmentally friendly catalyst for the synthesis of 5-ethoxymethylfurfural from 5-hydroxymethylfurfural and fructose. *Catalysis Science & Technology* 3(8): 2104–2112.

Wang, X.X., M.Y. Huang and Y.Y. Jiang. 1992. Hydrogenation catalytic behaviors of palladium complexes of chitin and chitosan. *Makromolekulare Chemie. Macromolecular Symposia* 59: 113.

Watson, K.J., J. Zhu, S.T. Nguyen and C.A.J. Mirkin. 1999. Hybrid nanoparticles with block copolymer shell structures. *American Chemical Society* 121: 462.

Yang, C., C.H. Choi, C.S. Lee and H. Yi. 2013. A facile synthesis-fabrication strategy for integration of catalytically active viral palladium nanostructures into polymeric hydrogel microparticles via replica molding. *ACS Nano* 7(6): 5032–5044.

Zhang, H., X. Lin, S. Chin and M.W. Grinstaff. 2015. Synthesis and characterization of poly(glyceric acid carbonate): A degradable analogue of poly(acrylic acid). *Journal of American Chemical Society* 1-8.

Zhang, Q., J. Kang, B. Yang, L. Zhao, Z. Hou and B. Tang. 2016. Immobilized cellulase on Fe_3O_4 nanoparticles as a magnetically recoverable biocatalyst for the decomposition of corncob. *Chinese Journal of Catalysis* 37(3): 389-397.

Zhao, H., W. Hao, Z. Xi and M. Cai. 2011. Palladium-catalyzed crosscoupling of PhSeSnBu3 with aryl and alkyl halides in ionic liquids: A practical synthetic method of diorganyl selenides. *New Journal of Chemistry* 35(11): 2661-2665.

Zhao, M., L. Sun and R.M.J. Crooks. 1998. Preparation of Cu nanoclusters within dendrimer templates. *American Chemical Society* 120: 4877.

Zhou, C.H.C., J.N. Beltramini, Y.X. Fan and G.Q.M. Lu. 2008. Chemoselective catalytic conversion of glycerol as a biorenewable source to valuable commodity chemicals. *Chemical Society Reviews* 37: 527.

Zhu, J., P.C. Wang and M. Lu. 2012. Synthesis of novel magnetic silica supported hybrid ionic liquid combining TEMPO and poly oxometalate and its application for selective oxidation of alcohols. *RSC Advances* 2(22): 8265–8268.

Polyether-supported Catalyst

Sapana Jadoun[1*], Juan Pablo Fuentes[2], Jorge Yáñez[3], Vaibhav Budhiraja[4] and Ufana Riaz[5,6]

[1] Departamento de Química, Facultad de Ciencias, Universidad de Tarapacá, Avda. General Velásquez, 1775, Arica, Chile

[2] Departamento de Ciencias Básicas, Facultad de Ciencias, Universidad Santo Tomás, Buena Vecindad #91, Puerto Montt, Chile

[3] Facultad de Ciencias Químicas, Departamento de Química Analítica e Inorgánica, Universidad de Concepción, 4070371 Edmundo Larenas 129, Concepción, Chile

[4] Department of Polymer Chemistry and Technology, National Institute of Chemistry, Hajdrihova 19, 1000 Ljubljana, Slovenia

[5] Material Research Laboratory, Department of Chemistry, Jamia Millia Islamia, New Delhi, India

[6] Department of Chemistry and Biochemistry, North Carolina Central University, NC 27707, USA

1. Introduction

Catalysis has always been a hot area in the domain of chemistry and in recent years many studies have been done; still, numerous distinguished challenges are yet to be explored (Bossion et al. 2019, Munirathinam et al. 2015). Improved tools for catalyst characterization and new approaches for catalyst screening might help to resolve many futuristic problems (Vogt and Weckhuysen 2022). New avenues have been opened for the discovery of completely new products because of catalyst development in the realm of polymer synthesis (Kiesewetter et al. 2010). The scientific community engaged in evolving developments is always striving for new catalyst discoveries. In the field of polymer science, catalysis is a key to many reactions (Shifrina et al. 2019). Any scientist or researcher is hardly ever aware of the precise methodology to achieve any novel product without knowing the extensive interpretation of catalysts. In

*Corresponding author: sjadoun022@gmail.com; sjadoun@academicos.uta.cl

general, the catalysts are usually put in service as an adolescent component in any reaction mixture enabling the desired chemical process to occur (Kwon et al. 2021).

Polymer-supported catalysts are a type of heterogeneous catalyst where the catalytic active site is anchored onto polymeric support (Gu et al. 2021). The polymeric support can be either a natural or a synthetic polymer, and it is chosen based on its compatibility with the reaction conditions and the nature of the catalytic active site (Nasrollahzadeh et al. 2019). Polyethers-supported catalysts are one of the types of polymer-supported catalysts widely used in various organic transformations including polymerization, hydrogenation, oxidation, and reduction reactions (Pirinen and Pakkanen 2015). They have several advantages over other types of catalysts including higher stability, lower toxicity, and ease of handling. The performance of polyethers-supported catalysts depends on the nature of the active site, the polyether polymer, and the method of preparation. The active site can be a metal complex, metal oxide, or a chiral ligand, depending on the specific reaction. New polyether structures can be more approachable with new catalysts making modern raw materials more attractive and lucrative (Gini et al. 2020, Hu et al. 2019).

There are various inborn attributes of ether linkage that drove the large-scale development and use of polyethers in a wide range of applications during the last several decades (Paraja et al. 2020). Talking about some of them highlights low polarity and low Van der Walls interaction characteristics, it provides fewer hurdles to the coiling and uncoiling of chains as the carbon-oxygen bond, and has lower barrier to rotation than the carbon-carbon bond. An important factor allowing for greater chain flexibility is that all its backbone units cumulate to the smallest "excluded volume" because ether oxygen has an even lower excluded volume than a methyl group (Wick and Theodorou 2004). Compared to ester, acetal, or amide links, it offers greater hydrolytic resistance because the carbon-oxygen bond has a stronger bond energy (85 kcal/mol) than the carbon-carbon bond (82 kcal/mol) (Price 1974).

2. Polyethers-supported catalysts

Polyether-supported catalysts have several advantages over other types of catalysts, including higher stability, easier handling, and recyclability (Dembinski 2004). This polymer provides stable and solid support for the active catalytic site, preventing aggregation and loss of activity. It allows the catalyst to be used for multiple reaction cycles, reducing the overall cost and waste generated by the reaction (Modak et al. 2021).

Polyether-supported catalysts can be used in a wide range of organic transformations, including hydrogenation, oxidation, reduction, and polymerization reactions. They are effective in both small-molecule and large-molecule synthesis, and can be used in both batch and continuous flow reactors. The choice of the active site, polyether polymer, and preparation method can greatly impact the performance of the polyether-supported catalyst. For example, the size and shape of the polyether polymer can influence the accessibility of the active site to the reactants, while the choice of metal complex or ligand can greatly affect the selectivity and efficiency of the reaction (Lu and Toy 2009).

One potential drawback of polyether-supported catalysts is that the polymer support can limit the diffusion of the reactants to the active site, leading to slower reaction rates. However, this can often be addressed through careful selection of the polymer and the preparation method (Fukunishi et al. 1981).

3.   Synthesis of polyether-supported catalyst

The synthesis of a polyether-supported catalyst typically involves several steps, including the preparation of the polyether, the attachment of the catalytic active site to the polymer, and the activation of the catalyst. Here is a general overview of the synthesis process:

(i)   Preparation of the polyether polymer: The first step is to prepare the polyether polymer that will serve as the support for the catalytic active site. The polymer can be either a natural or a synthetic polymer and is typically chosen based on its compatibility with the reaction conditions and the nature of the catalytic active site. The polymer is typically synthesized through a polymerization reaction, such as ring-opening polymerization (Wilms et al. 2010).

(ii) Attachment of the catalytic active site to the polymer: Once the polymer is prepared, the next step is to attach the catalytic active site to the polymer. This can be done through a variety of methods, such as covalent attachment, adsorption, or ion exchange. The choice of attachment method depends on the nature of the active site and the polymer support (Rudolf et al. 2015).

(iii) Activation of the catalyst: Once the active site is attached to the polymer, the catalyst needs to be activated to be effective. This typically involves a reduction or oxidation step, depending on the nature of the catalytic active site (Isono 2021).

The specific steps and conditions for synthesizing a polyether-supported catalyst will depend on the nature of the active site, the polymer support, and the desired reaction. Careful attention must be paid to optimizing the conditions to achieve the desired activity, selectivity, and stability of the catalyst.

Huang et al. developed a polyether sulfone (PES) based cyclitic non-permselective membrane reactor (CNMR) by basic blending process to achieve enhanced catalytic performance for cellulose for rapid reduction of phenolic pollutants in a continuous flow through the reaction from industrial wastewater (Huang et al. 2021). Polyepichlorohydrin (PECH), a functional polyether, was first synthesized in 1950's using catalytic ring opening polymerization of epichlorohydrin (ECH). ECH played a pioneering role in the discovery of epoxides. Being discovered by Edwin J. Vandenberg, it is popularly known as Vandenberg catalyst since then which empowered the synthesis of numerous brand new, economically viable, PECH-based rubber constituents. They reacted the alkyl aluminum–water catalyst with acetylacetone, to block a monomer coordination site to gain insight into some mechanistic aspects of epoxide polymerizations with the catalyst (Figure 1) (Shukla and Ferrier Jr 2021).

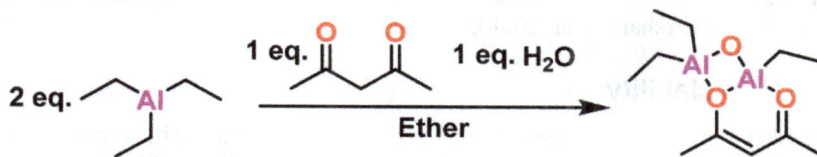

Figure 1. Vandenberg catalyst synthesis with suspected catalyst structure. Reprinted with permission of Wiley (Shukla and Ferrier Jr 2021)

Meticulous synthesis of well-defined hyperbranched polyglycerol (PG) using ring-opening multi-branching polymerization of glycidol to prepare hyperbranched and polyfunctional polyethers with measurable molar mass and low polydispersities (Mw/Mn=1.2-1.9) via various monomer addition protocols is rigorously used (Ahn 2018). Hu et al. studied various alkali and alkaline earth metal salts with various linear and branched polyether complexing agents/co-catalysts for the synthesis of cyclic carbonates from epoxides and CO_2, where polyethylene glycol dimethyl ethers appeared to be most promising ones (Hu et al. 2019).

4. Properties of polyether-supported catalysts

Polyether-supported catalysts have emerged as an important tool for a wide range of chemical transformations, offering many advantages over other types of catalysts. In this essay, we will discuss some of the key properties of polyether-supported catalysts, including their stability, selectivity, recyclability, and customizability (Dunjic et al. 1994).

4.1 High stability

One of the key properties of polyether-supported catalysts is their high stability. The polyether polymer provides a stable matrix for the catalytic active site, preventing aggregation and loss of activity. This allows the catalyst to be used for multiple reaction cycles, reducing the overall cost and waste generated by the reaction. In addition, the polyether support can help to reduce the leaching of the active site into the reaction mixture, which can be important for maintaining high activity and selectivity (Konieczynska et al. 2015).

4.2 Selectivity

Another important property of polyether-supported catalysts is their selectivity. The polymer support can help to increase the selectivity of the catalyst by allowing for more precise control over the reaction environment. For example, the size and shape of the polymer can influence the accessibility of the active site to the reactants, while the choice of metal complex or ligand can greatly affect the selectivity and efficiency of the reaction. This selectivity can be especially important in complex organic transformations where multiple reaction pathways are possible (Hirahata et al. 2008).

4.3 Recyclability

Polyether-supported catalysts are also highly recyclable, which can reduce the overall cost and waste generated by the reaction. The solid nature of the catalysts makes them easy to handle and store, and they can often be recovered and reused. This can be especially important in industrial-scale reactions, where the cost and waste generated by the reaction can be significant (Sawant et al. 2011).

4.4 Customizability

One of the most important properties of polyether-supported catalysts is their customizability. The choice of the active site and the polymer support can be customized to optimize the catalyst for a specific reaction. For example, the choice of polymer support can be influenced by the nature of the reaction and the reaction conditions, while the choice of active site can be tailored to the desired selectivity and efficiency of the reaction. This customizability has led to the development of a wide range of polyether-supported catalysts for a variety of chemical transformations (Harun et al. 2021).

4.5 Others

In addition to these key properties, polyether-supported catalysts also offer several other advantages. For example, they are compatible with a wide range of organic transformations, including hydrogenation, oxidation, reduction, and polymerization reactions. They are also often more environmentally friendly than other types of catalysts, as they can reduce the amount of waste generated by the reaction (Peighambardoust et al. 2011).

Polyether-supported catalysts are a promising area of research in the field of catalysis, offering many advantages over other types of catalysts. They are likely to continue to be an important tool for synthetic chemists and chemical engineers in the years to come (Yu et al. 2010).

5. Applications of polyether-supported catalyst

Polyether polyols are principal chemical raw materials with a versatile spectrum of applications. Their intense use in many industries like petroleum, sealants, foam, elastomers, adhesives, coatings, textiles, and paper proves the worth itself (Segura et al. 2005).

Polyether-supported catalysts have become increasingly important in the field of catalysis, offering a range of benefits over traditional catalysts. These benefits include high stability, selectivity, recyclability, and customizability, which make polyether-supported catalysts attractive for use in a wide range of chemical reactions. In this essay, we will discuss some of the major applications of polyether-supported catalysts.

5.1 Organic synthesis

Polyether-supported catalysts have been extensively used in organic synthesis reactions such as hydrogenation, oxidation, reduction, and condensation reactions. For example, polyether-supported palladium catalysts are commonly used for Suzuki-Miyaura coupling, a reaction used in the synthesis of complex organic molecules. Polyether-supported ruthenium and iridium catalysts have also been used in hydrogenation and transfer hydrogenation reactions, which are important in the synthesis of fine chemicals (Yu et al. 2010, Weber and Gokel 2012).

5.2 Environmental remediation

Polyether-supported catalysts have also found applications in environmental remediation, particularly in the removal of pollutants from wastewater and air. For example, polyether-supported copper catalysts have been used in the degradation of organic pollutants such as dyes and pesticides in water, while polyether-supported iron catalysts have been used in the removal of nitrogen oxides from the air (Ahmed and Jhung 2021).

5.3 Pharmaceuticals

Polyether-supported catalysts have emerged as a valuable tool in the synthesis of pharmaceuticals due to their high selectivity, low leaching, and recyclability. For example, polyether-supported chiral catalysts have been used in asymmetric hydrogenation reactions, which is an important step in the synthesis of chiral drugs. The selectivity of these catalysts can be fine-tuned by varying the polymer support, which can lead to higher yields and fewer side reactions (Thomas et al. 2014).

5.4 Polymerization

Polyether-supported catalysts have been used in a variety of polymerization reactions, including the polymerization of olefins and other monomers. Polyether-supported Ziegler-Natta catalysts are commonly used in the production of polyethylene and polypropylene, while polyether-supported metallocene catalysts have been used in the production of specialty polymers with unique properties, such as polyethylene with high strength and elasticity (Dookhith et al. 2022, Brocas et al. 2013).

5.5 Fuel cells

Polyether-supported catalysts have also been explored for use in fuel cells, which are an important technology for producing clean energy. Polyether-supported platinum catalysts are commonly used in fuel cells to catalyze the oxygen reduction reaction, which is an important step in the production of electricity (Peighambardoust et al. 2011).

5.6 Others

Additionally, polyether-supported catalysts have also been used in a range of other chemical reactions, including epoxidation, cycloaddition, and carbonylation reactions. The stability, selectivity, recyclability, and customizability of these catalysts make them attractive for use in a variety of chemical reactions, and they are likely to continue to be an important tool for synthetic chemists and chemical engineers in the years to come (Wilms et al. 2010, Homberg and Lacour 2020). They are also very much useful in nowadays booming aerospace industry applications, notably missile propellants. Gu et al. (2021) probed into the optimization of polyether polyol coming from propylene oxide using Single Factor Experiments and Response Surface Methodology by way of composite alkaline earth metal catalyst (BaO/MgO/γ-Al$_2$O$_3$). This is a highly stable and environmentally friendly catalyst favorable for steady production having very high catalytic activity, needs no post-treatment and most importantly can be reused (Li et al. 2020). ECH comprising polymers are utilized in a broad range of applications because of the alkyl chloride functional group and ether backbone. Some of the applications are Anion Exchange Membranes (AEM), Polymer Electrolytes (in lithium-ion batteries) and CO$_2$ separation (Noh et al. 2019). Hyperbranched polyglycerols have proved enormous potential to be used as soluble support for catalysts and organic synthesis, in biomineralization and biomedical applications (Noh et al. 2019).

6. Conclusion

Polyether-supported catalysts have emerged as a valuable tool in catalysis, offering a range of benefits over traditional catalysts. The stability, selectivity, recyclability, and customizability of these catalysts have made them attractive for use in a wide range of chemical reactions, including organic synthesis, pharmaceuticals, polymerization, environmental remediation, and fuel cells. The ability to tailor the polymer support and the active site of the catalyst has led to the development of a wide range of polyether-supported catalysts for specific applications. Polyether-supported catalysts have not only improved the efficiency of chemical reactions, but they have also reduced the environmental impact by providing a sustainable and eco-friendly solution to chemical processes. Moreover, the use of these catalysts has led to significant cost savings in the chemical industry due to their recyclability and long-term stability. The continued development of new polyether-supported catalysts and their applications is likely to play an increasingly important role in the field of catalysis in the coming years. As researchers continue to optimize these catalysts for specific applications, we can expect to see breakthroughs in the synthesis of complex organic molecules, the production of high-performance polymers, the removal of environmental pollutants, and the generation of clean energy. Overall, polyether-supported catalysts have the potential to revolutionize the field of catalysis and contribute to a more sustainable and efficient chemical industry.

Acknowledgement

The author Sapana Jadoun is grateful for the support of the National Research and Development Agency of Chile (ANID) and the projects, FONDECYT Postdoctoral 3200850, 3190383 and ANID/FONDAP/15110019.

References

Ahmed, Imteaz and Sung Hwa Jhung. 2021. Covalent organic framework-based materials: Synthesis, modification, and application in environmental remediation. *Coordination Chemistry Reviews* 441: 213989. Elsevier.

Ahn, Gyunhyeok. 2018. Design and Synthesis of Functional Polyamine Based on Novel Amino Glycidyl Ether. Graduate School of UNIST.

Bossion, Amaury, Katherine V. Heifferon, Leire Meabe, Nicolas Zivic, Daniel Taton, James L. Hedrick, Timothy E. Long and Haritz Sardon. 2019. Opportunities for organocatalysis in polymer synthesis via step-growth methods. *Progress in Polymer Science* 90: 164–210. Elsevier.

Brocas, Anne-Laure, Christos Mantzaridis, Deniz Tunc and Stephane Carlotti. 2013. Polyether synthesis: From activated or metal-free anionic ring-opening polymerization of epoxides to functionalization. *Progress in Polymer Science* 38(6): 845–873. Elsevier.

Dembinski, Roman. 2004. Recent advances in the Mitsunobu reaction: Modified reagents and the quest for chromatography-free separation. *European Journal of Organic Chemistry* 13: 2763–2772. Wiley Online Library.

Dookhith, Aaliyah Z., Nathaniel A. Lynd, Costantino Creton and Gabriel E. Sanoja. 2022. Controlling architecture and mechanical properties of polyether networks with organoaluminum catalysts. *Macromolecules* 55(13): 5601–5609. ACS Publications.

Dunjic, Branko, Alain Favre-Réguillon, Olivier Duclaux and Marc Lemaire. 1994. New polyether-based ionoselective materials. *Advanced Materials* 6(6): 484–486. Wiley Online Library.

Fukunishi, Koji, Bronislaw Czech and Steven L. Regen. 1981. Polyether-based triphase catalysts. A synthetic comparison. *The Journal of Organic Chemistry* 46(6): 1218–1221. ACS Publications.

Gini, Andrea, Miguel Paraja, Bartomeu Galmés, Celine Besnard, Amalia I. Poblador-Bahamonde, Naomi Sakai, Antonio Frontera and Stefan Matile. 2020. Pnictogen-bonding catalysis: Brevetoxin-type polyether cyclizations. *Chemical Science* 11(27): 7086–7091. Royal Society of Chemistry.

Gu, Yanlong, Seung Uk Son, Tao Li and Bien Tan. 2021. Low-cost hypercrosslinked polymers by direct knitting strategy for catalytic applications. *Advanced Functional Materials* 31(12): 2008265. Wiley Online Library.

Harun, Nur Ain Masleeza, Norazuwana Shaari and Nik Farah Hanis Nik Zaiman. 2021. A review of alternative polymer electrolyte membrane for fuel cell application based on sulfonated poly (ether ether ketone). *International Journal of Energy Research* 45(14): 19671–19708. Wiley Online Library.

Hirahata, Wataru, Renee M. Thomas, Emil B. Lobkovsky and Geoffrey W. Coates. 2008. Enantioselective polymerization of epoxides: A highly active and selective catalyst for the preparation of stereoregular polyethers and enantiopure epoxides. *Journal of the American Chemical Society* 130(52): 17658–17659. ACS Publications.

Homberg, Alexandre and Jérôme Lacour. 2020. From reactive carbenes to chiral polyether macrocycles in two steps – Synthesis and applications made easy? *Chemical Science* 11(25): 6362–6369. Royal Society of Chemistry.

Hu, Yuya, Johannes Steinbauer, Vivian Stefanow, Anke Spannenberg and Thomas Werner. 2019. Polyethers as complexing agents in calcium-catalyzed cyclic carbonate synthesis. *ACS Sustainable Chemistry & Engineering* 7(15): 13257–13269. ACS Publications.

Huang, Chiao-Ling, Yu-Ruei Kung, Yu-Jen Shao and Guey-Sheng Liou. 2021. Synthesis and characteristics of novel TPA-containing electrochromic poly(ether sulfone)s with dimethylamino substituents. *Electrochimica Acta* 368: 137552. Elsevier.

Isono, Takuya. 2021. Synthesis of functional and architectural polyethers via the anionic ring-opening polymerization of epoxide monomers using a phosphazene base catalyst. *Polymer Journal* 53(7): 753–764. doi:10.1038/s41428-021-00481-3.

Kiesewetter, Matthew K., Eun Ji Shin, James L. Hedrick and Robert M. Waymouth. 2010. Organocatalysis: Opportunities and challenges for polymer synthesis. *Macromolecules* 43(5): 2093–2107. ACS Publications.

Konieczynska, Marlena D., Xinrong Lin, Heng Zhang and Mark W. Grinstaff. 2015. Synthesis of aliphatic poly(ether 1,2-glycerol carbonate)s via copolymerization of CO_2 with glycidyl ethers using a cobalt salen catalyst and study of a thermally stable solid polymer electrolyte. *ACS Macro Letters* 4(5): 533–537. ACS Publications.

Kwon, Kitae, R. Thomas Simons, Meganathan Nandakumar and Jennifer L. Roizen. 2021. Strategies to generate nitrogen-centered radicals that may rely on photoredox catalysis: Development in reaction methodology and applications in organic synthesis. *Chemical Reviews* 122(2): 2353–2428. ACS Publications.

Li, Shuo, Zhenggui Gu, Kaijun Wang, Xiaoyan Cao, Yacheng Liu and Can Wang. 2020. Study on catalytic synthesis of low molecular weight polyether polyol by composite alkaline earth metal. *In: IOP Conference Series: Earth and Environmental Science*, 453: 12083. IOP Publishing.

Lu, Jinni and Patrick H. Toy. 2009. Organic polymer supports for synthesis and for reagent and catalyst immobilization. *Chemical Reviews* 109(2): 815–838. ACS Publications.

Modak, Arindam, Anindya Ghosh, Akshay R. Mankar, Ashish Pandey, Manickam Selvaraj, Kamal Kishore Pant, Biswajit Chowdhury and Asim Bhaumik. 2021. Cross-linked porous polymers as heterogeneous organocatalysts for task-specific applications in biomass transformations, CO_2 fixation, and asymmetric reactions. *ACS Sustainable Chemistry & Engineering* 9(37): 12431–12460. ACS Publications.

Munirathinam, Rajesh, Jurriaan Huskens and Willem Verboom. 2015. Supported catalysis in continuous-flow microreactors. *Advanced Synthesis & Catalysis* 357(6): 1093–1123. Wiley Online Library.

Nasrollahzadeh, Mahmoud, Mohaddeseh Sajjadi, Mohammadreza Shokouhimehr and Rajender S. Varma. 2019. Recent developments in palladium (nano) catalysts supported on polymers for selective and sustainable oxidation processes. *Coordination Chemistry Reviews* 397: 54–75. Elsevier.

Noh, Sangtaik, Jong Yeob Jeon, Santosh Adhikari, Yu Seung Kim and Chulsung Bae. 2019. Molecular engineering of hydroxide conducting polymers for anion exchange membranes in electrochemical energy conversion technology. *Accounts of Chemical Research* 52(9): 2745–2755. ACS Publications.

Paraja, Miguel, Andrea Gini, Naomi Sakai and Stefan Matile. 2020. Pnictogen-bonding catalysis: An interactive tool to uncover unorthodox mechanisms in polyether cascade cyclizations. *Chemistry – A European Journal* 26(67): 15471–15476. Wiley Online Library.

Peighambardoust, S.J., S. Rowshanzamir, M.G. Hosseini and M. Yazdanpour. 2011. Self-humidifying nanocomposite membranes based on sulfonated poly(ether ether ketone) and heteropolyacid supported Pt catalyst for fuel cells. *International Journal of Hydrogen Energy* 36(17): 10940–10957. Elsevier.

Pirinen, Sami and Tuula T. Pakkanen. 2015. Polyethers as potential electron donors for Ziegler–Natta ethylene polymerization catalysts. *Journal of Molecular Catalysis A: Chemical* 398: 177–183. Elsevier.

Price, Charles C. 1974. Polyethers. *Accounts of Chemical Research* 7(9): 294–301. ACS Publications.

Rudolf, Constantin, Irina Mazilu, Alexandru Chirieac, Brandusa Dragoi, Fatima Abi-Ghaida, Adrian Ungureanu, Ahmad Mehdi and Emil Dumitriu. 2015. Copper nanoparticles supported on polyether-functionalized mesoporous silica. Synthesis and application as hydrogenation catalysts. *Environmental Engineering & Management Journal (EEMJ)* 14(2).

Sawant, Dinesh, Yogesh Wagh, Kushal Bhatte, Anil Panda and Bhalchandra Bhanage. 2011. Palladium polyether diphosphinite complex anchored in polyethylene glycol as an efficient homogeneous recyclable catalyst for the heck reactions. *Tetrahedron Letters* 52(18): 2390–2393. Elsevier.

Segura, D.M., A.D. Nurse, A. McCourt, R. Phelps and A. Segura. 2005. Chemistry of polyurethane adhesives and sealants. *Handbook of Adhesives and Sealants* 1: 101–162. Elsevier.

Shifrina, Zinaida B., Valentina G. Matveeva and Lyudmila M. Bronstein. 2019. Role of polymer structures in catalysis by transition metal and metal oxide nanoparticle composites. *Chemical Reviews* 120(2): 1350–1396. ACS Publications.

Shukla, Geetanjali and Robert C. Ferrier Jr. 2021. The versatile, functional polyether, polyepichlorohydrin: History, synthesis, and applications. *Journal of Polymer Science* 59(22): 2704–2718. Wiley Online Library.

Thomas, Anja, Sophie S. Müller and Holger Frey. 2014. Beyond poly(ethylene glycol): Linear polyglycerol as a multifunctional polyether for biomedical and pharmaceutical applications. *Biomacromolecules* 15(6): 1935–1954. Wiley Online Library.

Vogt, Charlotte and Bert M. Weckhuysen. 2022. The concept of active site in heterogeneous catalysis. *Nature Reviews Chemistry* 6(2): 89–111. Nature Publishing Group UK London.

Weber, William P. and George W. Gokel. 2012. Phase Transfer Catalysis in Organic Synthesis. Vol. 4. Springer Science & Business Media.

Wick, Collin D. and Doros N. Theodorou. 2004. Connectivity-altering Monte Carlo simulations of the end group effects on volumetric properties for poly(ethylene oxide). *Macromolecules* 37(18): 7026–7033. ACS Publications.

Wilms, Daniel, Salah-Eddine Stiriba and Holger Frey. 2010. Hyperbranched polyglycerols: From the controlled synthesis of biocompatible polyether polyols to multipurpose applications. *Accounts of Chemical Research* 43(1): 129–141. ACS Publications.

Yu, Fengli, Ruili Zhang, Congxia Xie and Shitao Yu. 2010. Polyether-substituted thiazolium ionic liquid catalysts – A thermoregulated phase-separable catalysis system for the stetter reaction. *Green Chemistry* 12(7): 1196–1200. Royal Society of Chemistry.

Poly(ethylene imine) and Poly (2-oxazoline)-supported Catalyst

Rizwan Arif[1]*, Sapana Jadoun[2] and Anurakshee Verma[3]

[1] Department of Chemistry, School of Basic and Applied Sciences,
Lingaya's Vidyapeeth, Faridabad, Haryana, India
[2] Departamento de Química, Facultad de Ciencias, Universidad de Tarapacá,
Avda. General Velásquez, 1775, Arica, Chile
[3] Department of Applied Sciences, KCC Institute of Technology of Management,
Greater Noida, India

1. Introduction

The development of polymeric catalyst has attracted the researchers for the synthesis of new organic and inorganic macromolecules with control properties. The facilitation of these reactions in presence of visible light with respect to green chemistry is very interesting and promising approach. For the synthesis of polymer-supported catalysts, many significant techniques have been utilized and among them ring opening metathesis and atom transfer radical polymerization are the most widely utilized techniques. Catalyst required to synthesize the polymer is often more valuable and is always discarded after single polymerization and their use in this regard is nothing more than an academic curiosity. This renders the recovery and reuse of the less-precious supported catalyst inconsequential (Rossegger et al. 2013). Synthetic (co)polymer materials have been also investigated due to their wide range of properties and functions which can be modulated by tuning their structures' properties in biomedical sciences such as for controlled and triggered drugs delivery, responsive bio-interfaces mimicking natural surfaces' biosensors and in chemo-mechanical actuators; among them, stimuli-responsive (co)polymers have become a major polymer research area which shows significant applications in tissue engineering, bio-imaging and

---

*Corresponding author: arif.rizwan9@gmail.com

in environmental remediation (Rapoport 2007, Shim and Kwon 2012, Cheng et al. 2015, Traitel et al. 2008, Zhang et al. 2019, Mendes 2008, Stuart et al. 2010, Wei et al. 2017, Joseph and Jose 2021, Jeong et al. 2001, Liu and Urban 2010, Roy et al. 2010, Chen et al. 2010, Lu and Urban 2018). They undergo reversible and irreversible changes in their physical chemical properties after the modification and variations in pH, temperature, light, redox potential, ionic strength, ultrasound, electrical and magnetic fields as shown in Figure 1.

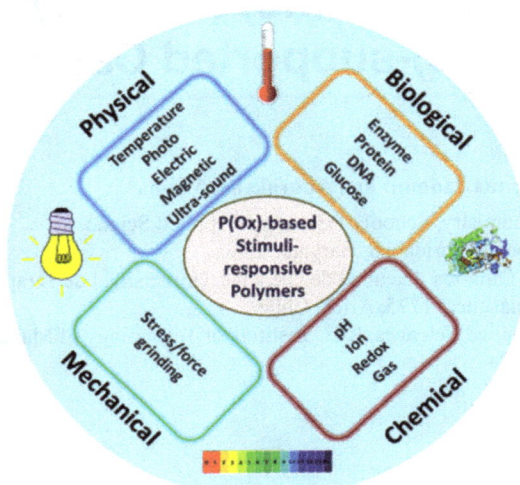

Figure 1. Schematic representation of different types of poly(2-oxazoline)s (POx)-based stimuli responsive (co)polymers. Licence Copyright: https://s100.copyright.com/CustomerAdmin/PLF.jsp?ref=7fd114d3-92eb-4b1e-bbb1-017832f40c68

Some natural stimuli-responsive polymers and bio-macromolecules like proteins and nucleic acids are also known as "environmentally sensitive" and "intelligent biomaterials" (Hoffman 1995, Kikuchi and Okano 2002, García-Fernández et al. 2019, Kocak et al. 2017, Kim and Matsunaga 2017, Bertrand and Gohy 2017, Sedighipoor et al. 2019, Huo et al. 2014, Yoshida et al. 2013, Magnusson et al. 2008, Hu et al. 2012, Shahriari et al. 2019, Ma and Shi 2014, Lin and Theato 2013, Manouras and Vamvakaki 2017).

But due to lack of some significant properties like non-toxicity, biocompatibility and biodegradability, they have been never considered as ideal biomedical materials. In last few years, many polymeric biomaterials have been prepared for biomedical applications using poly(ethylene glycol) (PEG) and various other significant polymeric properties (Kumar et al. 2007, Cabane et al. 2012, Berglund et al. 2016, Tian et al. 2012, Knop et al. 2010, D'souza and Shegokar 2016, De La Rosa 2014, Konradi et al. 2012, Hucknall et al. 2009, Banerjee et al. 2011). But PEG-containing polymers produce toxins in enzymatic oxidation in physiological environment; therefore, to overcome these limitations, new polymers have been synthesized, in which poly(2-oxazoline)s is considered to be a promising alternative to PEG-containing polymers. We can hydrolyze these P(Ox)s by acids and these are degradable in the presence of reactive oxygen species (ROS). The poly(2-oxazoline)s class of compounds attracted great attention of researchers due to its versatility in the synthesis of various materials with different properties. These polymeric 2-oxazoline materials are currently receiving much interest as potential biomaterials in microwave reactors which are designed for chemical synthesis (Bauer et al. 2012, Hoogenboom 2009, Hoogenboom and Schlaad 2011, Jana and Uchman 2020, Luxenhofer et al. 2012).

Now a days, polyoxazolines-based substituted polymers are in demand and considered as competitive as compared to other biocompatible polymer like PEG, poly(vinyl pyrrolidone), and poly(N-(2-hydroxypropyl) methacrylamide) for the synthesis and development of next-generation therapeutics (Van Kuringen et al. 2012, Ulbricht et al. 2014, Sedlacek et al. 2012). Schematic representation highlighting the research interest of researchers in P(Ox) for the development of polymer therapeutics is shown in Figure 2.

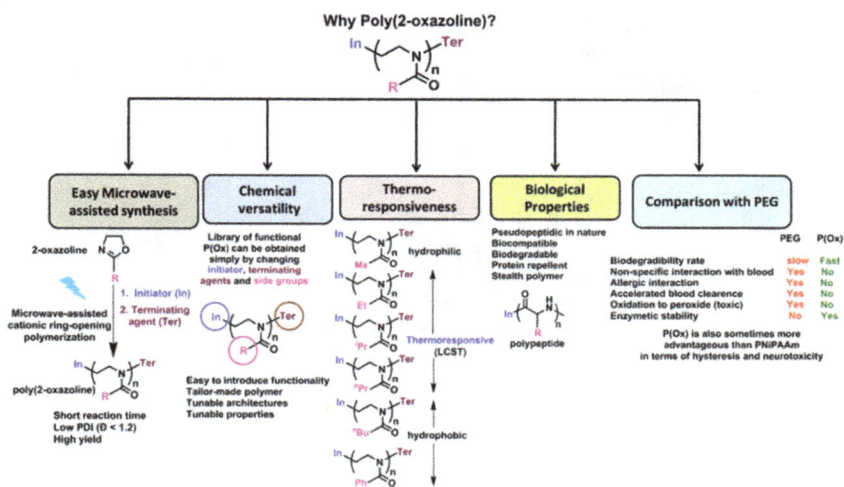

Figure 2. Schematic representation highlighting the underlying reasons for the increasing research interest in P(Ox) in the context of polymer therapeutics. Licence No: https://s100.copyright.com/CustomerAdmin/PLF.jsp?ref=7fd114d3-92eb-4b1e-bbb1-017832f40c68

2. Synthesis and applications of poly(ethylene imine) and poly(2-oxazoline)

Substituted poly(2-oxazoline)s were prepared by microwave assisted method in inert atmosphere. After weighing poly(2-methyl-2-oxazoline), poly(2-ethyl-2-5oxazoline), poly(2-phenyl-2-oxazoline) and poly(2-nonyl-2-oxazoline) in microwave vial, all these poly(2-oxazoline)s were heated with methyl tosylate and acetonitrile at 140 °C and quenching is carried out after the addition of water (Hwang et al. 2019). In the last few years, many investigations have been carried out by the researchers for the synthesis and development of poly(2-substituted 2-oxazolines) (POx) and other significant initiators and terminators via cationic ring-opening polymerization (CROP) and supercritical CO_2-assisted and microwave assisted green methodologies. Advancements in synthesis and development of POx made it a competitive alternative for polymer systems due to its easy synthetic methods and chemical diversity which in turn strengthened the liability of POx. Many functionalized POx products are part of current commercial market which increases the opportunity for the uptake of these promising materials. As POXs are considered as enormous potential material because of their structural properties and various biological and physicochemical properties, CROP can be considered to have a "living" character, which meets appropriate reaction conditions and has an integral characteristic (Hoogenboom 2017, Verbraeken et al. 2017, De Macedo et al. 2007, Bonifácio et al. 2012, Wang et al. 2007, Hoogenboom et al. 2005, Kempe et al. 2011). This living cationic ring-opening polymerization facilitates the incorporation of various functional groups by using appropriate initiators and terminating agents like 2-substituted 2-oxazolines (Figure 3).

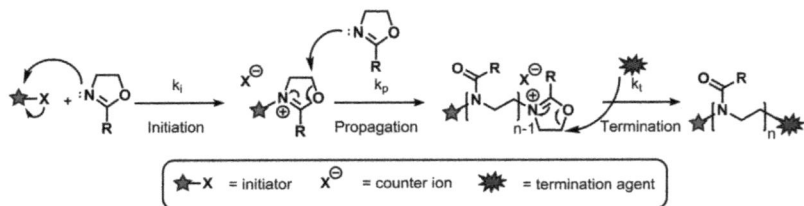

Figure 3. Schematic representation of the cationic ring-opening polymerization (CROP) mechanism: Demonstrating the initiation, propagation and termination steps. https://s100. copyright.com/CustomerAdmin/PLF.jsp?ref=487f80a7-0bb2-499d-b41b-b82141a286bb

Poly(2-oxazoline)s (POx) have been known for a little over 50 years and this family of polymers has seen very variable interest by researchers and industry. Poly(2-ethyl-2-oxazoline) (PEtOx) and hydrophilic poly(2-methyl-2-oxazoline) (PMeOx) are the actual chemical species investigated in the last few years. PEtOx is an amphiphilic polymer that exhibits excellent solubility in water and some other organic solvents and also very significant applications in the field of biomaterials like in organic electronics. Cationic ring-opening polymerization of 2-oxazoline monomers can be performed in a way to suppress all the side-reactions like termination of chain growth as this type of polymerization is also considered as living or quasi-living polymerization. Due to significant developments of polymerization and green methodologies like supercritical CO_2-assisted, microwave-assisted techniques and improved synthesis of initiators, monomers, and terminators, the synthesis of poly(2-substituted 2-oxazolines) (POx) via cationic ring-opening polymerization is the area of significant research during the last few years. These improvements have also strengthened the commercial viability of POx, and hence there are now a number of functionalized POx products available in the market, thereby increasing the opportunity for broad uptake of these promising materials. Specifically, care must be taken to ensure the absence of chain-transfer promoting impurities and reagents, as well as nucleophilic species that could cause unwanted chain termination (Figure 4).

Poly(ethylene imine) (PEI) containing primary, secondary and tertiary amino group, which is synthesized by the CROP of ethylene imine, is the best known polymer. We can modify the functionalities of PEI to obtain desired polymer by substituting various organic moieties (Kobayashi et al. 1986, Tomalia and Sheetz 1966). Poly(2-oxazoline)s was first synthesized by the research group of Tomalia and co-workers (1966). They studied the hydrolysis of poly(2-methyl-2-oxazoline) and PEI in the presence of sulphuric acid for the acetic acid formation. Hydrolysis of poly(2-oxazoline)s is a promising route for the preparation of linear PEI, and for the easy determination of structure-property relationships (SAR). Cationic ring-opening polymerization of 2-oxazolines is the best known synthetic procedure for the preparation of significant polymers with desired physiochemical properties (Tomalia and Sheetz 1966, Aoi and Okada 1996, Kobayashi et al. 1986, Hoogenboom 2007, Adams and Schubert 2007, Seeliger et al. 1966, Kagiya et al. 1966, Bassiri et al. 1967, Christova et al. 2002, 2003). Research group of Saegusa and other researchers studied the synthesis and characterization of linear PEI through the cationic ring-opening polymerization of 2-oxazoline and hydrolysis of the poly(2-oxazoline) in alkaline conditions. Poly(2-oxazoline)s can be also used as precursor for linear PEI upon hydrolysis under alkaline or acidic conditions (Saegusa et al. 1972, Tanaka et al. 1983, Jeong et al. 2001). In textile industry, PEI is used for the pre-treatment of fibres for a better adsorption or desorption of dyes, for the treatment of paper and for DNA and RNA delivery into cells due to its high charge density, which is due to the nitrogen atom present in it, which can be easily protonated and make PEI capable of condensing plasmid DNA and RNA into stable complexes through electrostatic interactions (Jones and Thompson 2009). As its contain variety of side chain (methyl, ethyle) and due to its versatile nature, poly(2-oxazoline) has been considered to be hydrophilic in nature while poly(2-oxazoline) with longer side chains has been considered to be hydrophobic in nature. Very few studies have been carried out by researchers on poly(2-oxazoline)s or poly(alkyl ethyleneimine)s (PalkylEI) due to very small difference in molecular structure in poly(2-oxazoline) and PalkylEI. We can prepare PalkylEI by two synthetic routes, i.e either by the alkylation of linear PEI and/or by the reduction of poly(2-oxazoline)s (Kobayashi et al. 1986, Jin 2002, Hoogenboom 2007, Aoi and Okada 1996, Wiesbrock et al. 2004, Sinnwell and Ritter 2005, Hu et al. 2017).

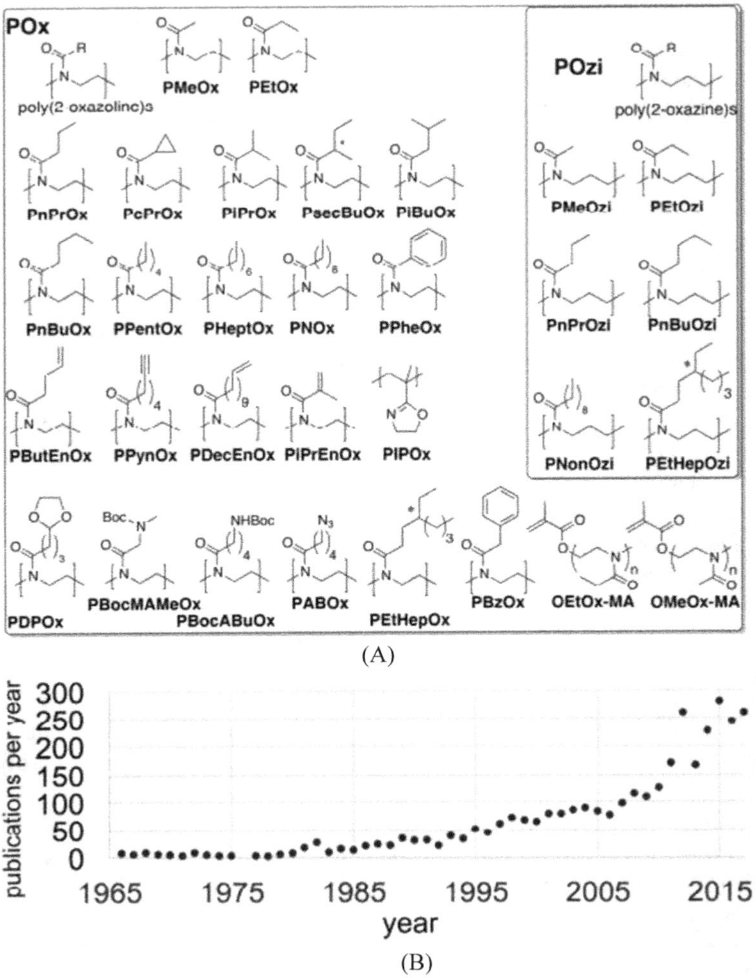

(A)

(B)

Figure 4. (A) Commonly used poly(2-oxazoline)s and some poly(2-oxazine)s relevant for this review. (B) Development of annual publications on poly(2-oxazoline)s. https://s100. copyright.com/CustomerAdmin/PLF.jsp?ref=25703466-ed95-4031-9147-7ede410a43b5

Frech and co-worker synthesized linear poly (methyl ethylene imine) (PMEI) and linear poly(ethyl ethylene imine) (PEEI) via the alkylation of linear PEI and Tanaka and research group synthesized PMEI by Eschweiler-Clarke reductive N-methylation method and reported its properties (Frech et al. 2005). PEtOx was also synthesized by cationic ring opening polymerization (CROP). 2-ethyl-2-oxazoline and 1,4-dibromo-2E-butene in 30:1 molar ratio were mixed in a flask in acetonitrile at 80 °C for 20 h in inert atmosphere. Quenching was carried out by adding methanolic KOH. Precipitation of prepared polymer was done in cold diethyl ether and dried under vacuum (Christova et al. 2002).

3. Conclusions

Researchers have been focused on polymer-supported catalyst for the synthesis and development of polymeric materials in the last few decades. Due to biomedical and pharmaceutical applications, family of polymers like poly(2-oxazoline)s (POX) and poly(ethylene imine) have gained much more interest as biomaterials. In view of these pharmaceutical and biomedical applications, we have tried to study the polymer-supported catalyst for the development of various polymers.

Abbreviations

Poly(ethylene glycol)	:	PEG
Poly(2-oxazoline)s	:	P(Ox)
Reactive oxygen species	:	ROS
Poly(vinyl pyrrolidone)	:	PVP
Poly(N-(2-hydroxypropyl) methacrylamide)	:	PHPMA
poly(2-methyl-2-oxazoline)	:	PMeOx
Poly(2-ethyl-2-5oxazoline)	:	PEtOx
Poly(2phenyl-2-oxazoline)	:	PPhOx
Poly(2-nonyl-2-oxazoline)	:	PNOx
Cationic ring-opening polymerization	:	CROP
Poly(ethylene imine)	:	PEI
Structure-property relationships	:	SAR
Ribonucleic acid	:	RNA
Deoxyribose nucleic Acid	:	DNA
Poly(alkyl ethyleneimine)s	:	PalkylEI
Poly (methyl ethylene imine)	:	PMEI
Poly(ethyl ethylene imine)	:	PEEI

Acknowledgement

Author Rizwan Arif is very much grateful to Department of Chemistry, LV. Author Sapana Jadoun is grateful for the support of National Research and Development Agency of Chile (ANID) for the project FONDECYT Postdoctoral 3200850.

References

Adams, Nico and Ulrich S. Schubert. 2007. Poly(2-oxazolines) in biological and biomedical application contexts. *Advanced Drug Delivery Reviews* 59(15): 1504–1520. doi:10.1016/j.addr.2007.08.018.

Aoi, Keigo and Masahiko Okada. 1996. Polymerization of oxazolines. *Progress in Polymer Science* 21(1): 151–208. Oxford. doi:10.1016/0079-6700(95)00020-8.

Banerjee, Indrani, Ravindra C. Pangule and Ravi S. Kane. 2011. Antifouling coatings: Recent developments in the design of surfaces that prevent fouling by proteins, bacteria, and marine organisms. *Advanced Materials* 23(6): 690–718. doi:10.1002/adma.201001215.

Bassiri, T.G., A. Levy and M. Litt. 1967. Polymerization of cyclic imino ethers. I. Oxazolines. *Journal of Polymer Science Part B: Polymer Letters* 5(9): 871–879. doi:10.1002/pol.1967.110050927.

Bauer, Marius, Christian Lautenschlaeger, Kristian Kempe, Lutz Tauhardt, Ulrich S. Schubert and Dagmar Fischer. 2012. Poly(2-ethyl-2-oxazoline) as alternative for the stealth polymer poly(ethylene glycol): Comparison of in vitro cytotoxicity and hemocompatibility. *Macromolecular Bioscience* 12(7): 986–998. doi:10.1002/mabi.201200017.

Bertrand, Olivier and Jean François Gohy. 2017. Photo-responsive polymers: Synthesis and applications. *Polymer Chemistry* 8(1): 52–73. doi:10.1039/c6py01082b.

Bonifácio, Vasco D.B., Vanessa G. Correia, Mariana G. Pinho, João C. Lima and Ana Aguiar-Ricardo. 2012. Blue emission of carbamic acid oligooxazoline biotags. *Materials Letters* 81: 205–208. doi:10.1016/j.matlet.2012.04.134.

Cabane, Etienne, Xiaoyan Zhang, Karolina Langowska, Cornelia G. Palivan and Wolfgang Meier. 2012. Stimuli-responsive polymers and their applications in nanomedicine. *Biointerphases* 7(1–4): 1–27. doi:10.1007/s13758-011-0009-3.

Chen, Tao, Robert Ferris, Jianming Zhang, Robert Ducker and Stefan Zauscher. 2010. Stimulus-responsive polymer brushes on surfaces: Transduction mechanisms and applications. *Progress in Polymer Science* 35(1–2): 94–112. Oxford. doi:10.1016/j.progpolymsci.2009.11.004.

Cheng, Weiren, Liuqun Gu, Wei Ren and Ye Liu. 2015. Stimuli-responsive polymers for anti-cancer drug delivery. *Materials Science and Engineering C* 45: 600–608. doi:10.1016/j.msec.2014.05.050.

Christova, Darinka, Rumiana Velichkova, Eric J. Goethals and Filip E. Du Prez. 2002. Amphiphilic segmented polymer networks based on poly(2-alkyl-2-oxazoline) and poly(methyl methacrylate). *Polymer* 43(17): 4585–4590. doi:10.1016/S0032-3861(02)00313-0.

Christova, Darinka, Rumiana Velichkova, Wouter Loos, Eric J. Goethals and Filip Du Prez. 2003. New thermo-responsive polymer materials based on poly(2-ethyl-2-oxazoline) segments. *Polymer* 44(8): 2255–2261. doi:10.1016/S0032-3861(03)00139-3.

D'souza, Anisha A. and Ranjita Shegokar. 2016. Polyethylene glycol (PEG): A versatile polymer for pharmaceutical applications. *Expert Opinion on Drug Delivery* 13(9): 1257–1275. doi:10.1080/17425247.2016.1182485.

Frech, Roger, Guinevere A. Giffin, Frank Yepez Castillo, Daniel T. Glatzhofer and Jördis Eisenblätter. 2005. Spectroscopic studies of polymer electrolytes based on poly(N-ethylethylenimine) and poly(N-methylethylenimine). *Electrochimica Acta* 50(19): 3963–3968. doi:10.1016/j.electacta.2005.02.051.

García-Fernández, Luis, Ana Mora-Boza and Felisa Reyes-Ortega. 2019. PH-responsive polymers: Properties, synthesis, and applications. *Smart Polymers and Their Applications*, 45–86. doi:10.1016/B978-0-08-102416-4.00003-X.

Hoffman, Allan S. 1995. 'Intelligent' polymers in medicine and biotechnology. *Artificial Organs* 19(5): 458–467. doi:10.1111/j.1525-1594.1995.tb02359.x.

Hoogenboom, Richard. 2007. Poly(2-oxazoline)s: Alive and kicking. *Macromolecular Chemistry and Physics* 208(1): 18–25. doi:10.1002/macp.200600558.

Hoogenboom, Richard. 2009. Poly(2-oxazoline)s: A polymer class with numerous potential applications. *Angewandte Chemie International Edition* 48(43): 7978–7994. doi:10.1002/anie.200901607.

Hoogenboom, Richard. 2017. 50 Years of poly(2-oxazoline)s. *European Polymer Journal* 88: 4448–4450. doi:10.1016/j.eurpolymj.2017.01.014.

Hoogenboom, Richard, Frank Wiesbrock, Mark A.M. Leenen, Michael A.R. Meier and Ulrich S. Schubert. 2005. Accelerating the living polymerization of 2-nonyl-2-oxazoline by implementing a microwave synthesizer into a high-throughput experimentation workflow. *Journal of Combinatorial Chemistry* 7(1): 10–13. doi:10.1021/cc049846f.

Hoogenboom, Richard and Helmut Schlaad. 2011. Bioinspired poly(2-oxazoline)s. *Polymers* 3(1): 467–488. doi:10.3390/polym3010467.

Hu, Jinming, Guoqing Zhang and Shiyong Liu. 2012. Enzyme-responsive polymeric assemblies, nanoparticles and hydrogels. *Chemical Society Reviews* 41(18): 5933–5949. doi:10.1039/c2cs35103j.

Hucknall, Angus, Srinath Rangarajan and Ashutosh Chilkoti. 2009. In pursuit of zero: Polymer brushes that resist the adsorption of proteins. *Advanced Materials* 21(23): 2441–2446. doi:10.1002/adma.200900383.

Huo, Meng, Jinying Yuan, Lei Tao and Yen Wei. 2014. Redox-responsive polymers for drug delivery: From molecular design to applications. *Polymer Chemistry* 5(5): 1519–1528. doi:10.1039/c3py01192e.

Hwang, Duhyeong, Jacob D. Ramsey, N. Makita, Clemens Sachse, Rainer Jordan, Marina Sokolsky-Papkov and Alexander V. Kabanov. 2019. Novel poly(2-oxazoline) block copolymer with aromatic heterocyclic side chains as a drug delivery platform. *Journal of Controlled Release* 307: 261–271. doi:10.1016/j.jconrel.2019.06.037.

Jana, Somdeb and Mariusz Uchman. 2020. Poly(2-oxazoline)-based stimulus-responsive (co)polymers: An overview of their design, solution properties, surface-chemistries and applications. *Progress in Polymer Science* 106(July): 101252. doi:10.1016/j.progpolymsci.2020.101252.

Jeong, Ji Hoon, Soon Ho Song, Dong Woo Lim, Haeshin Lee and Tae Gwan Park. 2001. DNA transfection using linear poly(ethylenimine) prepared by controlled acid hydrolysis of poly(2-ethyl-2-oxazoline). *Journal of Controlled Release* 73(2–3): 391–399. doi:10.1016/S0168-3659(01)00310-8.

Jin, R. 2002. Controlled location of porphyrin in aqueous micelles self-assembled from porphyrin centered amphiphilic star poly(oxazolines). *Advanced Materials* 14(12): 889–892. doi:10.1002/1521-4095(20020618)14:123.0.CO;2-6.

Jones, Russell G. and Craig B. Thompson. 2009. Tumor suppressors and cell metabolism: A recipe for cancer growth. *Genes & Development* 23(5): 537–548. Cold Spring Harbor Lab.

Joseph, Vincent and Jiya Jose. 2021. Stimulus-responsive polymers. *Nanohydrogels*, 109–125. Springer Singapore. doi:10.1007/978-981-15-7138-1_5.

Kempe, Kristian, C. Remzi Becer and Ulrich S. Schubert. 2011. Microwave-assisted polymerizations: Recent status and future perspectives. *Macromolecules* 44(15): 5825–5842. doi:10.1021/ma2004794.

Kikuchi, Akihiko and Teruo Okano. 2002. Intelligent thermoresponsive polymeric stationary phases for aqueous chromatography of biological compounds. *Progress in Polymer Science* 27(6): 1165–1193. Oxford. doi:10.1016/S0079-6700(02)00013-8.

Kim, Joo Ho, Yongseok Jung, Dajung Lee and Woo Dong Jang. 2016. Thermoresponsive polymer and fluorescent dye hybrids for tunable multicolor emission. *Advanced Materials* 28(18): 3499–3503. doi:10.1002/adma.201600043.

Kim, Young Jin and Yukiko T. Matsunaga. 2017. Thermo-responsive polymers and their application as smart biomaterials. *Journal of Materials Chemistry B* 5(23): 4307–4321. doi:10.1039/c7tb00157f.

Knop, Katrin, Richard Hoogenboom, Dagmar Fischer and Ulrich S. Schubert. 2010. Poly(ethylene glycol) in drug delivery: Pros and cons as well as potential alternatives. *Angewandte Chemie – International Edition* 49(36): 6288–6308. doi:10.1002/anie.200902672.

Kobayashi, Shiro, Toshio Igarashi, Yasuhiro Moriuchi and Takeo Saegusa. 1986. Block copolymers from cyclic imino ethers: A new class of nonionic polymer surfactant. *Macromolecules* 19(3): 535–541. doi:10.1021/ma00157a006.

Kocak, G., C. Tuncer and V. Bütün. 2017. PH-responsive polymers. *Polymer Chemistry* 8(1): 144–176. doi:10.1039/c6py01872f.

Konradi, Rupert, Canet Acikgoz and Marcus Textor. 2012. Polyoxazolines for nonfouling surface coatings – A direct comparison to the gold standard PEG. *Macromolecular Rapid Communications* 33(19): 1663–1676. doi:10.1002/marc.201200422.

Kumar, Ashok, Akshay Srivastava, Igor Yu Galaev and Bo Mattiasson. 2007. Smart polymers: Physical forms and bioengineering applications. *Progress in Polymer Science* 32(10): 1205–1237. Oxford. doi:10.1016/j.progpolymsci.2007.05.003.

Kuringen, Huub P.C. Van, Joke Lenoir, Els Adriaens, Johan Bender, Bruno G. De Geest and Richard Hoogenboom. 2012. Partial hydrolysis of poly(2-ethyl-2-oxazoline) and potential implications for biomedical applications. *Macromolecular Bioscience* 12(8): 1114–1123. doi:10.1002/mabi.201200080.

La Rosa, Victor R. De. 2014. Poly(2-oxazoline)s as materials for biomedical applications. *Journal of Materials Science: Materials in Medicine* 25(5): 1211–1225. doi:10.1007/s10856-013-5034-y.

Lin, Shaojian and Patrick Theato. 2013. CO_2-responsive polymers. *Macromolecular Rapid Communications* 34(14): 1118–1133. doi:10.1002/marc.201300288.

Liu, Fang and Marek W. Urban. 2010. Recent advances and challenges in designing stimuli-responsive polymers. *Progress in Polymer Science* 35(1–2): 3–23. Oxford. doi:10.1016/j.progpolymsci.2009.10.002.

Lu, Chunliang and Marek W. Urban. 2018. Stimuli-responsive polymer nano-science: Shape anisotropy, responsiveness, applications. *Progress in Polymer Science* 78: 24–46. doi:10.1016/j.progpolymsci.2017.07.005.

Luxenhofer, Robert, Yingchao Han, Anita Schulz, Jing Tong, Zhijian He, Alexander V. Kabanov and Rainer Jordan. 2012. Poly(2-oxazoline)s as polymer therapeutics. *Macromolecular Rapid Communications* 33(19): 1613–1631. doi:10.1002/marc.201200354.

Ma, Rujiang and Linqi Shi. 2014. Phenylboronic acid-based glucose-responsive polymeric nanoparticles: Synthesis and applications in drug delivery. *Polymer Chemistry* 5(5): 1503–1518. doi:10.1039/c3py01202f.

Macedo, Carlota Veiga De, Mara Soares Da Silva, Teresa Casimiro, Eurico J. Cabrita and Ana Aguiar-Ricardo. 2007. Boron trifluoride catalyzed polymerization of 2-substituted-2-oxazolines in supercritical carbon dioxide. *Green Chemistry* 9(9): 948–995. doi:10.1039/b617940a.

Magnusson, Johannes Pall, Adnan Khan, George Pasparakis, Aram Orner Saeed, Wenxin Wang and Cameron Alexander. 2008. Ion-sensitive 'Isothermal' responsive polymers prepared in water. *Journal of the American Chemical Society* 130(33): 10852–10853. doi:10.1021/ja802609r.

Manouras, Theodore and Maria Vamvakaki. 2017. Field responsive materials: Photo-, electro-, magnetic- and ultrasound-sensitive polymers. *Polymer Chemistry* 8(1): 74–96. doi:10.1039/c6py01455k.

Mendes, Paula M. 2008. Stimuli-responsive surfaces for bio-applications. *Chemical Society Reviews* 37(11): 2512–2529. doi:10.1039/b714635n.

O'Boussif, F. Lezoualc'h, M.A. Zanta, M.D. Mergny, D. Scherman, B. Demeneix, J.P. Behr. 1995. A versatile vector for gene and oligonucleotide transfer intocells in culture and in vivo: Polyethylenimine. *Proceedings of the National Academy of Sciences* 92(16): 7297-7301. http://www.ncbi.nlm.nih.gov/pubmed/25409609%0Ahttp://linkinghub. elsevier.com/retrieve/pii/B9780128001486000092%0Ahttp://www.pnas.org/cgi/ reprint/82/24/8658%5Cnpapers3://publication/uuid/DBA25972-E972-460A-AA67-50B259EFAAC6%0Ahttp://www.pnas.org/cgi/doi/10.

Rapoport, Natalya. 2007. Physical stimuli-responsive polymeric micelles for anti-cancer drug delivery. *Progress in Polymer Science* 32(8–9): 962–990. Oxford. doi:10.1016/j. progpolymsci.2007.05.009.

Richard Song, Maxwell Murphy, Chenshuang Li, Kang Ting, Chia Soo and Zhong Zheng. 2018. Current development of biodegradable polymeric materials for biomedical applications. *Drug Design, Development and Therapy* 12: 3117–3145.

Rossegger, Elisabeth, Verena Schenk and Frank Wiesbrock. 2013. Design strategies for functionalized poly(2-oxazoline)s and derived materials. *Polymers* 5(3): 956–1011. doi:10.3390/polym5030956.

Roy, Debashish, Jennifer N. Cambre and Brent S. Sumerlin. 2010. Future perspectives and recent advances in stimuli-responsive materials. *Progress in Polymer Science* 35(1–2): 278–301. Oxford. doi:10.1016/j.progpolymsci.2009.10.008.

Saegusa, Takeo, Hiroharu Ikeda and Hiroyasu Fujii. 1972. Crystalline polyethylenimine. *Macromolecules* 5(1): 108. doi:10.1021/ma60025a029.

Sedighipoor, Maryam, Ali Hossein Kianfar, Gholamhossein Mohammadnezhad, Helmar Görls, Winfried Plass, Amir Abbas Momtazi-Borojeni and Elham Abdollahi. 2019. Synthesis, crystal structure of novel unsymmetrical heterocyclic Schiff base Ni (II)/V (IV) complexes: Investigation of DNA binding, protein binding and in vitro cytotoxic activity. *Inorganica Chimica Acta* 488. Elsevier: 182–194.

Sedlacek, Ondrej, Bryn D. Monnery, Sergey K. Filippov, Richard Hoogenboom and Martin Hruby. 2012. Poly(2-oxazoline)s – Are they more advantageous for biomedical applications than other polymers? *Macromolecular Rapid Communications* 33(19): 1648–1662. doi:10.1002/marc.201200453.

Seeliger, W., E. Aufderhaar, W. Diepers, R. Feinauer, R. Nehring, W. Thier and H. Hellmann. 1966. Recent syntheses and reactions of cyclic imidic esters. *Angewandte Chemie International Edition in English* 5(10): 875–888. doi:10.1002/anie.196608751.

Shahriari, Mahsa, Mahsa Zahiri, Khalil Abnous, Seyed Mohammad Taghdisi, Mohammad Ramezani and Mona Alibolandi. 2019. Enzyme responsive drug delivery systems in cancer treatment. *Journal of Controlled Release* 308: 172–189. doi:10.1016/j. jconrel.2019.07.004.

Shim, Min Suk and Young Jik Kwon. 2012. Stimuli-responsive polymers and nanomaterials for gene delivery and imaging applications. *Advanced Drug Delivery Reviews* 64(11): 1046–1059. doi:10.1016/j.addr.2012.01.018.

Sinnwell, Sebastian and Helmut Ritter. 2005. Microwave accelerated polymerization of 2-phenyl-2-oxazoline. *Macromolecular Rapid Communications* 26(3): 160–163. doi:10.1002/marc.200400477.

Stuart, Martien A. Cohen, Wilhelm T.S. Huck, Jan Genzer, Marcus Müller, Christopher Ober, Manfred Stamm, Gleb B. Sukhorukov, Igal Szleifer, Vladimir V. Tsukruk, Marek Urban, Françoise Winnik, Stefan Zauscher, Igor Luzinov and Sergiy Minko. 2010. Emerging applications of stimuli-responsive polymer materials. *Nature Materials* 9(2): 101–113. doi:10.1038/nmat2614.

Tanaka, Ryuichi, Isao Ueoka, Yasuhiro Takaki, Kazuya Kataoka and Shogo Saito. 1983. High molecular weight linear poly(ethylenimine) and poly(IV-methylethylenimine). *Macromolecules* 16(6): 849–853. doi:10.1021/ma00240a003.

Tian, Huayu, Zhaohui Tang, Xiuli Zhuang, Xuesi Chen and Xiabin Jing. 2012. Biodegradable synthetic polymers: Preparation, functionalization and biomedical application. *Progress in Polymer Science* 37(2): 237–280. Oxford. doi:10.1016/j.progpolymsci.2011.06.004.

Tomalia, D.A. and D.P. Sheetz. 1966. Homopolymerization of 2-alkyl- and 2-aryl-2-oxazolines. *Journal of Polymer Science Part A-1: Polymer Chemistry* 4(9): 2253–2265. doi:10.1002/pol.1966.150040919.

Traitel, Tamar and Joseph Kost. 2008. Smart polymers for responsive drug-delivery systems. *Journal of Biomaterials Science, Polymer Edition* 19(6): 755–767. doi:10.1163/156856208784522065.

Tsutomu Kagiya, Shizuo Narisawa, Taneo Maeda and Kenichi Fukui. 1966. Ring-opening polymerization of 2-substituted 2-oxazolines. *A Bibliographical Catalogue of Seventeenth-Century German Books* (Published in Holland) 4: 441–445.

Ulbricht, Juliane, Rainer Jordan and Robert Luxenhofer. 2014. On the biodegradability of polyethylene glycol, polypeptoids and poly(2-oxazoline)s. *Biomaterials* 35(17): 4848–4861. doi:10.1016/j.biomaterials.2014.02.029.

Verbraeken, Bart, Bryn D. Monnery, Kathleen Lava and Richard Hoogenboom. 2017. The chemistry of poly(2-oxazoline)s. *European Polymer Journal* 88(March): 451–469. doi:10.1016/j.eurpolymj.2016.11.016.

Wang, Yuan Yuan, Wei Li and Li Yi Dai. 2007. Cationic ring-opening polymerization of 3,3-bis(chloromethyl)oxacyclobutane in ionic liquids. *Chinese Chemical Letters* 18(10): 1187–1190. doi:10.1016/j.cclet.2007.08.013.

Wei, Menglian, Yongfeng Gao, Xue Li and Michael J. Serpe. 2017. Stimuli-responsive polymers and their applications. *Polymer Chemistry* 8(1): 127–143. doi:10.1039/c6py01585a.

Wiesbrock, Frank, Richard Hoogenboom and Ulrich S. Schubert. 2004. Microwave-assisted polymer synthesis: State-of-the-art and future perspectives. *Macromolecular Rapid Communications* 25(20): 1739–1764. doi:10.1002/marc.200400313.

Yong Qiu and Kinam Park. 1998. Environment-sensitive hydrogels for drug delivery. *Psychiatry and Clinical Neurosciences* 52(2): 145–147.

Yoshida, Takayuki, Tsz Chung Lai, Glen S. Kwon and Kazuhiro Sako. 2013. pH- and ion-sensitive polymers for drug delivery. *Expert Opinion on Drug Delivery* 10(11): 1497–1513. doi:10.1517/17425247.2013.821978.

Zhang, Aitang, Kenward Jung, Aihua Li, Jingquan Liu and Cyrille Boyer. 2019. Recent advances in stimuli-responsive polymer systems for remotely controlled drug release. *Progress in Polymer Science* 99. doi:10.1016/j.progpolymsci.2019.101164.

Zhang, Ning, Robert Luxenhofer and Rainer Jordan. 2012. Thermoresponsive poly(2-oxazoline) molecular brushes by living ionic polymerization: Modulation of the cloud point by random and block copolymer pendant chains. *Macromolecular Chemistry and Physics* 213(18): 1963–1969. doi:10.1002/macp.201200261.

Polyisobutylene and Polynorbornenes-supported Catalysts

Deepshikha Rathore* and Diksha Verma

Department of Chemistry, University of Delhi, Delhi - 110 007, India

1. Introduction

Catalysis is a prime area of synthetic chemistry which can be used to perform typical reactions under mild conditions with significantly higher yields in shorter reaction times. The reuse and recycling of catalysts is currently an important focus area in organic synthesis (Holladay and Albrecht 2012). In this endeavour, polymeric supported catalysts have become increasingly popular and are widely used, particularly since the rapid development of contemporary chemistry that idealises the concepts of "green" and "sustainable" (Ley and Baxendale 2002). The primary benefits of these solid support catalysts are their ease of synthesis, effortless fabrication, physical separability, recyclability, and high environmental stability. Starting with the Merrifield and Letsinger's ground-breaking discovery of cross-linked polystyrene in peptide synthesis, insoluble polymeric systems have achieved a lime light in the catalyst world (Benaglia 2007, Benaglia et al. 2003). However, there are various shortcomings in the use of these insoluble resins due to their heterogeneous nature in the reaction conditions. Therefore, replacing the insoluble heterogeneous resins with soluble homogeneous polymeric supports have been employed and are considered preferably due to their ease of characterization, high reactivity, selectivity, polydispersity, and flexibility. There are many polymeric support systems available that have been widely used as catalyst, such as atactic-polypropylene (PP), or isotactic-poly(propylene-co-hexene) (iPPH), poly(ethylene glycol) (PEG), poly(N-alkylacrylamide) and polystyrene derivatives etc. (Anna et al. 2013). PEG is the most common soluble

*Corresponding author: shikha.rathore4@gmail.com

support polymer used for the synthesis and catalytic activity due to its wide commercial availability and history of being easily modified (Bandaiphet and Kennedy 2004, Dickerson et al. 2003). The soluble Polyisobutylene (PIB) and Poly(norbornene) (PNB) based supports have recently been widely involved in the catalytic reactions, and are regarded as two important pillars of the polymer supported catalyst (Bergbreiter et al. 2011a & b, Sommer and Weck 2006, Su et al. 2010).

PIB is a vinyl polymer made by cationic polymerization of isobutylene (IB) monomer. Despite the fact that it has a linear structure, it is classified as a synthetic rubber or elastomer. They have unique properties such as low moisture permeability, chemical resistivity, oxidative and thermal stability (Boyd and Pant 1991, Kunal et al. 2008). PIB and its derivatives have the advantage of a structurally simple and chemically inert alkane backbone. It is considered as an analogue of PEG with a commercially available average molecular weight of 1000 or 2300 Da. It is mostly soluble in aliphatic and aromatic hydrocarbon solvents but insoluble in polar solvents due to their non-crystalline, paraffinic, and nonpolar nature. This phase soluble selectivity of PIB and its derivatives is appealing, as any polymer or species attached to it can be easily removed from the reaction mixture simply by extraction. PIB-attached ligands and metal complexes can be easily characterized by NMR spectroscopy because they have only two major protons appearing in the range of $\delta 2.00\text{-}1.00$ ppm (Li et al. 2005). Furthermore, PIB contains complex heterogeneous morphology, good swelling properties, and extremely high mechanical strength. Therefore, the polymer based on IB with branched architectures has the potential to be used in a wide range of applications such as microencapsulation, dispersion of chemicals and catalysts, production of reactive polymers, modifications of rheological properties of solutions and suspensions, and additive polymer processing operations. The major advantage of using PIB as a catalyst is its easy alteration in carbon-carbon double bond by standard reactions of alkene. Also, PIB supported ligands behave like a low molecular weight counterpart in metal coordination and are useful in catalysis. Thus, commercially available PIB derivatives have been frequently modified terminally to prepare a wide variety of soluble polymeric reagents and ligands such as PIB-bounded phosphine ligands, salen catalysts, olefins metathesis catalysts, organocatalysts, Cu(I) catalyst for atom transfer radical polymerization and azide-alkene cyclization, Rh(II) and Cu(II) cyclopropanation catalyst, and vinyl ether sequestering agents (Bergbreiter and Li 2004, Liang and Bergbreiter 2016).

On the other hand, PNB which was first developed by CdF-Chimie in France in 1976, is notable for being the first ROMP polymer to be used for commercial purposes by Nippon Zeon Company of Japan under the trade name Norsorex®. The monomer for PNB is synthesised by Diels-Alder reaction of cyclopentadiene and ethylene. The PNB is made up of a cyclohexene ring with a methylene bridge that restricts rotation around the main chain. An exceptionally high molecular weight ($>3,000,000$ g mol^{-1}) material with a predominance of *trans* olefins is produced as a result of polymerization in the presence of a

RuCl$_3$/HCl/BuOH catalyst system. The material has remarkable qualities, acting as both an elastomeric and thermoplastic material due to its large molecular weight and unsaturated microstructure (Blank and Janiak 2009). Norsorex®'s exceptional damping qualities in applications for shock, vibration, and sound control are a result of its Tg of 37 °C. It can absorb up to ten times its dry weight in hydrocarbons due to its low polarity and large molecular weight, generating a gel that retains acceptable mechanical qualities. Because of this, it has specific uses for removing oil spills. In the field of catalysis, solubility/swellability profile, level of functionalization, and potential participation of the polymer backbone are just a few of the various characteristics of polymeric supports that need to be taken into account (Kann 2010). PNB is a good option for catalyst support due to its distinctive solubility selectivity, which allows its removal from the reaction mixture by straightforward filtration. PIB supported catalyst such as salen catalysts, C–C coupling-catalyst, olefins metathesis catalysts, metal organocatalysts etc. have been used widely (Fredlund et al. 2017, Karimi et al. 2013, Madhavan et al. 2011).

Here, a brief survey on recent developments in the field of PIB and PNB supported catalyst system is given. We have considered the progressive efforts made in the area of catalyst preparation, properties, catalytic applications including its catalytic performance and recyclability, since 2005. In terms of content, we chose to include a few representative articles published on the topic of PIB and PNB supported catalyst in addition of research available so as to give an overview.

2. Modes of isobutylene and norbornene polymerization

PIB can be classified on the basis of their molecular weight as low molecular weight PIB (Mn <5000 g/mol), medium molecular weight PIB (Mn = 10,000-100,000 g/mol), and high molecular weight PIB (Mn >100,000 g/mol). Amongst them, low molecular weight PIB is the most demanding in the market and is available as highly reactive (HR) PIB and "conventional" PIB. HR PIB is primarily composed of terminal *exo*-olefin groups while the "conventional" PIB holds predominantly internal tri- and tetra-substituted olefinic end groups and is thus found to be less reactive (Alves et al. 2021). For the mode of polymerization, different cationic polymerization approaches have been accomplished in the last few decades. Particularly, the chain transfer polymerization (CCTP) in non-polar solvents using Lewis acid-ether complexes (LA-ether) has gained interest for the polymerization method (Figure 1) (Rajasekhar et al. 2017, 2020).

There are three ways to polymerize bicyclo [2.2.1] hept-2-ene, often known as norbornene (NB), and its derivatives (Figure 2). Each path results in a particular form of polymer and may be distinguished from one another by the catalyst used. The ring opening metathesis polymerization (ROMP), which is technically used

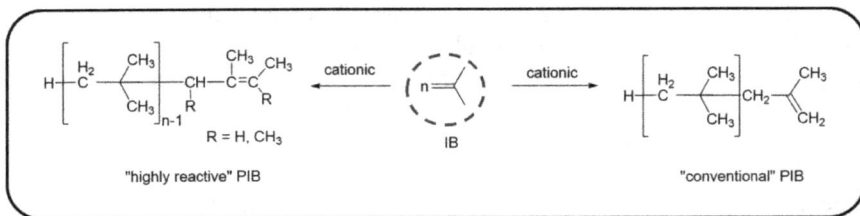

Figure 1. Modes of isobutylene (IB) polymerization

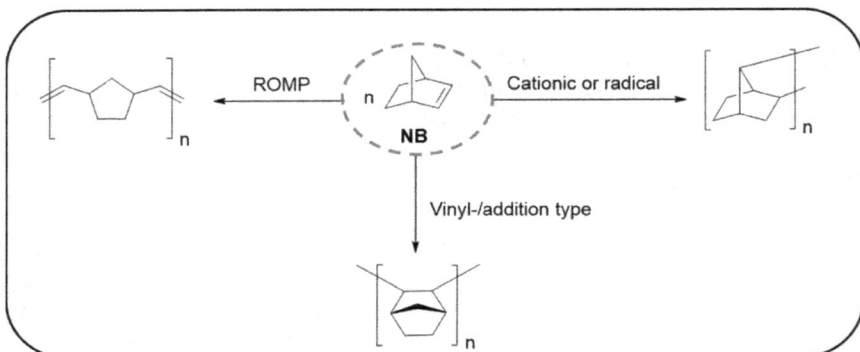

Figure 2. Modes of norbornene (NB) polymerization

in the Norsorex method, is the most well-known NB polymerization technique (Ohm and Vial 1978). This highly controlled polymerization method enables tuning of the metal loading as well as the solubility of the produced catalysts. The majority of these catalysts are, in fact, soluble in the reaction media, but they are easily recoverable by adding diethyl ether or ethyl acetate to the catalyst solution. Figure 2 shows a schematic illustration of the three distinct NB polymerization processes (Frank et al. 2018, Xu et al. 2017). Other technical ROMP methods usually utilize metal halides, tungsten, molybdenum, rhenium or ruthenium and metal oxides or metal oxo-chlorides along with alkylating agents (e.g. R^4Sn, $EtAlCl_2$) and promoting agents (e.g. O_2, EtOH, PhOH). Academic ROMP research has concentrated on molecular single-component catalysts with metal-carbene complexes playing a key role, whereas commercial catalysts are typically heterogeneous. The complexes of tungsten, molybdenum, and ruthenium with carbenes are a few examples. There have been reports of the potential for ring-opening metathesis polymerization of NB using catalyst systems based on titanium, zirconium, hafnium, vanadium, niobium, tantalum, osmium, and rhenium. The cationic and radical polymerizations of NB are poorly understood. The outcome is a low molecular mass oligomeric material with 2,7-connectivity of the monomer. tert-Butyl peracetate, tert-butyl perpivalate, or azoisobutyronitrile (AIBN) were some examples of initiators for the radical polymerization. Ethylalmunium dichloride (EtAlCl$_2$) was used to launch the cationic polymerization. Additionally, NB can be polymerized while preserving the bicyclic structural unit, i.e. by just opening the double bond of the component. Vinyl or addition polymerization is the term used for this type of polymerization, which is similar to the traditional olefin polymerization. There are no longer any double bonds in the vinyl addition product. NB's vinyl polymerization can be either a homo- or a copolymerization (Alentiev and Bermeshev 2022, Blank and Janiak 2009).

3. Polyisobutylene-supported catalysts

Utilising polymer-supported catalysts to remove metal species from a reaction mixture is a successful approach. The fascinating terminal vinyl group is a key characteristic of PIB, which can easily be modified and utilized for numerous organic syntheses as catalyst support. A number of researchers have made efforts in this direction e.g. Hongfa and his co-researchers reported PIB as a catalyst support in thermomorphic system and used it for Pd catalyzed cross-coupling reactions (Hongfa et al. 2007). The PIB terminus was attached to the sulfur-carbon-sulfur ligand and phosphine ligand which are used to prepare Pd complexes 1 and 2 (Bergbreiter et al. 2000, Bergbreiter and Sung 2006). The obtained catalysts were used to perform Heck, Sonogashira, and allylic substitution reactions with good recyclability. In their study, PIB is also used to support chiral bisoxazoloine ligands for copper-catalyzed cyclopropanation reactions (Bergbreiter and Tian 2007).

In subsequent work, a PIB-supported RCM catalyst (3) and a chromium-based polycarbonate polymerization catalyst (4) have been reported (Hongfa et al. 2008). Both of these catalysts exhibited similar activity to their unanchored counterparts, and can easily be recovered for reuse used without any appreciable loss of activity. Another report also mentioned the use of PIB-supported Cu(I) complex (5) for ATRP polymerization of styrene. Additionally, β-diketone ligands were also prepared in only two steps using PIB methyl ketone enolate as a nucleophile and readily available low molecular weight carbonyl compounds (ester or acid chloride) as the electrophile. Further, the ligand was subjected to undergo an exchange reaction with Ni(OAc)$_2$ or Co(OAc)$_2$ for the formation of PIB-supported Ni and Co complex (6) which is used as a catalyst for the epoxidation reaction. Bergbreiter et al. prepared PIB-bound 10-*N*-phenylphenothiazine (PIB-PTH) photoredox catalyst (7) (Liang and Bergbreiter 2016). This organocatalyst bounded PIB is effective in the light-mediated radical polymerization reactions of the acrylate monomers, has good catalytic activity, is recyclable, and provides modest control over the polymer molecular weight and polydispersity.

PIB oligomers can bind to *N*-heterocyclic carbene (NHC) in various ways and form metal complexes that are phase selective soluble in heptane of thermomorphic mixtures of heptane and polar solvents. The prepared PIB-NHC complex is known as Hoveyda-Grubbs-type catalysts (8/9). The first synthesis

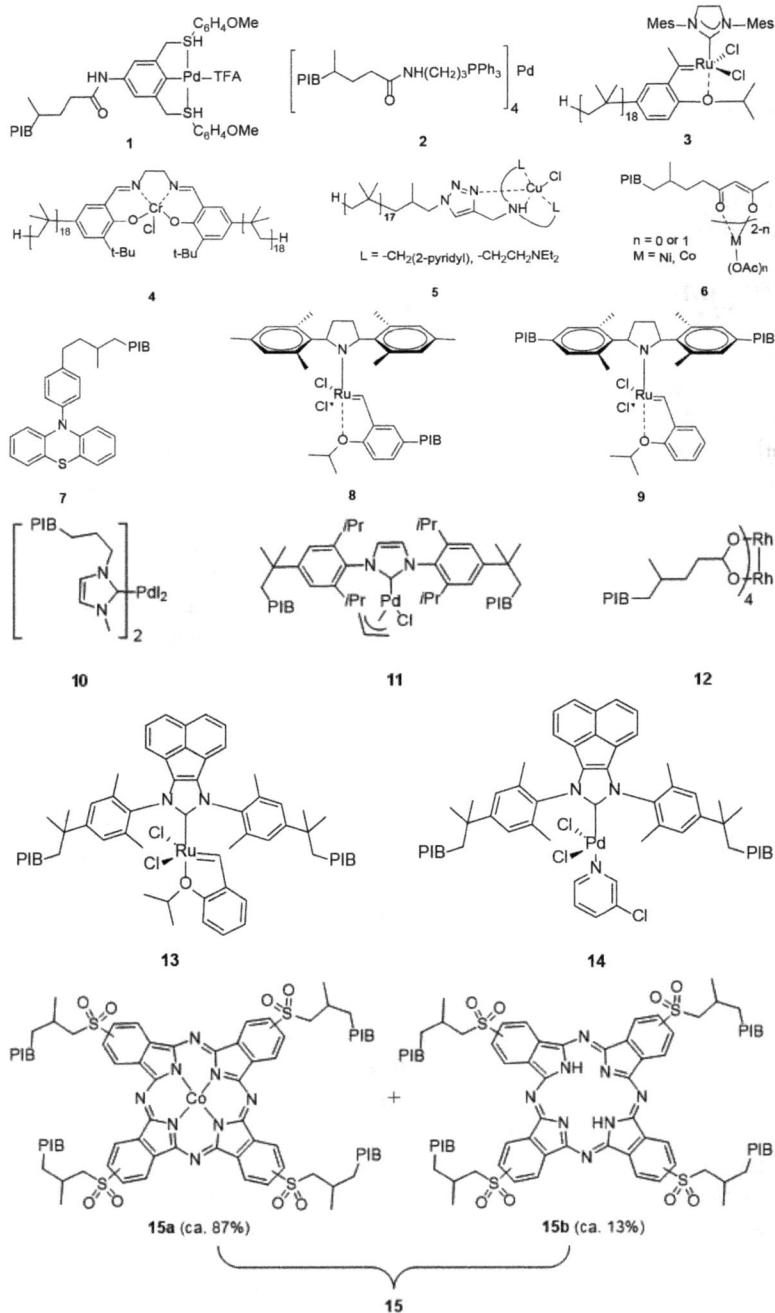

Scheme 1. Important examples of polyisobutylene-supported catalysts

and application of the Hoveyda-Grubbs catalyst was described in 2007 (Hongfa et al. 2007, Su et al. 2010). These catalysts are used for ring-closing metathesis reactions of dienes affording >90% average yield. In the catalytic reactions, the recoverability/recyclability was executed by ICP-MS analysis (Bano et al. 2022). In 2011, PIB-supported N-heterocyclic carbene catalyst, N,N-dialkyl and N,N-bis(2,6-diisopropylphenyl) (10/11) were prepared and successfully employed to form palladium cross-coupling catalysts (Scheme 1).

The ligand exchange of PIB-COOH with $Rh_2(OAc)_4$ in refluxing toluene produced the PIB-supported Rh(II) carboxylate catalyst 12. The cyclopropanation of octene with ethyl diazoacetate in cyclohexane or heptane worked well with this catalyst 12. In 2007, Bergbreiter and Tian performed a post-reaction extraction with ethylene glycol diacetate (EGDA), the polymer-supported catalyst 12 was easily separated and recovered (Bergbreiter and Tian 2007). The synthesis of the poly isobutylene-derived BIAN-IMes-based H-G2 catalyst 13 (Scheme 1) was reported by Bazzi and colleagues in 2016. The benefit of polyisobutylene tether is the ability to recycle the catalyst through the use of a non-polar solvent for the reaction and acetonitrile for the extraction process. Furthermore, catalyst 13 was used to produce high molecular weight polymers by polymerizing norbornene derivatives with a ring-opening reaction (Suriboot et al. 2016). A C–N cross-coupling utilizing catalyst 14 immobilized on PIB was recently described by Suriboot and colleagues as the Buchwald-Hartwig amination process (Chen et al. 2021). Chao and Bergbreiter reported an efficient recyclable homogeneous catalyst (15) using a cobalt phthalocyanine (CoMPc) that contains covalently linked PIB groups as phase anchors in a semithermomorphic system (Chao and Bergbreiter 2016).

4. Polynorbornenes-supported catalysts

Pd(II) pincer complexes (SCS, NCN, PCP) are among the most promising and well-defined transition metal catalysts for carbon-carbon coupling processes (Scheme 2). Through the use of either amide or urea linkages, PCP and SCS-pincer Pd(II) complexes have been covalently immobilized on the soluble PIB and PNB substrates (Yu et al. 2005).

In this context, the Jones research group developed a five step process to get desired product as PNB-immobilized SCS-*N* pincer Pd(II) complex (16) using the SCS-NH$_2$ complex and cyclobutadiene as starting material in 82% yield. Because of the presence of amide bond in this polymer, the supports are readily soluble under reaction conditions (Yu et al. 2004). Sommer et al. developed the soluble PNBs PCP pincer Pd(II) complexes 17 and 18 starting from palladium containing monomers (Anna et al. 2013, Yu et al. 2005). In 2006, NHC was used as the ligand rather than a SCS pincer by Weck and Sommer when they developed a PNB-supported Pd catalyst (19/20) using the same ROMP method (Sommer and Weck 2006). In another approach, Weck and co-workers reported the development and applications of a number of PNB supported metal (Mn/Co/

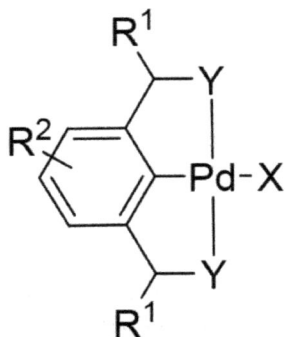

$$Y = NR_2, PR_2, SR, etc.$$
$$R^1, R^2 = H, alkyl, aryl, etc.$$
$$X = Cl, Br, I, OTf, OAc, etc.$$

Scheme 2. Pincer palladium (II) complexes.

A1) salen ligands and catalysts (21-25) shown in scheme 3 that support hydrolytic kinetic resolutions (HKR), asymmetric epoxidations, and conjugate additions of cyanide (Madhavan et al. 2008). Based on a sequential click/ROMP approach, a nanomagnetic cobalt/carbon core-shell-based NB polymeric tag ligand was developed by Reiser and colleagues in 2010 (catalyst 26) (Schätz et al. 2010).

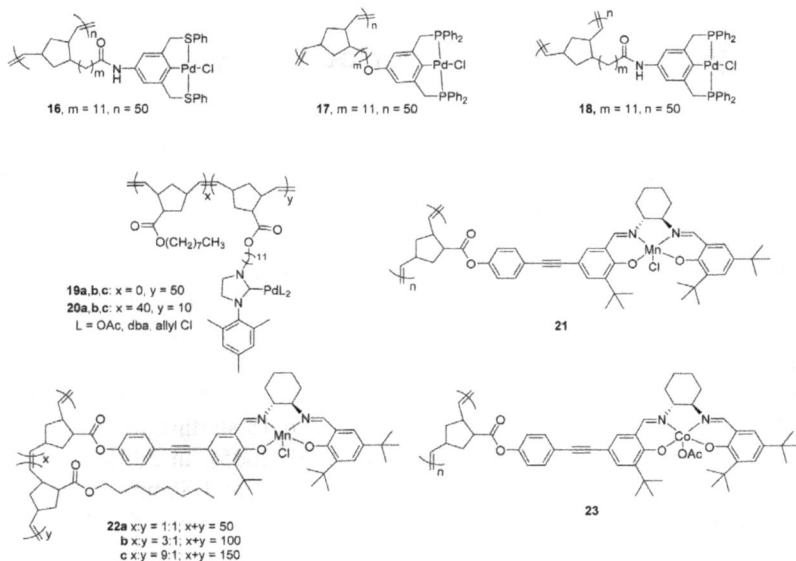

Scheme 3. (*Contd.*)

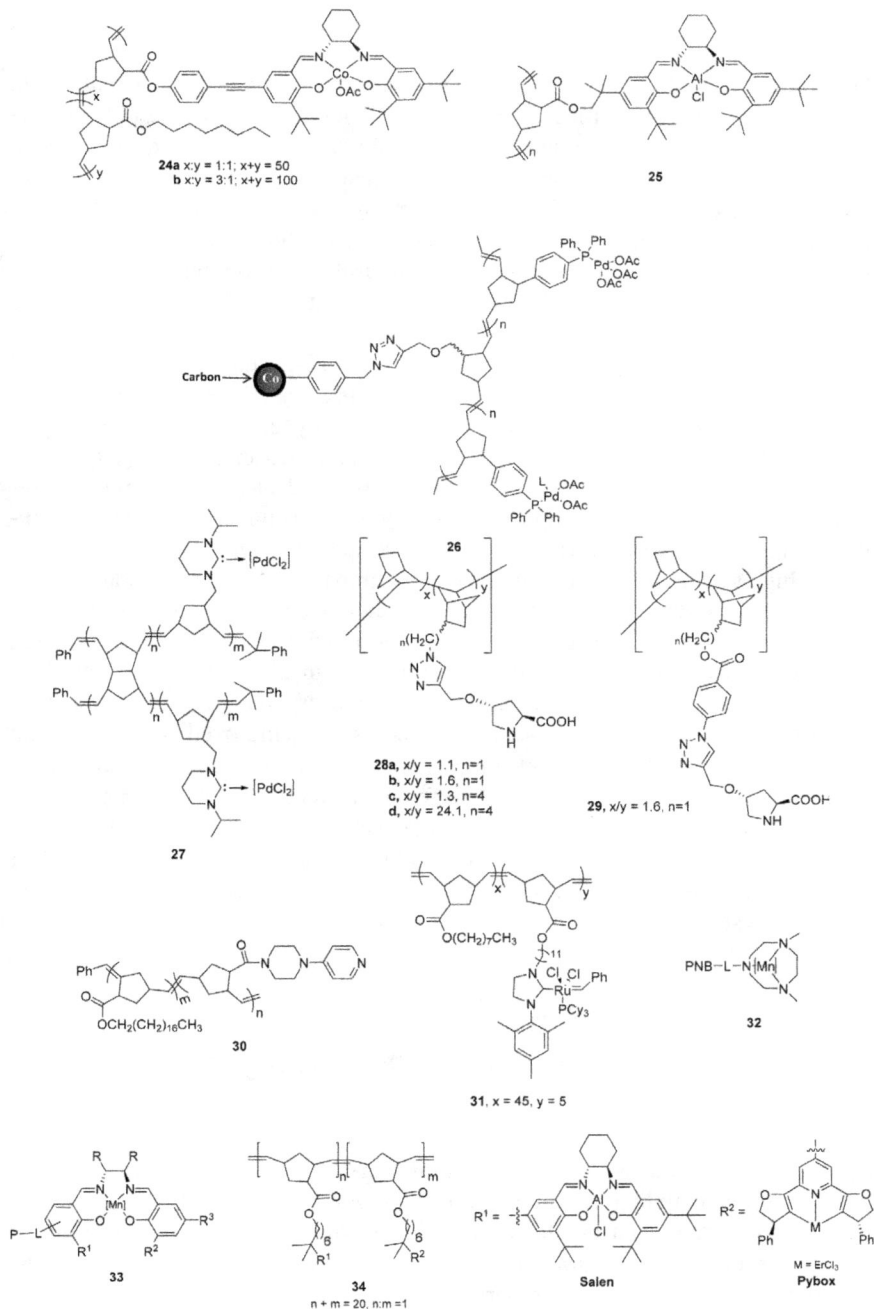

Scheme 3. Important examples of polynorbornenes-supported catalysts

Although NHCs serve as effective ligands for metals such as palladium, the free carbenes are unstable substances that can be challenging to work with. However, carbon dioxide can act as a protective group for carbenes (Bantu et al. 2009). In a solid phase setting, Pawar and Buchmeiser used this useful technique to create a carbon dioxide adduct of a tetrahydropyrimidine-derived carbene attached to a polymeric support. This in situ preparation involved the ring-opening metathesis of a pendant NB functionality attached to the protected carbene. The protected polymer-bound ligand was subsequently changed into a supported metal complex, where the metal might be palladium, iridium, or rhodium but palladium supported catalyst 27 shows effective catalytic activity in carbon-carbon coupling reactions (Pawar and Buchmeiser 2010).

Sagamanova et al. reported the preparation of PNB-supported proline 28a–d and 29 and their application as catalysts in the direct asymmetric aldol reaction. These catalysts (28a–d and 29) were prepared by using the ω-bromoalkyl functionalized PNB as starting reactant (Sagamanova et al. 2015). In 2017, a nonpolar phase tag was incorporated by Fredlund et al. to develop phase-selectively soluble PNB support (30) in 82% yield. In contrast to existing ROMP-based supports that rely on wasteful solvent precipitations, these catalytic supports have a high selectivity for nonpolar solvents compared to their polar counterparts and can be recovered and reused five times in their analysis (Fredlund et al. 2017).

An influential, PNB-supported ruthenium based catalyst 31 for ring closing metathesis was introduced by Sommer and Weck in 2006. In order to produce polymers having 1,4,7-triazacyclononane moieties, manganese-loaded PNB based support 32 was investigated and successfully employed for epoxidation of alkene using H_2O_2 as oxidant. Further for comparison study for epoxidation of PNB supported salen, catalysts 33 were developed using a modular approach, much like supported triazacyclononane catalysts (32) (Ju et al. 2019). A different approach utilized by the researchers to produce an effective PNB based catalyst 34, the R-salen and R-pybox monomers were copolymerized. In this case, two salen catalysts attached to the same polymer backbone were cooperating in a bimetallic transition state.

5. Applications of polyisobutylene and polynorbornenes-supported catalysts

5.1 Coupling reactions

Heck, Suzuki, and Sonogashira's transition-metal-catalyzed processes for carbon-carbon coupling reactions have undergone extensive development for their various significant synthetic applications such as versatile precursors in the total synthesis of natural products, pharmaceuticals, functional materials, conducting polymers, among others (Sutthasupa et al. 2010, Yu et al. 2005). However, coupling reactions driven by polymer-supported catalysts have revealed a growing interest in the design and development of environmentally benign and economically advantageous catalytic systems in the form of supported metal nanoparticles or supported metal complexes. In this section, some of the following soluble PIB and PNB based metal complexes as catalysts are summarized below.

5.1.1 The Heck coupling

In the Heck reaction, an organohalide (aryl or vinyl halides) is cross-coupled with an alkene in the presence of a base to produce a substituted alkene as shown in Scheme 4. Metal-catalyzed reactions are hazardous to the environment and challenging to control. Also, leaching of Pd catalyst is a very tedious task. To overcome these issues, various polymer-supported catalysts have been used in this coupling reaction. In light of this, (PIB)-supported NHC-Pd complexes 10 and 11 reported by Bergbreiter et al. exhibited high activity in the Heck coupling reaction. Catalyst 10 could be recycled upto 10 runs (1 mol% of Pd, 75°C) and 6 runs (130°C) with high yields; however, catalyst 11 showed a sharp decline in yield in the second run. The breakdown of 11 under the described reaction circumstances is the key factor for such an observation (Bergbreiter et al. 2011b, Suriboot et al. 2016). In the Heck coupling of iodobenzene and acrylic acid or methyl acrylate, the PIB-supported SCS complex 1 was utilized. The highly active aryl iodides were the sole compounds with which this catalyst worked, as was the case with earlier supported SCS-Pd(II) species. Better results were while working with the phosphine complexed 2 species with 90-92% yields.

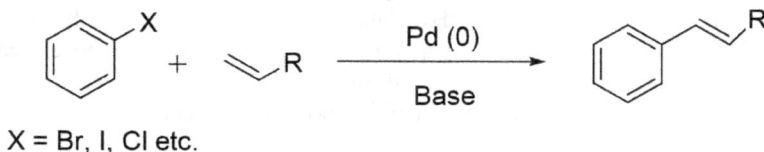

X = Br, I, Cl etc.

Scheme 4. Representation of Heck coupling reaction

In 2005, Jones and his co-researchers' soluble PNB supported SCS-pincer Pd^{II} catalyst (16) was reported for Heck coupling reaction of iodobenzene with butyl acrylate in 99% yield within one hour. However, pincer complexes only served as precatalysts for soluble Pd species, according to the results of poisoning studies using Hg(0) and poly(4-vinylpyridine) (PVPy) instead of leaching of the complex. A year later, Weck and Sommer utilized the same methodology to synthesize a PNB-supported Pd catalyst (19/20) but they substituted SCS pincer with a NHC as the ligand. In the Heck, Suzuki and Sonagashira coupling reactions, all three synthesized PNB-supported Pd-NHC catalysts (20a-c) had the same activity as their tiny monomer predecessors.

It has been established that the Heck reaction could be successfully carried out using the most effective PNB-based catalyst (19a). In 2010, Pawar and Bucheister reported carbon dioxide adduct of a tetrahydropyrimidine-derived carbene attached to a polymeric support. The complex catalyst's protected polymer-bound ligand contains PNB as a polymer support in addition to palladium metal species. Furthermore, the Heck reaction of styrene and butyl acrylate with various aryl bromides was found to be particularly well-catalyzed by the palladium-supported complex 27, with turnover numbers as high as 100,000 (TOF 25,000). Herein, PIB and PNB based catalysts utilized for Heck coupling are summarized in Table 1.

5.1.2 Suzuki–Miyaura cross-coupling

The Suzuki Coupling process, which is the cross-coupling of boric acid and its esters with organohalides in the presence of solvent, is depicted in Scheme 5. This coupling process was discovered in 1979 by Akira Suzuki, who was awarded the Nobel Prize in 2010. In following work, PIB-supported chromium-based polycarbonate (4) catalyzed the reaction of iodobenzene and phyenylboronic acid to obtain Suzuki coupling product. This catalyst exhibited similar activity to their unanchored counterparts, and can easily be recovered for reuse without appreciable loss of activity. The Suzuki-Miyaura cross-coupling reaction of haloarenes with arylboronic acid in THF/water in the presence of Na_2CO_3 was carried out at 65°C using PNB supported hybrid magnetic nanoparticles (26). According to Reiser et al., the yield for the coupling product of a few iodo- and bromoarenes was high (86–96%) but the yield for chlorobenzene was significantly lower (38%). Catalyst recycling can be accomplished with ease by using an external magnetic field due to the magnetic moment of Co/C nanoparticles (Schätz et al. 2010). In 2006, Weck and Sommer reported synthesized PNB-supported Pd-NHC catalysts (20a-c) for remarkable activity with chloroarenes, including sterically hindered substrates and the extremely challenging 2-bromopyridines. One of the catalysts' key features is that it produces the relevant coupling products in good to excellent yields.

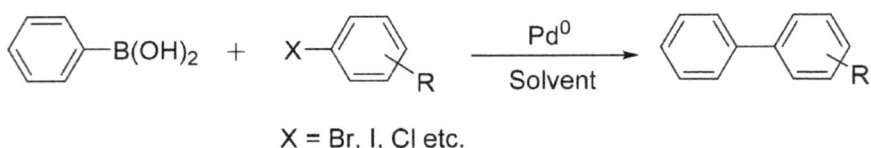

X = Br, I, Cl etc.

Scheme 5. Demonstration of Suzuki coupling reaction

It has also been shown that palladium precursors can induce polymer-supported NHC-based metal catalysts through ring-opening metathesis. Polymers 19a-c and 20a-c were produced by copolymerizing substituted NB co-monomer with palladium-based monomers in various ratios. Each of the six supported palladium catalysts was used to perform the Suzuki coupling reactions (19a-c and 20a-c). Isolated yields of 80–99% for all the substrates were attained (Karimi et al. 2013). The examples provided in this coupling reaction (Table 2), which is based on PIB and PNB-based catalysts, clearly demonstrate how crucial and

Table 1. Heck coupling reaction performed by PIB and PNB supported catalyst

Catalyst	Substrate	Solvent/Temp. (°C)	Catalyst loading (mol%)	Time (h)	Ref.
1	(structure) R = H, CH$_3$	Heptane–DMA/100	1	-	Bergbreiter et al. 2004
2	(structure) X = I, Br, R = Ac, H R' = H, Me, Bu	Heptane/DMF/100	1	6–12	Priyadarshani et al. 2013
10	(structure) R = CH$_3$, OCH$_3$	Heptane or DMF/130	1	0.5	Bergbreiter 2011
11	(structure)	Decane or DMF/65	2.5	10	Bergbreiter 2011
16	(structure)	DMF/120	10	1	Yu et al. 2005
17	(structure)	DMF/120	10	1.5	Anna et al. 2013
18	(structure)	DMF/120	10	1	Anna et al. 2013
19a	(structure)	DMF/120	5	0.5	Sommer and Weck 2006
27	(structure)	DMF/150	0.001	4	Pawar and Buchmeiser 2010

Table 2. Suzuki coupling reaction performed by PIB and PNB supported catalyst

Catalyst	Substrate	Solvent/Temp. (°C)	Catalyst loading (mol%)	Time (h)	Ref.
2	X = Br, R = H, Ac, Cl R' = H, OMe	Heptane/DMF/100	1	6-12	Priyadarshani et al. 2013
19(a-c)	R = H, CH₃	Dioxane/80	1	0.5-2	Karimi et al. 2013
20(a-c)	R = H, CH₃	Dioxane/80	1	0.5-2	Sommer and Weck 2006
26	X = Cl, I, Br	THF/Water/65	1.1	2-12	Schätz et al. 2010

even inevitable it is to use this technique as a significant tool in the process of organic synthesis of biologically important compounds.

5.1.3 Sonogashira Coupling

The Sonogashira Coupling was proposed by N. Hagihara, Y. Tohda and Kenkichi Sonogashira in 1975. Sonogashira Coupling reaction is an extention of Cassar and Dieck and Heck reactions. It is a process of cross-coupling reaction of a terminal alkyne and an aryl or vinyl halide by utilizing palladium catalyst as well as copper co-catalyst to form carbon–carbon bonds in organic synthesis (Scheme 6). In the Sonagashira alkyne-arene coupling chemistry, the PIB-PPh$_2$-ligated Pd(0) complex 2 had good activity. In a heptane DMA thermomorphic system, Sonagashira reactions were conducted, and recycling was accomplished by the system's simple cooling and phase separation (Table 3). Weck and Sommer in 2006 utilized the PNB-supported Pd catalyst with an NHC as the ligand. All three of the synthesized PNB-anchored Pd-NHC catalysts (20a-c) have the same activity as their small monomer for the Sonagashira coupling process (Sommer and Weck 2006).

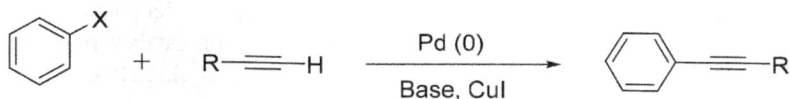

X = Br, I, Cl, OTf etc.

Scheme 6. Illustration of Sonogashira coupling reaction

5.2 Epoxidation

The epoxidation reaction of the C=C is one of the most powerful and versatile synthetic chemistry methods, from the laboratory to the industrial scale. It is used to produce bulk compounds like ethylene oxide and propylene oxide as well as fine chemicals with high added value (i.e. epoxidized terpenes). A range of reagents, including as air oxidation, hypochlorous acid, hydrogen peroxide, and organic peracids, are used in the chemical process known as epoxidation to convert the carbon-carbon double bond into oxiranes (epoxides). The choice is based on the substrate type, the features of the catalyst being used, as well as the process's safety, efficacy, and long-term viability. Several efforts have been made to develop polymer-supported epoxidation catalysts and comprehensive reviews of polymer-supported epoxidation catalysts in general have been published. The most recent chapter concerning polymer-supported metal complex epoxidation catalysts in particular appeared in 2008 (Arnold 2008). It should be emphasized that only three polymer types account for around 80% of the documented epoxidation catalysts. About 65% of the researched catalytic systems are based on polystyrene, and remaining 15% are made of polymethacrylate, and poly (ethylene glycol). The trend in recent research toward developing soluble polymer-supported catalysts has grown in popularity. Using a range of thermosetting supports, such as PNB, polybenzimidazole, and polyimide, intriguing results have already been obtained.

In order to produce polymers having 1,4,7-triazacyclononane moieties, NB connected to the azacycle was subjected to ring-opening metathesis polymerization (ROMP). For the epoxidation of alkene using H_2O_2 as the oxidant, manganese-loaded PNB based support (32) was investigated. High activity under mild reaction conditions comparable to or even superior to that of the monomeric complex was reported. Additionally, for comparing the reusability and recyclability of the catalyst, manganese (II) Schiff base complexes supported on a styrene-DVB copolymer were studied, but their activity in the epoxidation of cyclooctene and NB with tert-butyl hydroperoxide was modest and decreases with recycling.

According to a report, PNB supported salen catalysts [33] were developed using a modular approach, much like supported triazacyclononane catalysts [32]. For this, ring-opening metathesis polymerization was used to create a manganese-salen complex coupled to a NB monomer. When m-CPBA/N-methylmorpholine-N-oxide (NMO) was used as the oxidant, the resultant polymers and copolymers demonstrated strong catalytic activity and enantioselectivity; however, a notable drop in activity and selectivity was noted upon catalyst recycling in the epoxidation of 1,2-dihydronaphthalene (Table 4) (Ju et al. 2019).

5.3 Miscellaneous

In 2013, Anna et al. reported the activity of polymer-supported Al salen-complex (25) for the enantioselective 1,4-addition reaction of cyanide to various α,β-unsaturated imides obtaining high yields in each case (between 88 and 96%) along with high ee (98 or 99%). In only one case, the yield was low (22%) when tert-butyl-substituted imide was used as substrates. With the exception of the tert-butyl-substituted imide substrate, compound 25 was both more active and selective than its unsupported analogue, even with a metal loading of just 5 mol% (as compared to 15 mol% for the unsupported analogue). They proposed that the flexible backbone increased the concentration of local catalysts, hence enhancing catalytic activity. Additionally, the enantioselectivity and activity of catalyst 25 may be recycled up to five times without any significant loss of activity.

In another study, two salen catalysts mounted onto the same polymer backbone were cooperating in a bimetallic transition state. The R-salen monomer and the R-pybox monomer were copolymerized to create catalyst 34. Seven carbon atom linkers were chosen for both monomers because they allowed the best interactions between the two catalytic PNB-supported salen centres that were required for the transformation. To evaluate the catalytic activity of polynorborene catalyst 34, cyanide addition to alpha-beta unsaturated imide was deployed, and 88% of the isolated yield with 80% ee was obtained. Without

Table 3. Sonogashira coupling reaction performed by PIB and PNB supported catalyst

Catalyst	Substrate	Solvent/Temp. (°C)	Catalyst loading (mol%)	Time (h)	Ref.
2	Alkyne-arene	Heptane-DMA/70	1	2	Bergbreiter et al. 2004
19a	R—Br + R'—≡—Si(Me)$_3$ R = H, CH$_3$, CHO R' = H, Ph	THF/80	2	2-2.5	Karimi et al. 2013
20(a-c)	R—Br + R'—≡—Si(Me)$_3$ R = H, CH$_3$, CHO R' = H, Ph	THF/DMAc/80	3	2.5-3	Sommer and Weck 2006

Table 4. Epoxidation reaction of alkene performed by PIB and PNB supported catalyst

Catalyst	Alkene	Oxidant	Time (h)	Temp. (°C)	Yield (%)	TON[a]	ee (%)	Runs[b]	Ref.
32	Styrene	H$_2$O$_2$	3	0	80	80	-	1	Arnold 2008
33	Styrene	m-CPBA/NMO	0.083	-20	100	25	33	3	Arnold 2008

[a] TON, turnover number (moles of epoxide per moles of catalyst).
[b] Maximum number of reactions (initial run β recycling experiments) carried out with the catalyst systems.

any significant conversion losses, the catalyst remained in the reaction flask and was reported to be utilized for three cycles. With each consecutive cycle, ee decreased, showing a reduction in catalytic selectivity that could be attributable to the degradation of chiral pybox ligand. Compound 30 was employed in DMAP assisted Boc protection of 2,6-dimethylphenol yielding 92% of the product in less than 20 minutes. Through copper-catalyzed azide-alkyne cycloaddition (CuAAC) reactions, L-proline derivatives have been immobilized onto PNB support to create PNB supported catalysts (28a-d and 29), which are employed for the direct asymmetric aldol reaction of cyclohexanone with p-nitrobenzaldehyde in aqueous medium (Sagamanova et al. 2015, Salvo et al. 2016).

An efficient recyclable homogeneous catalyst (15) for the reduction of nitroarene utilising hydrazine hydrate as a reducing reagent was demonstrated using a cobalt phthalocyanine (CoMPc) that contains covalently linked PIB groups as phase anchors in a semithermomorphic system at 110°C and is highly soluble in both nonpolar and polar organic solvents. Several solvents were used, but the best results were obtained with an equivolume mixture of ethylene glycol and heptane (Chao and Bergbreiter 2016).

Furthermore, allylic substitution of cinnamyl acetate by secondary amines in a latent biphasic system was successfully accomplished using a PIB-supported Pd catalyst 25. In this instance, the separation was accomplished by adding 10 vol% water to a mixture of EtOH and heptane solvent that was utilized at room temperature. This catalyst for allylic substitution was successfully recycled five times. The Pd(0) complex 2 with PIB-PPh$_2$-ligation displayed good activity in the allylic substitution process. In order to produce biphasic conditions and phase separation, the allylic substitution reactions were carried out in heptane-EtOH latent biphasic systems. The catalyst was reused and recycled in this setting (Priyadarshani et al. 2013).

In Buchwalde-Hartwig aryl amination process, the PIB-bound catalyst 11 was initially tested. At 80°C in either heptane or 1,2-dimethoxyethane, 4-bromotoluene underwent full conversion to *N*-(4-methylphenyl)morpholine in 20 minutes with a 1 mol% catalyst. Ring opening of epoxides with thiols catalysed by PIB-supported salen Cr(III) 4 at room temperature for 24 h gives the desired product (Bergbreiter et al. 2011a). Recently, Bazzi and colleagues described the Buchwald-Hartwig amination process as a Palladium-BIAN-NHC-catalyzed C–N cross-coupling using catalyst 14 immoblized on PIB. Easy catalyst separation in a biphasic heptane/acetonitrile combination is a practical benefit of PIB-tagging. However, in comparison to the similar imidazolylidene-based catalyst, Pd-BIAN-NHC catalyst 14 shows somewhat reduced reactivity under the studied reaction conditions (Chen et al. 2021). PNB-supported complex 31 (3 mol% loading) was reported for the ring closing metathesis of Diethyl malonate (DEM) in the presence of dichloromethane (DCM) solvent at 45°C to get the desired product (Table 5) (Sommer and Weck 2006).

Table 5. Miscellaneous reactions performed by PIB and PNB supported catalysts

Catalyst	Reaction	Time (h)	Temp. (°C)	Yield (%)	ee (%)	Runs	Ref.
2	Allylic substitution	6	80	65-90	-	8	Priyadarshani et al. 2013
4	Ring opening of epoxides with thiols	24	25	34-99	-	4-5	Bergbreiter 2011
11	Buchwalde-Hartwig aryl amination process (C–N coupling reaction)	0.3	80	99	-	2	Bergbreiter 2011
14	Buchwald-Hartwig amination (C–N coupling reaction)	17	100	79	-	2	Chen et al. 2021
15	Reduction of nitroarene	24	110	65-86	-	10	Chao and Bergbreiter 2016
25	Enantioselective 1,4-additon reaction of cyanide to α,β-unsaturated imide	36	45	88-96	98-99	5	Anna et al. 2013
25	Allylic substitution of cinnamyl acetate by secondary amines	-	25	99-100	-	5	Bergbreiter et al. 2004
28(a-d)	Direct asymmetric aldol reaction	23	25	63-98	90-97	7	Sagamanova et al. 2015, Salvo et al. 2016
29	Direct asymmetric aldol reaction	22	25	55-99	69-98	7	Sagamanova et al. 2015, Salvo et al. 2016
30	Boc protection of 2,6-dimethylphenol	0.3	-	92	98	5	Yolsal et al. 2020
34	Cyanide addition to α,β-unsaturated imide	18	45	88	80	3	Madhavan et al. 2011

6. Summary and future prospects

In conclusion, this literature review's findings indicate that PIB and PNB oligomers are effective non-polar phase-selective supports that can serve as both a catalytic support and a ligand used in catalytic chemistry. These catalysts' molecular building blocks are adaptable by nature, making it possible to systematically alter their catalytic performance. Another interesting feature of these catalysts was shown to be separable from the product by straightforward filtration or by precipitation after addition of a small amount of solvent, in the case where the supported catalyst is soluble in the reaction medium. These catalysts demonstrated high efficiency and/or excellent selectivity in a variety of reactions. Without losing activity or selectivity, the recovered catalyst is frequently able to be employed in additional cycles. This would not only make it easier to understand the catalyst's structure and conduct additional mechanistic research on pertinent reactions, but it might also result in the creation of brand-new single-site catalysts for highly selective synthesis.

The supporting catalyst's solubility characteristics are also highly significant. While heterogeneously supported catalysts are typically thought to be more stable, manageable, and easily recovered and recycled, a homogenous catalytic system is expected to be more reactive, stereochemically more efficient, and more dependable in reproducibility. Although there is no perfect support, one must choose the right one for each unique catalytic system. In this regard, substantial advancement may result from the multidisciplinary expertise being developed with the assistance of organic and inorganic material chemists, particularly in the area of polymeric materials.

Furthermore, recently reported PIB/PNB based catalysts, which combine organic polymers with inorganic components, offer new opportunities for the development of hybrid and effective polymer-support for carbon-carbon coupling reactions, carbon-nitrogen coupling reactions, epoxidation catalysts, and other catalysts. It can be predicted that catalytic systems based on alternative polymers still have a lot of potential.

Acknowledgement

We are thankful to University of Delhi for financial assistance. The award of Junior and Senior Research Fellowship to DV by CSIR (India) is gratefully acknowledged.

References

Alentiev, D.A. and M.V. Bermeshev. 2022. Design and synthesis of porous organic polymeric materials from norbornene derivatives. *Polymer Reviews* 62(2): 400-437. https://doi.org/10.1080/15583724.2021.1933026

Alves, J.B., M.K. Vasconcelos, L.H.R. Mangia, M. Tatagiba, J. Fidalgo, D. Campos, P.L. Invernici, M.V. Rebouças, M.H.S. Andrade and J.C. Pinto. 2021. A bibliometric survey on polyisobutylene manufacture. *Processes* 9(8): 1315. https://doi.org/10.3390/pr9081315

Anna, M., G. Romanazzi and P. Mastrorilli. 2013. Polymer supported catalysts obtained from metal-containing monomers. *Current Organic Chemistry* 17(12): 1236-1273. https://doi.org/10.2174/1385272811317120003

Arnold, U. 2008. Metal species supported on organic polymers as catalysts for the epoxidation of alkenes. pp. 387-411. *In:* Oyama, S.T. (ed.). Mechanisms in Homogeneous and Heterogeneous Epoxidation Catalysis. Elsevier. ISBN 9780444531889. https://doi.org/10.1016/B978-0-444-53188-9.00015-8

Bandaiphet, C. and J.F. Kennedy. 2004. Polymeric materials in organic synthesis and catalysis. *Carbohydrate Polymers* 56(2): 1-110. https://doi.org/10.1016/j.carbpol.2004.01.003

Bano, T., A.F. Zahoor, N. Rasool, M. Irfan and A. Mansha. 2022. Recent trends in Grubbs catalysis toward the synthesis of natural products: A review. *Journal of the Iranian Chemical Society* 19(6): 2131-2170. https://doi.org/10.1007/s13738-021-02463-x

Bantu, B., G.M. Pawar, U. Decker, K. Wurst, A.M. Schmidt and M.R. Buchmeiser. 2009. CO_2 and Sn^{II} adducts of N-heterocyclic carbenes as delayed-action catalysts for polyurethane synthesis. *Chemistry – A European Journal* 15(13): 3103-3109. https://doi.org/10.1002/chem.200802670

Benaglia, M. 2007. Organocatalysis: Recoverable, soluble polymer-supported organic catalysts. pp. 76-96. *In:* M. Reetz, B. List, S. Jaroch and H. Weinmann (eds.). Ernst Schering Foundation Symposium Proceedings Vol. 2. Springer, ISBN 978-3-540-73495-6. https://doi.org/10.1007/2789_2007_067

Benaglia, Maurizio, A. Puglisi and F. Cozzi. 2003. Polymer-supported organic catalysts. *Chemical Reviews* 103(9): 3401-3430. https://doi.org/10.1021/cr010440o

Bergbreiter, D.E., P.L. Osburn, A. Wilson and E.M. Sink. 2000. Palladium-catalyzed C–C coupling under thermomorphic conditions. *Journal of the American Chemical Society* 122(38): 9058-9064. https://doi.org/10.1021/ja001708g

Bergbreiter, D.E. and J. Li. 2004. Terminally functionalized polyisobutylene oligomers as soluble supports in catalysis. *Chemical Communications* 4(1): 42-43. https://doi.org/10.1039/b312368e

Bergbreiter, D.E. and S.D. Sung. 2006. Liquid/liquid biphasic recovery/reuse of soluble polymer-supported catalysts. *Advanced Synthesis and Catalysis* 348(12–13): 1352-1366. https://doi.org/10.1002/adsc.200606144

Bergbreiter, D.E. and J. Tian. 2007. Soluble polyisobutylene-supported reusable catalysts for olefin cyclopropanation. *Tetrahedron Letters* 48(26): 4499-4503. https://doi.org/10.1016/j.tetlet.2007.04.147

Bergbreiter, D.E., C. Hobbs and C. Hongfa. 2011a. Polyolefin-supported recoverable/reusable Cr(III)-salen catalysts. *Journal of Organic Chemistry* 76(2): 523-533. https://doi.org/10.1021/jo102044m

Bergbreiter, D.E., H.L. Su, H. Koizumi and J. Tian. 2011b. Polyisobutylene-supported N-heterocyclic carbene palladium catalysts. *Journal of Organometallic Chemistry* 696(6): 1272-1279. https://doi.org/10.1016/j.jorganchem.2010.10.058

Blank, F. and C. Janiak. 2009. Metal catalysts for the vinyl/addition polymerization of norbornene. *Coordination Chemistry Reviews* 253(7–8): 827-861. https://doi.org/10.1016/j.ccr.2008.05.010

Boyd, R.H. and P.V.K. Pant. 1991. Molecular packing and diffusion in polyisobutylene. *Macromolecules* 24(23): 6325-6331. https://doi.org/10.1021/ma00023a040

Chao, C.G. and D.E. Bergbreiter. 2016. Highly organic phase soluble polyisobutylene-bound cobalt phthalocyanines as recyclable catalysts for nitroarene reduction. *Catalysis Communications* 77: 89-93. https://doi.org/10.1016/j.catcom.2016.01.022

Chen, C., F.S. Liu and M. Szostak, 2021. BIAN-NHC ligands in transition-metal-catalysis: A perfect union of sterically encumbered, electronically tunable N-heterocyclic carbenes. *Chemistry – A European Journal* 27(14): 4478-4499. https://doi.org/10.1002/chem.202003923

Dickerson, T.J., N.N. Reed and K.D. Janda. 2003. Soluble polymers as scaffolds for recoverable catalysts and reagents. *Chemical Reviews* 102(10): 3325-3344. https://doi.org/10.1002/chin.200303259

Flid, V.R., M.L. Gringolts, R.S. Shamsiev and E.S. Finkelshtein. 2018. Norbornene, norbornadiene and their derivatives: Promising semi-products for organic synthesis and production of polymeric materials. *Russian Chemical Reviews* 87(12): 1169-1205. https://doi.org/10.1070/rcr4834

Frank, D., P. Espeel, N. Badi and F.D. Prez. 2018. Structurally diverse polymers from norbornene and thiolactone containing building blocks. *European Polymer Journal* 98: 246-253. https://doi.org/10.1016/j.eurpolymj.2017.11.023

Fredlund, A., V.A. Kothapalli and C.E. Hobbs. 2017. Phase-selectively soluble polynorbornene as a catalyst support. *Polymer Chemistry* 8(3): 516-519. https://doi.org/10.1039/c6py02041k

Holladay, J.E. and K.O. Albrecht. 2012. Catalysis of organic reactions. *Topics in Catalysis* 55(7–10): 419-420. https://doi.org/10.1007/s11244-012-9814-2

Hongfa, C., J. Tian, H.S. Bazzi and D.E. Bergbreiter. 2007. Heptane-soluble ring-closing metathesis catalysts. *Organic Letters* 9(17): 3259-3261. https://doi.org/10.1021/ol071210k

Hongfa, C., J. Tian, J. Andreatta, D.J. Darensbourg and D.E. Bergbreiter. 2008. A phase separable polycarbonate polymerization catalyst. *Chemical Communications* 8: 975-977. https://doi.org/10.1039/b711861a

Ju, P., S. Wu, Q. Su, X. Li, Z. Liu, G. Li and Q. Wu. 2019. Salen-porphyrin-based conjugated microporous polymer supported Pd nanoparticles: Highly efficient heterogeneous catalysts for aqueous C–C coupling reactions. *Journal of Materials Chemistry A* 7(6): 2660-2666. https://doi.org/10.1039/c8ta11330k

Kann, N. 2010. Recent applications of polymer supported organometallic catalysts in organic synthesis. *Molecules* 15(9): 6306-6331. https://doi.org/10.3390/molecules15096306

Karimi, B., S. Abedi and A. Zamani. 2013. Coupling reactions induced by polymer-supported catalysts. pp. 141-200. A. Molnar (ed.), Palladium-catalyzed Coupling Reactions: Practical Aspects and Future Developments. Wiley. ISBN 9783527648283. https://doi.org/10.1002/9783527648283.

Palladium-catalysed coupling reactions: practical aspects and future developments

Kunal, K., M. Paluch, C.M. Roland, J.E. Puskas, Y. Chen and A.P. Sokolov. 2008. Polyisobutylene: A most unusual polymer. *Journal of Polymer Science, Part B: Polymer Physics* 46(13): 1390-1399. https://doi.org/10.1002/polb.21473

Ley, S.V. and I.R. Baxendale. 2002. New tools and concepts for modern organic synthesis. *Nature Reviews Drug Discovery* 1(8): 573-586. https://doi.org/10.1038/nrd871

Li, J., S. Sung, J. Tian and D.E. Bergbreiter. 2005. Polyisobutylene supports – A non-polar hydrocarbon analog of PEG supports. *Tetrahedron* 61(51): 12081-12092. https://doi.org/10.1016/j.tet.2005.07.119

Liang, Y. and D.E. Bergbreiter. 2016. Recyclable polyisobutylene (PIB)-bound organic photoredox catalyst catalyzed polymerization reactions. *Polymer Chemistry* 7(12): 2161-2165. https://doi.org/10.1039/c6py00114a

Madhavan, N., C.W. Jones and M. Weck. 2008. Rational approach to polymer-supported catalysts: Synergy between catalytic reaction mechanism and polymer design. *Accounts of Chemical Research* 41(9): 1153-1165. https://doi.org/10.1021/ar800081y

Madhavan, N., W. Sommer and M. Weck. 2011. Supporting multiple organometallic catalysts on poly(norbornene) for cyanide addition to α,β-unsaturated imides. *Journal of Molecular Catalysis A: Chemical* 334(1–2): 1-7. https://doi.org/10.1016/j.molcata.2010.10.023

Oh, J.K. 2008. Recent advances in controlled/living radical polymerization in emulsion and dispersion. *Journal of Polymer Science, Part A: Polymer Chemistry* 46(21): 6983-7001. https://doi.org/10.1002/pola.23011

Ohm, R.F. and T.M. Vial. 1978. A new synthetic rubber Norsorex® polynorbornene. *Journal of Elastomers and Plastics* 10(2): 150. https://doi.org/10.1177/009524437801000205

Pawar, G.M. and M.R. Buchmeiser. 2010. Polymer-supported, carbon dioxide-protected n-heterocyclic carbenes: Synthesis and application in órgano- and organometallic catalysis. *Advanced Synthesis and Catalysis* 352(5): 917-928. https://doi.org/10.1002/adsc.200900658

Priyadarshani, N., J. Suriboot and D.E. Bergbreiter. 2013. Recycling Pd colloidal catalysts using polymeric phosphine ligands and polyethylene as a solvent. *Green Chemistry* 15(5): 1361-1367. https://doi.org/10.1039/c3gc36932c

Rajasekhar, T., U. Haldar, J. Emert, P. Dimitrov, R. Severt and R. Faust. 2017. Catalytic chain transfer polymerization of isobutylene: The role of nucleophilic impurities. *Journal of Polymer Science, Part A: Polymer Chemistry* 55(22): 3697-3704. https://doi.org/10.1002/pola.28751

Rajasekhar, T., G. Singh, G.S. Kapur and S.S.V. Ramakumar. 2020. Recent advances in catalytic chain transfer polymerization of isobutylene: A review. *RSC Advances* 10(31): 18180-18191. https://doi.org/10.1039/d0ra01945c

Sagamanova, I.K., S. Sayalero, S. Martínez-Arranz, A.C. Albéniz and M.A. Pericàs. 2015. Asymmetric organocatalysts supported on vinyl addition polynorbornenes for work in aqueous media. *Catalysis Science and Technology* 5(2): 754-764. https://doi.org/10.1039/c4cy01344a

Salvo, A.M.P., F. Giacalone and M. Gruttadauria. 2016. Advances in organic and organic-inorganic hybrid polymeric supports for catalytic applications. *Molecules* 21(10): 1288-1340. https://doi.org/10.3390/molecules21101288

Schätz, A., T.R. Long, R.N. Grass, W.J. Stark, P.R. Hanson and O. Reiser. 2010. Immobilization on a nanomagnetic Co/C surface using ROM polymerization: Generation of a hybrid material as support for a recyclable palladium catalyst. *Advanced Functional Materials* 20(24): 4323-4328. https://doi.org/10.1002/adfm.201000959

Sommer, W.J. and M. Weck. 2006. Poly(norbornene)-supported N-heterocyclic carbenes as ligands in catalysis. *Advanced Synthesis and Catalysis* 348(15): 2101-2113. https://doi.org/10.1002/adsc.200606135

Su, H.L., C. Hongfa, H.S. Bazzi and D.E. Bergbreiter. 2010. Polyisobutylene phase-anchored ruthenium complexes. *Macromolecular Symposia* 297(1): 25-32. https://doi.org/10.1002/masy.200900092

Suriboot, J., H.S. Bazzi and D.E. Bergbreiter. 2016. Supported catalysts useful in ring-closing metathesis, cross metathesis, and ring-opening metathesis polymerization. *Polymers* 8(4): 140. https://doi.org/10.3390/polym8040140

Sutthasupa, S., M. Shiotsuki and F. Sanda. 2010. Recent advances in ring-opening metathesis polymerization, and application to synthesis of functional materials. *Polymer Journal* 42(12): 905-915. https://doi.org/10.1038/pj.2010.94

Xu, Y.M., K. Li, Y. Wang, W. Deng and Z.J. Yao. 2017. Mononuclear nickel(II) complexes with schiff base ligands: Synthesis, characterization, and catalytic activity in norbornene polymerization. *Polymers* 9(3): 105-115. https://doi.org/10.3390/polym9030105

Yolsal, U., T.A.R. Horton, M. Wang and M.P. Shaver. 2020. Polymer-supported Lewis acids and bases: Synthesis and applications. *Progress in Polymer Science* 111: 101313. https://doi.org/10.1016/j.progpolymsci.2020.101313

Yu, K., W. Sommer, M. Weck and C.W. Jones. 2004. Silica and polymer-tethered Pd-SCS-pincer complexes: Evidence for precatalyst decomposition to form soluble catalytic species in Mizoroki–Heck chemistry. *Journal of Catalysis* 226(1): 101-110. https://doi.org/10.1016/j.jcat.2004.05.015

Yu, K., W. Sommer, J.M. Richardson, M. Weck and C.W. Jones. 2005. Evidence that SCS pincer Pd(II) complexes are only precatalysts in Heck catalysis and the implications for catalyst recovery and reuse. *Advanced Synthesis and Catalysis* 347(1): 161-171. https://doi.org/10.1002/adsc.200404264

Polymer-supported Phosphine Reagents

Girdhar Pal Singh* and Narendra Pal Singh Chauhan

Department of Chemistry, Faculty of Science, Bhupal Nobles' University, Udaipur - 313002, Rajasthan, India

1. Introduction

Philippe Gengembre (1783) observed phosphine, an inorganic compound, by heating phosphorus in an aqueous solution of K_2CO_3. The unique element phosphorus plays crucial roles in both organic and inorganic chemistry. Its ability to readily transition into +3 and +5 oxidation states and form three and five covalent bonds with carbon, hydrogen, nitrogen, and oxygen is largely attributable to its versatility in synthetic applications (Methot et al. 2004). It is a colorless gas with a rotten-fish odor, highly poisonous, sparingly soluble in water, soluble in organic solvents, and it acts as a Lewis base by donating its lone pair of electrons when it reacts with hydrogen iodide to function as a Lewis base. PH_3 is a major component of the Holme signal. Phosphine fumigants are a typical household product used to limit the invasion of pests, rats, and rabbits in a wide variety of stored grains. In addition, phosphine is utilized as a doping agent for n-type semiconductors, a polymerization initiator, a condensation catalyst, and a polymer-supported reagent in the fabrication of flame retardants for cotton fabrics. (Salmeia et al. 2016). The majority of the molecules used in the polymer chemistry of phosphorus have stable C-P bonds, or they are inorganic acids or their derivatives. These phosphorus-containing compounds are few in number compared to other organic polymers (Strasser and Teasdale 2020).

*Corresponding author: girdharpal@gmail.com

It can be prepared using following:

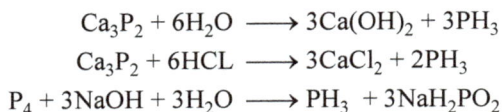

$$Ca_3P_2 + 6H_2O \longrightarrow 3Ca(OH)_2 + 3PH_3$$
$$Ca_3P_2 + 6HCL \longrightarrow 3CaCl_2 + 2PH_3$$
$$P_4 + 3NaOH + 3H_2O \longrightarrow PH_3 + 3NaH_2PO_2$$

Polyphosphazene, also referred to as an inorganic rubber, was the first synthetic rubber that was recorded by Stokes in 1897. Its technological potential was first overlooked. However, in recent years, a growing interest in organophosphine polymers has been noticed, which is certainly related to their significant contribution to science and industry. These polymers have unique properties such as medical sector, anticorrosive compounds, flame retardant, metal ions removal from wastewater etc. (Chauhan and Chundawat 2019). Furthermore, it has been demonstrated that phosphorus-containing compounds can improve material solubility (hydrophilicity), adhesion, biocompatibility, and biodegradability. Different chemical properties of phosphine based compounds are depicted in Figure 1.

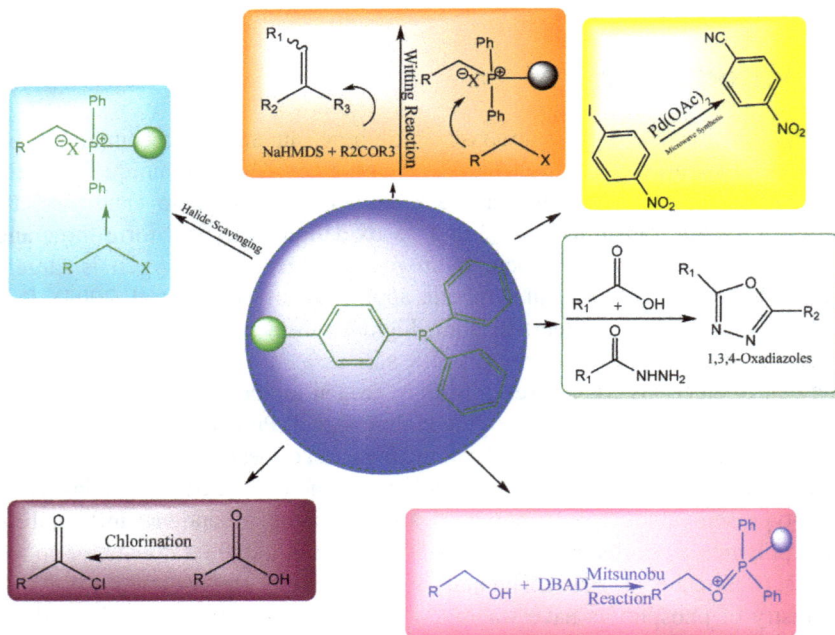

Figure 1. Different chemical properties based on phosphine based compounds

Non-cross-linked polystyrene was used in 1983 as the base polymer for making poly(styryldiphenylphosphine), which had a phosphine loading of 2.7–3.0 mmol per gram of polymer. This was the first report of a soluble polymer-supported phosphine. It's interesting to note that the polymer-supported phosphine's efficiency in producing alkyl chlorides from an alcohol and carbon tetrachloride was found to be comparable to poly(styryldiphenylphosphine) cross-linked with divinylbenzene. Initially, the reaction's starting linear polystyrene was soluble, but once the reaction was complete, the polymer precipitated, allowing for filtration-based purification; however, after long-term storage, the polymeric reagent appeared to cross-link and become insoluble. It has been reported that a triphenylphosphine reagent on NCPS has been further improved for a Staudinger/aza-Wittig process, with better outcomes than the reagent's insoluble-polymer variant.

In a subsequent study, this second-generation reagent was used in the highly regioselective SN_2' Mitsunobu reaction to synthesize E-tri-substituted alkenes from Baylis-Hillman adducts. Interestingly, the soluble polymer-bound reagent significantly outperformed triphenylphosphine in this reaction in terms of regioselectivity.

2. Synthetic methods

In Suzuki reactions, palladium-supported triphenylphosphine-functionalized microporous knitting aryl network polymers (Pd@KAPs(Ph-PPh$_3$)) have been employed (Li et al. 2012). It has demonstrated a number of benefits, including the ability to prevent the contamination of products with palladium and ligand residue, good reusability of the expensive metal and phosphine ligand, and an abundance of open micropore and macropore structures that are advantageous for catalysis. Palladium supported on triphenylphosphine functionalized porous organic polymer (Pd@KAPs(Ph-PPh$_3$)) is reported as a catalyst in an effective method for the alkoxycarbonylation of aryl iodides. When different aryl iodides are carbonylated with alcohols and phenols under CO balloon pressure, the corresponding products are produced in moderate to excellent yields (74–96%) (Lei et al. 2015).

By incorporating [(Me$_2$S)AuCl] into a phosphine-based hyper-cross-linked polymer, Shen and coworkers developed a neutral gold(I) complex, Au@HCPs-PPh$_3$, which has been shown to be an effective and versatile heterogeneous catalyst for the regioselective hydration of a number of alkynes to ketones (Shen et al. 2022). For their high catalytic activity and stability, Rh-based catalysts supported on porous vinyl triphenylphosphine (3V-PPh$_3$) polymers have been recommended. Density functional theory (DFT) was used to examine the formation process of catalyst active species after CO insertion as well as the mechanism of the reaction of ethylene hydroformylation to propionaldehyde. However, the effect of the carrier-ligand bifunctional interaction of 3V-PPh$_3$ on the mechanism of heterogeneous ethylene developed two types of crosslinked 3V-PPh$_3$ supported Rh-based catalysts (Xie et al. 2022).

By reacting 2,2-bis(4-hydroxyphenyl)propane and bis(4-fluorophenyl) phenylphosphine, phosphine-containing poly(arylene ether)s with three covalent bonds phosphorus are developed. In order to prevent partial phosphine derivative oxidation during polycondensation, the reaction conditions are changed. In contrast, copolymers with high molar contents of phosphine oxide (up to 45 mol-% P = O repeating units) and 85 mol-% repeating units containing phosphine are produced by solvent-free melt polycondensation of bis(4-fluorophenyl) phenylphosphine with bis(trimethylsilyl)-2,2-bis(4-oxyphenyl)propane and CsF as catalyst (Satpathi et al. 2020).

An important part of metal catalysis for chemical transformations at the industrial and laboratory levels is the hybridization of porous synthetic polymers and sophisticated ligands. The high permeability, ease of modification, rapid mass transfer properties and high stability of a monolithic porous polymer, which is a single piece with continuous macropores, are desirable. A monolithic porous polystyrene with three-fold cross-linked PPh_3 (M-PS-TPP) was developed by Matsumoto and coworkers for transition-metal catalysis (Matsumoto et al. 2020). By using a porogenic solvent to induce phase separation during polymerization, the monolithic and macroporous structure of M-PS-TPP was achieved. Suzuki-Miyaura cross-coupling of chloroarenes, a difficult Pd-catalyzed reaction, was made easier by the macroporous characteristics and controlled mono-P-ligating behavior of M-PS-TPP.

By using azodiisobutyronitrile (AIBN) as an initiator in solution polymerization, 4-(diphenylphosphino) styrene (DPPS), di(ethylene glycol) methyl ether methacrylate (DEGMA), and oligo(ethylene glycol) methyl ether methacrylate (OEGMA$_{300}$), a copolymer with an average molecular weight of 300, was developed. Different lower critical solution temperatures (LCST) were used to synthesize a series of P(DEGMA-co-DPPS-co-OEGMA$_{300}$) copolymers. The Suzuki-Miyaura reaction was successfully carried out in water in the presence of the thermoresponsive copolymer-supported palladium catalyst (Chen et al. 2019). The Suzuki-Miyaura coupling reaction has been used to develop a "one-pot" method for synthesizing polymer-supported phosphine-Pd catalyst.

A "one-pot" strategy has been developed for the synthesis of polymer-supported phosphine-Pd catalyst using Suzuki–Miyaura coupling reaction. On the skeleton of the polymer containing phosphine, the fine and uniform Pd nanoparticles were evenly distributed. In the Suzuki-Miyaura coupling reaction with the mild conditions and water-ethanol mixed media, this Pd catalyst demonstrated excellent catalytic activity and recycling performance (Chen et al. 2017).

In order to catalyze the homogeneous enantioselective hydroarylation of aryl allyl ethers and produce the benzene-fused cyclic skeletons that are frequently found in pharmaceuticals, natural alkaloids, and fascinating building blocks in organic synthesis, Xu and coworkers developed the Pd/Xu-Phos system (Zhang et al. 2018). In heterogeneous asymmetric catalysis, the immobilization of Pd/Xu-Phos catalyst can also demonstrate exceptional catalytic performance. However, the Xu-Phos chiral ligand has a greater steric hindrance than the previous polymer-bound chiral gold catalyst, which makes it more difficult to immobilize, and the ligand with more electron density is more susceptible to oxidation (Zhang et al. 2023). Furthermore, the coordination becomes more complicated and difficult to control due to the Pd catalyst with bidentate chelation (Figure 2).

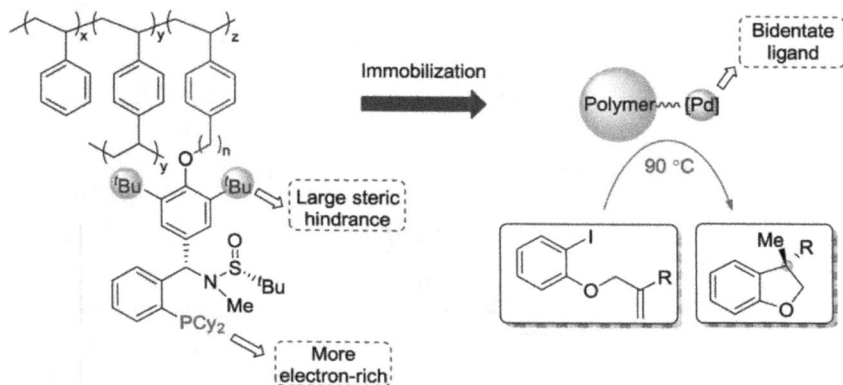

Figure 2. Immobilization of polymer-supported Xu-Phos to form polymer-supported palladium based complexes having excellent catalytic activity, higher cross linking degree and good recyclability (Zhang et al. 2023)

Phosphine-based conjugated hyper crosslinked polymer (PPh$_3$-CHCP) photocatalyst and its use in the development of the first large-scale, persistent Cu-ATRP (atom transfer radical polymerization) powered by sunlight with a limited O$_2$ tolerance (without a deoxygenation procedure) (Fang et al. 2023). Reversible addition-fragmentation transfer (RAFT) polymerization was used to produce a polymer with the bis(hydroxymethyl)phosphine oxide pendant group from a phosphine-containing monomer, 4-vinylbenzyl-bis-hydroxymethyl phosphine oxide (VBzBHPO) (Sun et al. 2021). The structure and ^{1}H NMR spectrum is depicted in Figure 3.

Dichloromethane has been used as a solvent in the synthesis of coordination polymers based on diethyl-1-(pyridine-2-yl)phosphine and copper (I) iodide. According to Scheme 1, by the previously described method, diethylpyridylphosphine was produced in a two-step reaction that was set up with primary phosphine PyPH$_2$ (Enikeeva et al. 2023).

The electronic and steric influences on phosphine, amine, and phosphine oxide moieties are systematically investigated for terpolymerization in a series of [P,O]-type cationic Pd and Ni complexes developed by Li and coworkers and supported by a diphosphazane monoxide (PNPO) platform (Li et al. 2023).

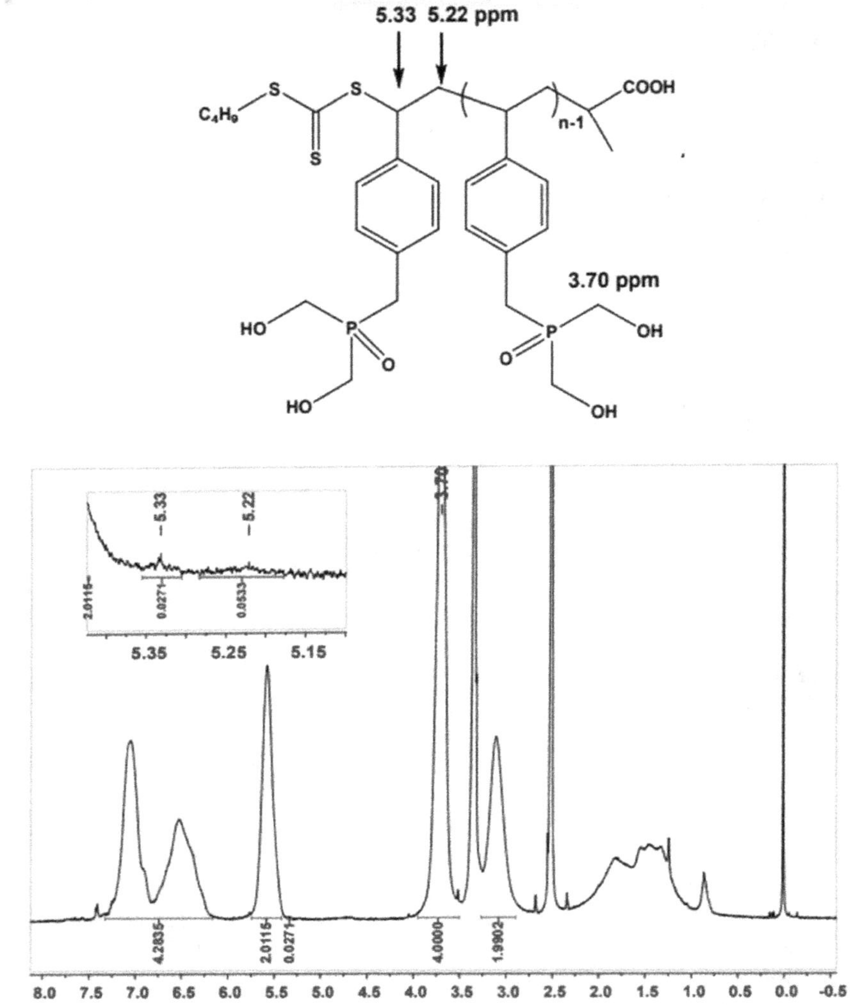

Figure 3. Structure and ^{1}H NMR spectrum of VBzBHPO (Sun et al. 2021)

3. Applications

Applications for polymer-supported reagents in synthetic organic chemistry are numerous. As a result, Ley has effectively demonstrated the use of supported reagents in medicinal chemistry by the multi-step synthesis of pharmacological targets and natural products like sildenafil (Viagra), epimaritidine, and epibatidine (Baxendale et al. 2000). Using polymer-supported reagents is an increasingly common approach in the field of organic synthesis, particularly in

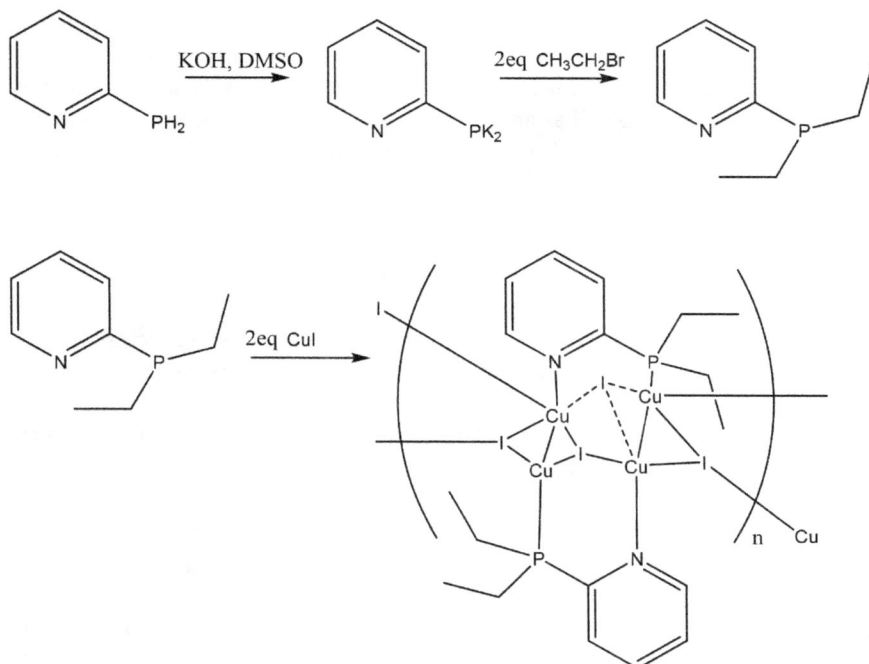

Scheme 1. Preparation of diethylpyidylphosphine and its coordination polymers with copper.

the pharmaceutical industry. As a result of the catalysts being immobilized on polymers via covalent or coordination interactions in recent years, a significant number of polymer-supported Lewis acids have been developed. These acids are frequently utilized in synthetic organic chemistry.

Phosphine-functionalized polymer reagents are frequently used in organic synthesis, particularly in Mitsunobu, Wittig, Aza-Wittig and, Staudinger reaction-type processes (Xie et al. 2021, 2022). Functional group transformations are frequently carried out under mild reaction conditions and pH neutrality by taking advantage of P(III)'s strong affinity for oxygen and nitrogen. Because the phosphorus atom is rarely incorporated into the product during organic synthesis, the production of phosphorus by-products is unavoidable. These are notoriously difficult to remove, especially if the desired product is very polar. Immobilizing organophosphorus reagents on insoluble solid supports is therefore particularly appealing because it makes it simple to remove and recover the phosphorus reagents (and their by-products) via simple filtration.

Polymer-supported polystyryldiphenylphosphine (PS-PPh$_2$), an analogue of the common triphenylphosphine (PPh$_3$), is one of the most useful reagents that may be credited for various chemical reactions. The latter is a frequently used chemical reagent that, in numerous instances, is oxidized to triphenylphosphine oxide (Ph$_3$PO). Ph$_3$PO must be removed from the product via expensive chromatographic separation and/or crystallization procedures, making PPh$_3$ worthless. Triphenylphosphine supported by polystyrene was made available as an alternative. This reagent has an advantage over its soluble equivalent in that it may be filtered to remove it from any oxidation byproducts. The reagent can be prepared in a single step and is conveniently supplied by a number of chemical suppliers (100–200 mesh, labelling extent: ~3 mmol g^{-1} triphenylphosphine loading).

To bind transition metals and explore potential electrocatalytic uses, porous organic polymers with phosphine oxide groups have been developed. As mono- and di-phosphine monomers with numerous phenyl substituents were cross-linked, the Friedel-Crafts reaction and the oxidation process were used to develop phosphine oxide porous polymers with up to 0.92 cm^3/g of pore capacity and about 990 m^2/g of surface area (Bonfant et al. 2022). Different porous polymers having phosphine oxide units are depicted in Fig. 4.

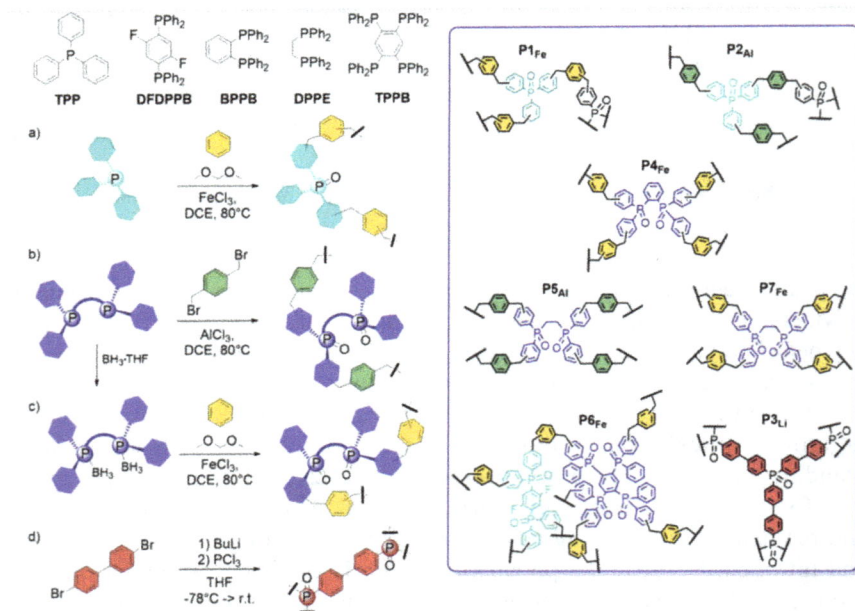

Figure 4. Porous polymers having phosphine oxide moieties (Bonfant et al. 2022)

4. Conclusion

Phosphine based polymer-supported catalysts are very promising candidates for organic functional group conversions. These catalysts have high catalytic activity and superior cycling stability. It exhibits various catalytic activities towards various reactions including reduction, cross-coupling (Suzuki-Miyaura, Heck coupling), hydrogenation of alkynes to alkenes etc.

References

Baxendale, I.R. and Steven V. Ley. 2000. Polymer-supported reagents for multi-step organic synthesis: Application to the synthesis of sildenafil. *Bioorganic & Medicinal Chemistry Letters* 10(17): 1983-1986.

Bonfant, G., D. Balestri, J. Perego, A. Comotti, S. Bracco, M. Koepf, M. Gennari and L. Marchiò. 2022. Phosphine oxide porous organic polymers incorporating cobalt (II) ions: Synthesis, characterization, and investigation of H2 production. *ACS Omega* 7(7): 6104-6112.

Chauhan, N.P.S. and N.S. Chundawat. 2019. Inorganic and organometallic polymers. *De Gruyter* 1(1): 1-144.

Chen, J., Zhang Ju, D. Zhu and T. Li. 2017. Novel polymer-supported phosphine palladium catalyst: One-pot synthesis from and application in Suzuki–Miyaura coupling reaction. *Journal of Porous Materials* 24: 847-853.

Chen, T., S. Zhang, L. Hua, Z. Xu, L. Zhou and J. Wang. 2019. Triphenylphosphine-containing thermo-responsive copolymers: Synthesis, characterization and catalysis application. *Macromolecular Research* 27: 931-937.

Enikeeva, K.R., A.V. Shamsieva, A.G. Strelnik, R.R. Fayzullin, D.V. Zakhrrychey, I.E. Kolesnikov, I.R. Dayanova, T.P. Gerasimova, I.D. Strelnik, E.I. Musina, A.A. Karasik and O.G. Sinyashin. 2023. Green emissive copper (I) coordination polymer supported by the diethylpyridylphosphine ligand as a luminescent sensor for overheating processes. *Molecules* 28(2): 706.

Fang, W.-W., G.Y. Yang, Z.H. Fan, Z.C. Chen, X.L. Hu, Z. Zhan, I. Hussain, Y. Lu, T. He and B.E. Tan. 2023. Conjugated cross-linked phosphine as broadband light or sunlight-driven photocatalyst for large-scale atom transfer radical polymerization. *Nature Communications* 14(1): 2891.

Lei, Y., L. Wu, X. Zhang and H. Mei. 2015. Palladium supported on triphenylphosphine functionalized porous organic polymer: A highly active and recyclable catalyst for alkoxycarbonylation of aryl iodides. *Journal of Molecular Catalysis A: Chemical* 398: 164-169.

Li, B., Guan, Z., Wang, W., Yang, X., Hu, J., Tan, B. Li, T. 2012. Highly dispersed Pd catalyst locked in knitting aryl network polymers for Suzuki–Miyaura coupling reactions of aryl chlorides in aqueous media. *Advanced Materials* 24(25): 3390-3395.

Li, S.-H., S.-Y. Chen, X.-B. Lu and Y. Liu. 2023. Favorable propylene-incorporated terpolymerization of ethylene with CO mediated by cationic [P, O]-Pd and Ni complexes. *ACS Inorganic Chemistry* 62(5): 2228-2235.

Matsumoto, H., H.Y. Hoshino, T. Iwai and Y. Miura. 2020. Polystyrene-supported PPh3 in monolithic porous material: Effect of cross-linking degree on coordination mode and catalytic activity in Pd-catalyzed C–C cross-coupling of aryl chlorides. *Chemistry Select* 12(16): 4034-4037.

Methot, J.L. and W.R. Roush. 2004. Nucleophilic phosphine organocatalysis. *Advanced Synthesis & Catalysis* (9-10): 1035-1050.

Salmeia, K.A., S. Gaan and G. Malucelli. 2016. Recent advances for flame retardancy of textiles based on phosphorus chemistry. *Polymers* 8(9): 319.

Satpathi, H., D. Pospiech, S. Banerjee and B. Voit. 2020. New trivalent phosphorus containing poly (arylene ether)s as alternative reactants for the Mitsunobu reaction. *European Polymer Journal* 140: 110045.

Shen, L., X. Han, B. Dong, Y. Yang, J. Yang and F. Li. 2022. Phosphine-based hyper-cross-linked polymer-supported neutral gold (I) complex as a recyclable catalyst for the regioselective hydration of alkynes to ketones. *ACS Applied Polymeric Materials* 4(10): 7408-7416.

Strasser, P. and I.J.M. Teasdale 2020. Main-chain phosphorus-containing polymers for therapeutic applications. *Molecules* 25(7): 1716.

Sun, J., C. Wang, Y. Hong, Z. Tan and C.M. Liu. 2021. Phosphine oxide-containing multifunctional polymer via RAFT polymerization and its high-density post-polymerization modification in water. *ACS Applied Polymeric Materials* 3(6): 3214-3226.

Xie, C., A.J. Smaligo, X.R. Song and O. Kwon. 2021. Phosphorus-based catalysis. *ACS Central Science* 7(4): 536-558.

Xie, C., J. Kim, B.K. Mai, S. Cao, R. Ye, X.Y. Wang, P. Liu and O. Kwon. 2022. Enantioselective synthesis of quaternary oxindoles: Desymmetrizing Staudinger–Aza-Wittig reaction enabled by a bespoke HypPhos oxide catalyst. *Journal of American Chemical Society* 144(46): 21318-21327.

Zhang, H., B. Xu, Z.M. Zhang and J. Zhang. 2023. Polymer-supported chiral palladium-based complexes as efficient heterogeneous catalysts for asymmetric reductive Heck reaction. *Green Synthesis and Catalysis* (In Press).

Zhang, Z.M., B. Xu, Y. Qian, L. Wu, Y. Wu, L. Zhou, Y. Liu and J. Zhang. 2018. Palladium-catalyzed enantioselective reductive heck reactions: Convenient access to 3,3-disubstituted 2,3-dihydrobenzofuran. *Angewandte Chemie* 57(32): 10373-10377.

Ruthenium and Iridium Containing Polymer-supported Catalyst

Avinash Kumar Srivastava*, Yachana Upadhyay and Raj Kumar Joshi*

Department of Chemistry, Malaviya National Institute of Technology Jaipur, Jaipur - 302017, Rajasthan, India

1. Introduction

Catalysis is an indispensable technology, and since the idea of catalyst-mediated chemical transformation has been established, it has been evolving continuously. Catalysis has become a critical pursuit for chemistry, and the improvements in catalyst-based synthesis methodologies increase the efficiency and sustainability of chemical synthesis while reducing byproducts and wastes (Chorkendorff and Niemantsverdriet 2017). The trending modern chemical processes usually do not feature several green synthesis principles. A careful process design including several vital strategies, methodologies, and techniques can develop a chemical process that would be safe, customarily dependent on renewable resources, and efficient in energy and waste generation, which is evident through various examples available in the literature. A catalyst can alternate to every required parameter to perform chemical transformations. It can play a central role by aiming to use safer and cost-effective chemicals, increasing the atom economy of the reaction, and reducing the number of steps for the synthesis. It is always a necessity to innovate catalytic technologies.

The fundamental categories of catalysts, i.e. homogeneous and heterogeneous catalysts, have pros and cons. The heterogeneous catalysts are of more importance to the industries, although homogeneous catalysts can be fine-tuned to better selectivity by using suitable ligands. However, it is rarely seen that the presence of ligands can provide substantial thermal stability and insensitivity towards air

*Corresponding author: aircmd.avi@gmail.com, rkjoshi.chy@mnit.ac.in

and moisture (Joshi et al. 2019, 2020, 2021). This is the foremost reason that restricts homogeneous catalysts' applications in commercial applications.

The heterogeneous catalysts refer to the type of catalysis in which catalysts are not soluble in the reaction media, and because of that, such catalysts can be easily separated (Ross 2012, Rase 2016, Urakawa et al. 2021). In addition, the heterogeneous catalysts are more tolerant to the extreme reaction conditions of heat, pressure and mostly, these catalysts do not show air and moisture sensitivity (Astruc et al. 2005). The major drawback the heterogeneous catalysts suffer is the desired products' selectivity. For heterogeneous catalysts, fine-tuning of the catalytic properties to achieve a better extent of selectivity requires a significant amount of effort and an intelligent design (Somorjai et al. 2009).

Based on the pros and cons of the known homogeneous and heterogeneous catalysts, there is a need for an intermediary alternate that can address the issues of both types of catalysis. It is well-known that there are several types of intramolecular interactions, which are broadly termed as covalent, non-covalent, and encapsulation. Using these properties, there is a possibility to partially heterogenise a homogeneous catalyst. Taking this idea of partial heterogenization seriously, many research groups are developing transition metal-based catalysts by immobilizing them on a support. The covalent interactions, i.e. the ligands of any homogeneous complexes, are covalently bound to an insoluble support, ultimately producing an immobilized catalyst, while in the non-covalent interactions, physisorption, intermolecular hydrogen bonding, electrostatic interactions, or Vander Waal forces can induce heterogeneity in the catalyst. Encapsulation is a different approach for heterogenization, which involves a physical entrapment of the catalyst in zeolites, porous polymers, and metal-organic frameworks or multitopic ligands.

Polymers are the most versatile material; since their discovery, their applications have increased exponentially. Polymers are an excellent material for catalyst support because polymeric supports are nontoxic, nonvolatile, and reusable. Polymers as catalyst support can play various roles, depending upon the catalyst available. Polymers are the only type of catalytic support which falls under all the catalytic supports, i.e. covalent, non-covalent and encapsulation. Covalently bonded polymeric supports are most commonly known, while encapsulation of nanoparticles in porous via impregnation is often reported. Non-covalent interaction between metal complexes and polymer support is rarely seen.

In 1963, an American biochemist Prof. Robert B. Merrifield, introduced the concept of polymer-supported catalyst (Merrifield 1963). This proof of concept includes synthesizing chloromethylated cross-linked resin by copolymerizing styrene and divinylbenzene. This copolymer was used as a support for the peptides. Since then, several strategies have been developed and commercialized for organic synthesis using metal complexes supported on styrene polymers (McNamara et al. 2002, Fan et al. 2002, Thomas et al. 2005, Haag et al. 2004). Based on the available literature in the past, several approaches have been explored for the heterogenization of the catalyst. Some recent reports show the use of organic polymers (Buchmeiser 2005, Lu et al. 2009, Altava et al. 2018), organic-inorganic polymers (Li et al. 2014), inorganic bulk supports (Regalbuto 2017), magnetic nanoparticles, and nanoparticles (Freire et al. 2012, Rossi et al. 2012). All of these reports embrace some excellent research related to catalyst heterogenization.

Metal complexes as catalysts have made countless contributions to organic synthesis. Most metals present in the 3d, 4d, and 5d transition series are excellent catalysts, and the main reason is the various achievable oxidation states. Besides palladium and rhodium, the most widely used transition metals are used for catalysis. Ruthenium and iridium have shown the most comprehensive scope of catalytic applications. Ruthenium especially shows stability in the broad range of oxidation states (–2 to +8), which is why it is suitable for catalysis (Griffith 1967). In addition, ruthenium is relatively inexpensive compared to most of the 4d and 5d transition metals, having considerable catalytic potential. While iridium-based complexes also show excellent potential for catalytic application. However, the iridium-based catalysis advancement is relatively slow because it is more valuable than gold, silver, and platinum. Nevertheless, the investigations show the significant stability of Ir-H and Ir-C bonds, enabling better isolation of iridium complexes (Vaska 1968).

2. Ruthenium containing polymer-supported catalysts

Pittman and coworkers reported the first illustration of an organic synthesis carried over a cross-linked polymer-supported catalyst in 1974 (Pittman and Smith 1974). This sequential catalytic reaction was performed by two catalysts anchored over a mixture of two resins. The ruthenium complex was immobilized on PPh3 functionalized resins using a standard methodology. Various research groups well accepted this methodology during the 1970s because of its straightforward synthetic approach and efficiency. In recent years, phosphine-based polymers and linkers have been used for catalytic support. Palkovits and coworkers in 2019 reported a synthesis of solid phosphine-based polymers 1,2-bis(diphenylphosphino)-ethane (pDPPE), 1,3,5-tris(4-bromophenyl)benzene (pTPPB), and triphenylphosphine (pTPP) and via simple wet impregnation of $[RuCl_2(\text{p-cymene})]_2$ to phosphine polymer in methanol, which offered an excellent catalyst (Palkovits et al. 2019).

Bergbreiter and coworkers during the 1980s reported some different methodologies to use polymers in the reactions, not as a support but as an ion exchange material, which increases the activity of catalysts by removing those byproducts, which were poisoning these catalysts (Bergbreiter et al. 1987). However, this method increases reaction productivity by increasing the catalytic activity but does not produce a reusable catalyst. The same group in 1989 ended up with the development of a polyethylene-based Ru catalyst, using an earlier known methodology, which requires a PPh_3 functionalized polymer as support (Phelps & Bergbreiter 1989). Another modification of their previous work was reported in 1990, where Bergbreiter and coworkers used polyethylene carboxylate as a support material. This polyethylene carboxylate works as a support and recovers the Ru(II) present in the reaction mixture; in other words, this catalyst regenerated itself during the catalysis (Bergbreiter and Treadwell 1990).

Support materials were combined to induce selectivity in the reaction; charcoal and anionic polymer composite with ruthenium particles show excellent hydrophilicity. As the reaction was performed in the aqueous media, the hydrophilicity of the support material became an important factor controlling the selectivity (Hronec et al. 1996). Cross-linking polymers usually show enhanced stability as compared to linear chain polymers. In the earlier reports, polystyrene, especially the PPh_3 functionalized polystyrene, was frequently used as catalyst support. Ram and coworkers in 1998 reported an observation that the polystyrene and divinylbenzene copolymers are more stable as catalyst support. However, the extent of cross-linking is also one factor that controls the catalytic activity; as observed, 2% cross-linked polymer was found best as catalytic support while more than 5% cross-linking reduces the overall catalyst activity. It is a common phenomenon that ligands usually show a significant effect on the selectivity of product formation. In this series, Ram and coworkers used chloro-methylated polystyrene-divinylbenzene copolymer to support and further functionalize it with the Schiff base as a ligand (Antony et al. 1998). It was observed that the type of Schiff base and extent of cross-linking significantly affects the catalytic activity (Scheme 1).

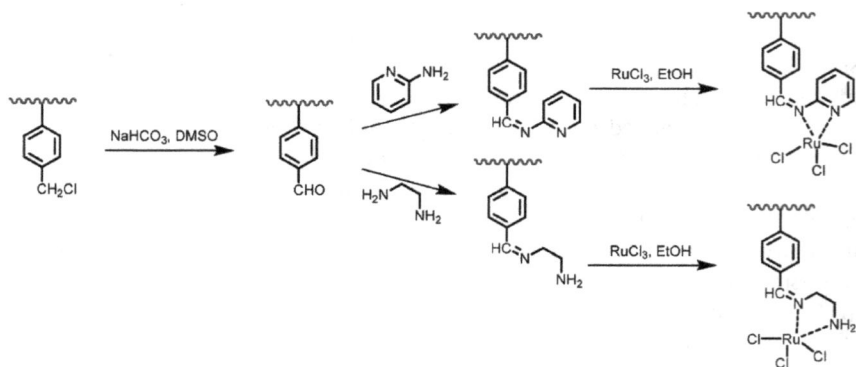

Scheme 1. Schiff base functionalized chloro-methylated polystyrene-divinylbenzene copolymer as catalyst support (Antony et al. 1998)

Barrett and coworkers synthesized a release-return (Boomerang) type polymer-supported catalyst and reported catalytic activity in 1999. Boomerang type catalyst has a substantial benefit in that it is a heterogeneous catalyst, but in the reaction media, it behaves as a homogeneous catalyst (Barrett et al. 1999). This report includes an exciting outcome showing the increased efficiency of a polymer-supported catalyst. Here, Grubbs catalyst was used with vinyl resins, and a simple shaking of a mixture produces desired catalyst. It was observed that the extended time of shaking reduces the catalytic activity, which may be due to the binding of active sites with the polymer. However, this catalyst is highly active and reusable up to three catalytic cycles. Because of a bulky polymer matrix, polymer-supported catalysts always suffered low enantio- and stereo-selectivity.

Homogeneous asymmetric catalysis is always considered a distinctive approach because of its wide applications. During the 1990s, several reports came up with polymer-supported asymmetric catalysis. However, most polymer-supported asymmetric catalysts lose their efficiency and selectivity significantly. As an alternative, Chan and coworkers reported an idea of "one phase catalysis and two-phase separation," which retains the catalytic activity and selectivity while remaining as a polymer-supported heterogeneous catalyst. Polymer-supported BINAP ligand was prepared by polycondensation of (*S*)-**1** or (*R*)-**1**, terephthaloyl chloride, and (2*S*,4*S*)-pentanediol. These polyester-supported ligands are usually soluble in non-polar solvents like benzene and toluene and can be precipitated by adding methanol in the mixture (Chan et al. 1999).

Porphyrins ligand-based complexes are well known for catalysis as some unusual selectivity was observed with such ligand systems. A ruthenium porphyrin complex was synthesized and covalently immobilized on Merrifield Peptide Resin (MPR) (Scheme 2). MPR resin is a chloromethylated styrene-divinylbenzene copolymer, which is not soluble in benzyl chloride but soluble in DCM and DMF. This property of solid support allows the metal porphyrin complex to remain intact with the solid surface (Che et al. 2000). A similar approach of using chloromethylated styrene-divinylbenzene copolymer was used with EDTA (Dalal and Ram et al. 2001) (Scheme 3).

Other than EDTA, Schiff bases are also reported as a ligand for complexations and an anchor to the polymer support. Yusuff and coworkers prepared a Schiff base functionalized polystyrene bound aldehyde and 1,2-phenylenediamine (PS-opd), 2-aminophenol (PS-ap), or 2-aminobenzimidazole (PS-ab) condensation polymers (Yusuff et al. 2007). This is the only report showing condensation polymer as catalyst support.

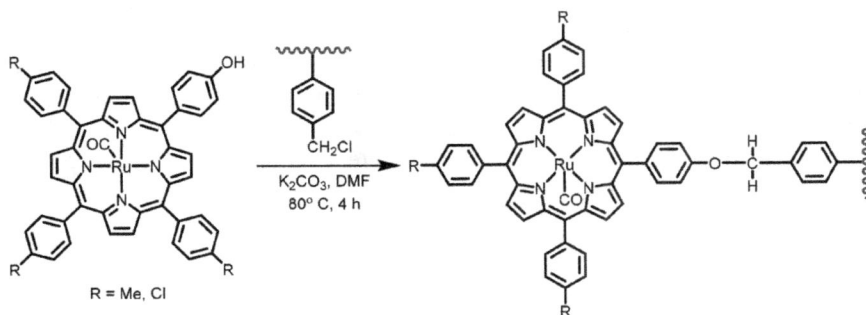

Scheme 2. Ruthenium porphyrin complex covalently immobilized to chloromethylated styrene-divinylbenzene or MPR (Merrifield Polymer Resins) (Yu et al. 2000)

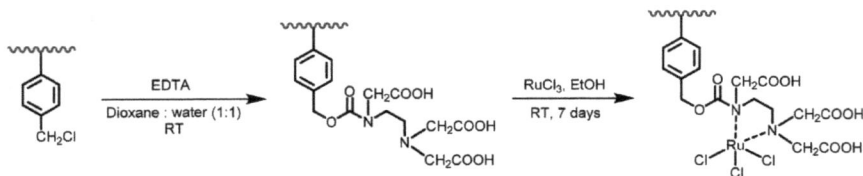

Scheme 3. Ruthenium EDTA complex covalently immobilized to chloromethylated styrene-divinylbenzene or MPR (Merrifield Polymer Resins) (Dalal and Ram et al. 2001)

The work of Ram and coworkers during the 1990s exploring the extent of cross-linking, the presence of ligand as a covalent anchor, and the type of ligand significantly affected the catalytic potential of immobilized catalysts. In this series, Brown and coworkers added a new hierarchy based on the type of cross-linking functionality. This report involved sulfide and ether cross-linked polystyrene as catalyst support (Brown et al. 2001). It was observed that the sulfide cross-linking had reduced the catalytic activity because of low polarity and the high-density polymer matrix. While with the ether cross-linking, better results were observed when alcohols were used as a solvent. However, water and other polar aprotic solvents were also not suitable because of swelling in the polymer matrix.

Before discussing further methodological advancements, let us detour and understand the limitations observed in the previous polymer-supported Ru catalysis. Ruthenium-based catalysis has always received considerable attention from researchers and industries. It is because it is comparatively cheaper than other 4d transition metals. In addition, we are well aware of the work of some prominent scientists like Chauvin, Grubbs, Schrock, and Dixneuf. Heterogenization of Grubbs and Grubbs type catalysts was much explored by various research groups. As we discussed at the beginning of this section, the phosphine-functionalized polymers were excellent support materials. However, these phosphine-based polymer matrices reduce the catalytic activity, and hence the amount of metal loading required for transformation is usually increased to two times compared to their homogenous counterpart. Hoveyda and coworkers used a dendrimer-supported Ru catalyst (Hoveyda et al. 2000). Yes, it is also an example of an immobilized catalyst but not in the scope of this chapter. Various other researchers reported using polystyrenes and polystyrenes copolymers with or without cross-linking; also, there are some examples of ligand-functionalized polymer matrices as catalyst support. There are some more examples reported by Yao, Blechert, and coworkers using Grubbs-type immobilized catalysts. Yao reported polyethylene glycol (PEG) as a support material; compared to phosphine functionalized and halide functionalized polystyrenes, PEG offers remarkable chemical stability and is cost effective and support material (Yao 2000). A slight modification in Yao's approach has been observed in the work of Lamaty and coworkers. Here, Polyethylene glycol (PEG) or Diethylene glycol (DEG) was used as polymer support (Lamarty et al. 2003) (Scheme 4).

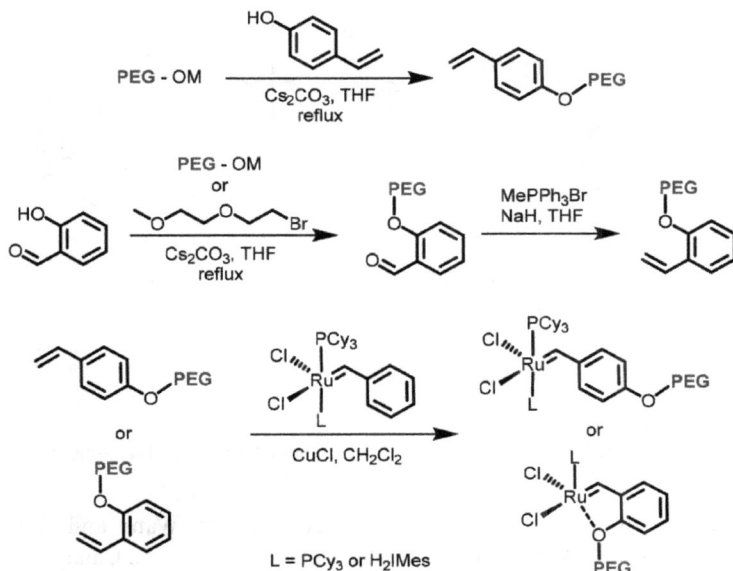

Scheme 4. Synthesis of PEG supported Grubbs type ruthenium catalyst
(Lamarty et al. 2003)

While Blechert and coworkers utilize the same Merrifield resin (polystyrene-divinylbenzene copolymer) as catalyst support (Blechert et al. 2000, 2001), there is one standard limitation, which has been observed in almost all the previous reports: several synthetic steps are required to develop polymer-supported catalyst. Iwasa and coworkers reported a two-step synthesis of polymer-supported ruthenium complex in 2013. The idea was significantly considerable as compared to earlier reports. It was made possible by changing the synthetic protocol; most of the earlier known methodologies are based on polymerization, followed by the immobilization of the metal complexes. However, Iwasa and coworkers slightly changed the order of this synthetic protocol. Using starting material p-chloromethyl benzoyl chloride and 2-amino-2-methyl propane-1-ol, Ru(II)-*dm*-Pheox complex was prepared further by reaction of acrylic acid, which produces a ruthenium complex having an acrylate functionality, which is used as one of the reactants in styrene-divinylbenzene copolymerization (Iwasa et al. 2013) (Scheme 5).

Even though multistep synthesis of polymer-supported rthenium catalysts was explored until the previous decade, Chen and coworkers reported using monodisperse polystyrene as a support material to immobilize ruthenium nanoparticles (Chen et al. 2009) (Figure 1). We have also discussed chiral synthesis using polymer-supported catalysts. Most polymer matrices reduced the catalytic activity to a certain extent but performed asymmetric catalysis using immobilized catalysts. It is because asymmetric catalysis requires specific ligands to induce

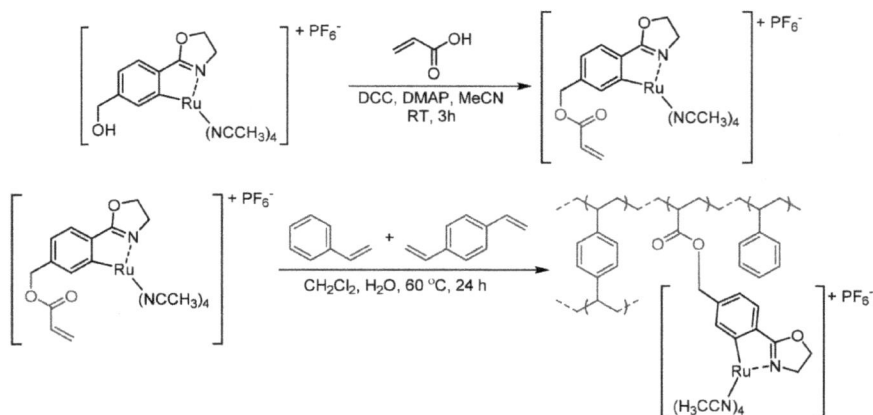

Scheme 5. Synthesis of porous-polymer supported Ru(II)-*dm*-pheox catalyst (Iwasa et al. 2013)

stereo and enantio-selectivity in the final compounds. Wang and coworkers reported the polymer-supported Ru-TsDPEN catalyst (Noyori's Catalyst) using a similar approach by using polystyrene as a support material (Wang et al. 2005).

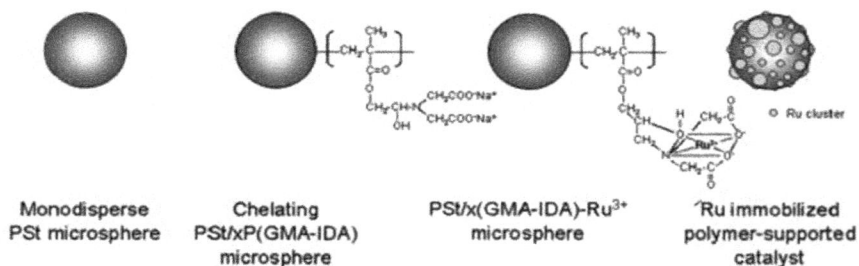

| Monodisperse PSt microsphere | Chelating PSt/xP(GMA-IDA) microsphere | PSt/x(GMA-IDA)-Ru³⁺ microsphere | Ru immobilized polymer-supported catalyst |

Figure 1. A Schematic explaining the preparation of ruthenium immobilized polymer-supported catalyst (Chen et al. 2009)

The idea of multistep synthesis and copolymerization of the polymer matrix can be an expensive approach because most of the earlier methodologies were based on the covalent anchoring of the ruthenium complexes or the complexation of functionalized polymer matrix with the ruthenium. An alternate methodology reported by Nolan and coworkers, despite covalently anchoring ruthenium complex to the polymer support, this idea was based on the impregnation of ruthenium complex to the porous polymer support. A simple polydivinylbenzene and ruthenium complex in toluene produce the desired catalyst (Nolan et al. 2002) (Scheme 6). Lau and coworkers established quite a similar idea in 2011; cation exchange resins Dowex-50W and Chelex-100 have used catalyst support polypyridyl ruthenium(II) complex, cis-[RuII(2,9-Me$_2$phen)$_2$(H$_2$O)$_2$]$^{2+}$

(Lau et al. 2011). Another example of catalyst synthesis by polymer impregnation was recently reported in 2016. Here, amine-functionalized nanoporous polymer (AFPS)-supported Ru nanoparticle-based catalysts were prepared by a simple impregnation-chemical reduction method (Hwang et al. 2016) (Figure 2). In all the cases, the porosity of the support material enables ruthenium complexes to impregnate into the porous sites while leaving the active catalytic sites available for catalysis.

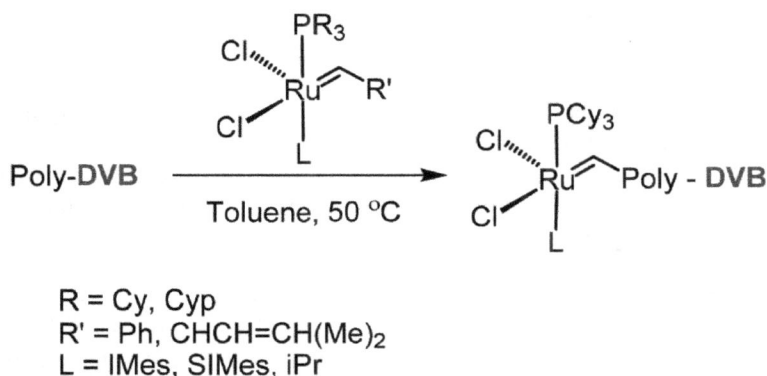

Poly-DVB $\xrightarrow{\text{Toluene, 50 °C}}$

R = Cy, Cyp
R' = Ph, CHCH=CH(Me)$_2$
L = IMes, SIMes, iPr

Scheme 6. Synthesis of poly – DVB supported Ru catalyst by impregnation (Nolan et al. 2002)

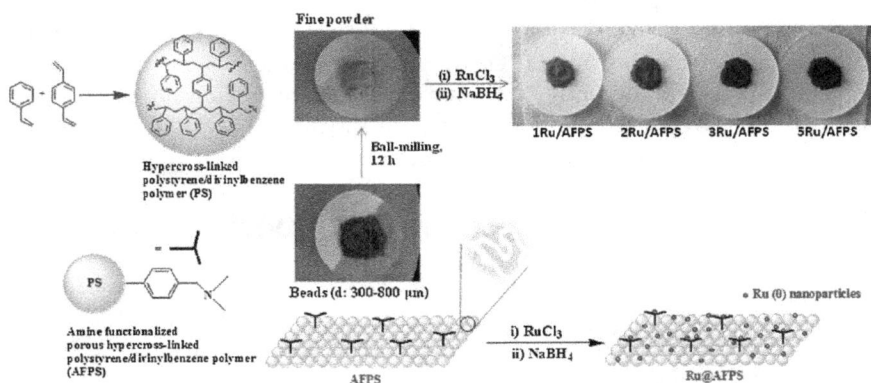

Figure 2. Amine functionalized nanoporous polymer (AFPS) supported Ru nanoparticle-based catalyst prepared by impregnation (Hwang et al. 2016)

As we know, the extent of heterogenization of catalyst has an inverse relation with the catalytic activity; however, it cannot be explained through any mathematical equation. It is understood that as the heterogenization increases, the homogeneity of the reaction mixture usually decreases and hence the catalytic activity decreases. All the examples we had discussed in this section focused entirely on the ruthenium complexes' heterogenization by immobilizing it on the polymer support. However, it is essential as immobilization increases thermal stability, reusability, and ease of handling. It is to mention that such immobilization reduces the catalytic activity and shows several inherent issues like such complexes are challenging to characterize. Because most of the catalyst supports are crystalline, these materials have very low mechanical strengths, which reduces the long-term application of these materials. Soluble polymers are an alternative that can address these issues. We have discussed one such example in the previous paragraphs; PEG is a soluble polymer that is readily available and cost-effective. It can be noted that even though polymers are well explored as catalyst support, soluble polymers are yet an uncharted territory. Polyisobutylene (PIB) has been widely used for various applications, but it recently came under the limelight and got recognition as catalyst support. It is non-crystalline and chemically inert; moreover, the characterization of PIB is comparatively easy because its 1H NMR spectrum shows only two types of protons. Above all, this compound has a vinyl terminus, compared to polyether in the case of PEG. The presence of vinyl groups offers accessibility to further functionalization.

In 2010, Bergbreiter and coworkers reported using PIB as soluble polymer and support for ruthenium complex (Bergbreiter et al. 2010) because PIB has the vinyl terminal, converted to an NHC by electrophilic aromatic substitution that phase immobilize a benzylidene Ru catalyst. The formed complex is soluble in various solvents and can be used as a homogeneous or biphasic catalyst. Again in 2016, the same group reported using polyisobutylene and poly(4-dodecylstyrene) copolymer as soluble polymer support for transition metal catalysts (Liang et al. 2016). Porous organic polymers (POPs) are another class of polymers categorized as soluble polymers. Recently, Yoon and coworkers reported a novel POP that incorporates NNN pincer type of ligand, i.e. 2,6-Bis(pyrazol-3-yl)pyridine, (3-bpp-POP) was prepared via solvent-knitting method.

3. Iridium containing polymer-supported catalysts

A study was reported in the year 1975, which comprises the controlled addition of triphenyl phosphines to $Rh_6(CO)_{16}$. This study was further extended to add Rh carbonyls on polystyrenes functionalized with PPh_2 (Watters et al. 1975). This idea was further implemented with $Ir_4(CO)_{12}$ using a similar methodology (Stuntz and Shapley 1976). However, the synthesized complexes were not used for any catalytic applications. In 1978, Gates and coworkers used a similar synthetic approach of PPh2 functionalized styrenes and referred to it as a novel catalyst (Gates et al. 1978).

After a decade in 1989, sulphoxide functionalized polystyrene was used as polymer support to produce Ir(III) complexes. This is the first report in which polymer-supported Ir(III) complex has been used for catalysis. The formed complex has been used for its potential application as a catalyst for hydrogen transfer reaction from isopropanol to ketones (Davies and Sood 1989). Since 2001, Ir-based polymer-supported complexes were not established as an option for catalysis. It is because no continuous development in synthetic approaches has been done. A report by Ley and coworkers in 2001 attracted the research groups by highlighting the catalytic application of polymer-supported Ir catalyst (Ley et al. 2001). Using earlier known Ir-based Felkin's catalyst (Felkin et al. 1976, 1978), Ley and coworkers immobilized the Felkin's catalyst to a PPh$_2$ functionalized polymer support.

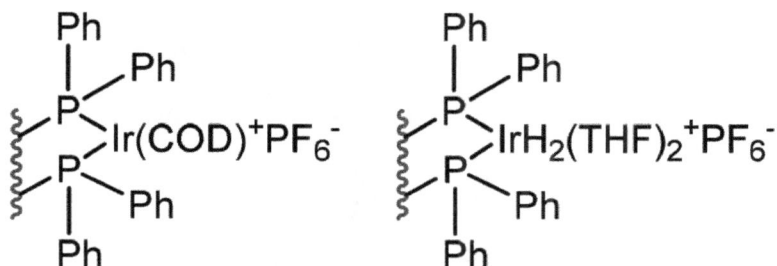

Figure 3. Polymer-supported iridium catalyst (Felkin et al. 1976, 1978)

Until 2016, a minimal number of polymer immobilized Ir-based heterogeneous catalysts were known for catalytic application. Boisson and coworkers elaborated a simple way to synthesize a polyethylene-supported Ir catalyst (Boisson et al. 2016). The catalyst stability was further improved by the NHC ligand used for the complexation. The synthetic protocol involves the imidazole-based NHC ligand attached to the polyethylene by iodine-functionalized polyethylene with NHC ligand (Scheme 7).

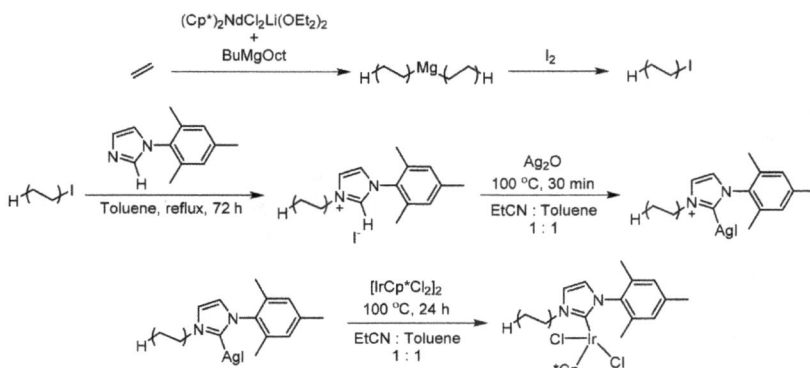

Scheme 7. Synthesis of polyethylene-supported Ir(III)-NHC complex
(Boisson et al. 2016)

Single-atom catalysis is a relatively new area of interest in heterogeneous catalysis, and in recent times, such catalysis has gained significant attention from various research groups. More importantly, supported single atom catalysis is an excellent approach with polymer. An amino pyridine functionalized porous organic polymer was used to develop an anatomically dispersed Ir catalyst (Shao et al. 2019) (Figure 4). This catalyst was highly stable and did not require any ligand to anchor the polymer support. This methodology is straightforward and robust in catalytic activity compared to catalyst loading and recyclability.

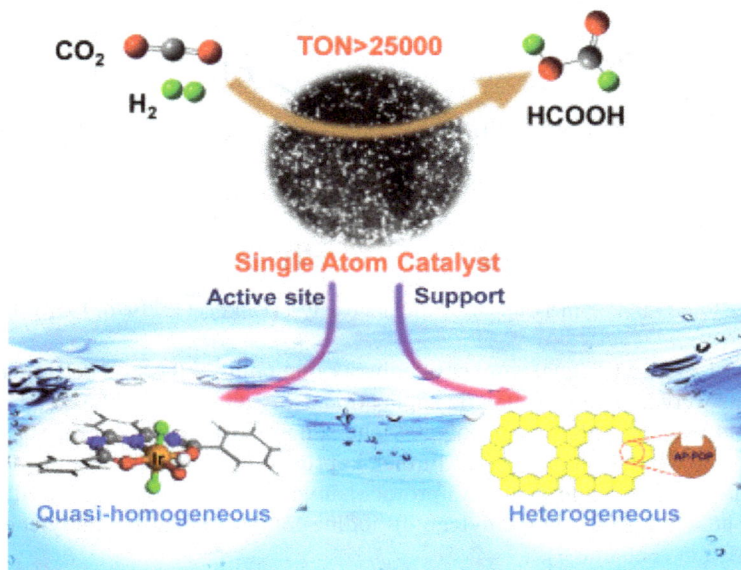

Figure 4. Polymer-supported iridium-based single-atom catalyst (Shao et al. 2019)

Sun and coworkers reported the only example of polymer-supported heterogeneous Ir photocatalyst (Sun et al. 2020). In here, organic porous polymers were used to support Ir complexes. These complexes were synthesized through a Sonogashira coupling reaction. Heterogeneous visible-light photocatalysts were prepared through a reaction between two $[Ir(ppy)_2(dtbbpy)]^+$-based biopic linkers and triphenylmethane tetraborate. Such polymer-conjugated complexes provide promising platforms for organic synthesis. In the previous year, Wang and coworkers reported a four-step synthetic protocol for synthesizing divinylbenzene polymer functionalized with benzothiazole ligated Ir catalyst (Wang et al. 2021). 4-vinyl benzyl boronic acid was used to bridge benzothiazole and divinylbenzene polymer. This catalyst was heterogeneous with excellent reproducibility up to several catalytic cycles, the synthetic pathway was relatively easy, and a very common Ir dimer $[Ir_2(Cp^*Cl)_2]$ has been used as a precursor.

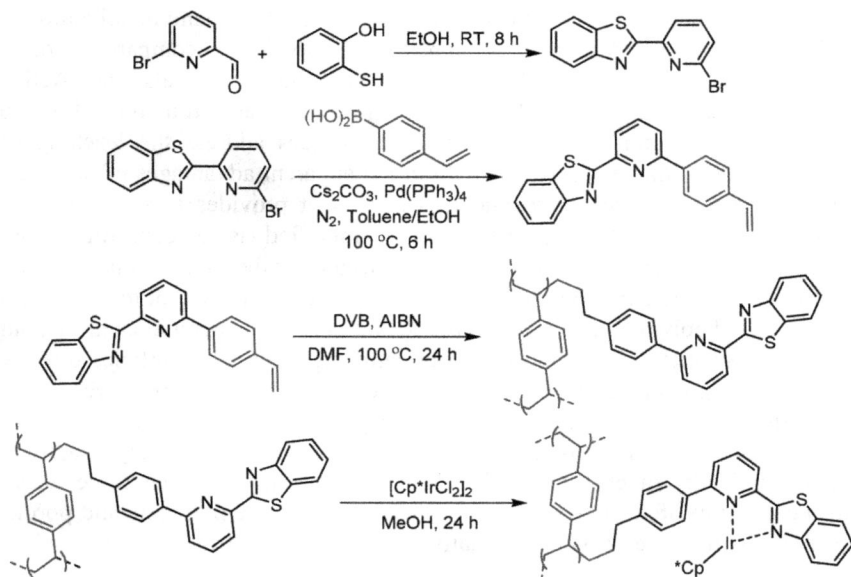

Scheme 8. Synthesis of POP-MBTS-Ir (Wang et al. 2021)

Li and coworkers have used a slightly different approach; here, a coordinative immobilization has been used to anchor a half sandwich Ir complex. $Ir_2(Cp*Cl)_2$ dimer has been used as a precursor with poly(4-vinyl pyridine) polymer (Li et al. 2021). $Cp*Ir@P4VP$ catalyst was synthesized through a single-step reaction of P4VP and Ir precursor. This Ir catalyst was supported on a linear polymeric chain, unlike previous polymer-supported complexes, where cross-linked polymer was used as support (Scheme 9).

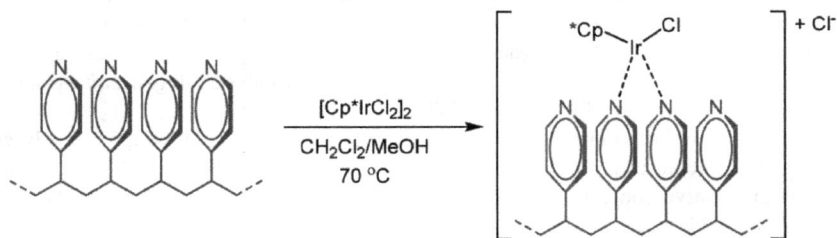

Scheme 9. Synthetic route for $Cp*Ir@P4VP$ immobilized catalyst (Li et al. 2021)

4. Conclusion

Polymer-bound catalysts are getting considerable interest in academic and industrial research in the last five decades. These supported catalysts provide an alternate paradigm in catalytic synthesis protocols. An organometallic moiety anchored to polymer support shows considerable advantages compared to other heterogenization techniques. Significantly, the pharmaceuticals and Active Pharmaceutical Industries (API) industries greatly benefit from this idea. No other heterogenization or immobilization techniques address the leaching of metal atoms into the products. Apart from the apparent advantages of polymer-supported catalysts like heterogeneous catalysts, it provides a comprehensive option of support materials, which can be modified as needed. Ruthenium complexes have been well explored for many decades in the case of ruthenium and iridium-based polymer-supported catalysts. However, there is a lot to study with iridium-based polymer-supported catalysts. Not only for ruthenium catalysts, but there are also methods available for impregnating other nanoparticles directly to the porous polymer supports, which can be considered a comparatively more straightforward method to immobilize catalysts. Since the establishment of this idea, few limitations have been observed, like the loss of catalytic activity and the difficulty of characterization of prepared catalyst, which some recent advances in this area can overcome. For example, soluble polymers and porous organic polymers are possible alternatives to address these issues.

References

Ahmed, M., A.G.M. Barrett, D.C. Braddock, S.M. Cramp and P.A. Procopiou. 1999. A recyclable 'boomerang' polymer-supported ruthenium catalyst for olefin metathesis. *Tetrahedron Letters* 40: 8657-8662.

Altava, B., M.I. Burguete, E. García-Verdugo and S.V. Luis. 2018. Chiral catalysts immobilized on achiral polymers: Effect of the polymer support on the performance of the catalyst. *Chemical Society Reviews* 47: 2722-2771.

Antony, R., G.L. Tembe, M. Ravindranathan and R.N. Ram. 1998. Polymer supported Ru(lll) complexes, synthesis and catalytic activity. *Polymer* 39: 4327-4333.

Astruc, D., F. Lu and J.R. Aranzaes. 2005. Nanoparticles as recyclable catalysts: The frontier between homogeneous and heterogeneous catalysis. *Angewandte Chemie International Edition* 44(48): 7852-7872.

Baudry, D., M. Ephritikhine and H. Felkin. 1978. Isomerisation of allyl ethers catalysed by the cationic iridium complex $[Ir(cyclo-octa-1,5-diene)(PMePh_2)_2]PF_6$. A highly stereoselective route to *trans*-propenyl ethers. *Journal of the Chemical Society, Chemical Communications* 694-695.

Bergbreiter, D.E., M.S. Bursten, G.L. Parsons and K. Cook. 1987. Polymeric cofactors which accelerate homogeneous rhodium(I) and ruthenium(II) catalyzed hydrogenations of alkenes. *Applied Organometallic Chemistry* 1: 65-71.

Bergbreiter, D.E. and D.R. Treadwell. 1990. Polyethylene carboxylate-bound triruthenium clusters as alcohol oxidation catalysts. *Reactive Polymers* 12: 291-295.

Bergbreiter, D.E., H. Su, C. Hongfa and H. Bazzi. 2010. Polyisobutylene phase-anchored ruthenium complexes. *Macromolecular Symposia* 297: 25-32.

Boisson, C., I. Romanenko, S. Norsic, L. Veyre, R. Sayah, F. D'Agosto, J. Raynaud, B. Lacote and C. Thieuleux. 2016. Active and recyclable polyethylene-supported iridium-(Nheterocyclic carbene) catalyst for hydrogen/deuterium exchange reactions. *Advanced Synthesis and Catalysis* 358: 2317-2323.

Brown, J.M., P.G. Breed and J.A. Ramsden. 2001. Scope and limitations of ruthenium-catalyzed metathesis of simple polymer-bound alkenes. *Canadian Journal of Chemistry* 79: 1049-1057.

Buchmeiser, M.R. (ed.) (2005). *Polymeric Materials in Organic Synthesis and Catalysis.* Wiley-VCH Weinheim.

Chen, C., C. Chen and Y. Huang. 2009. Method of preparing Ru-immobilized polymer-supported catalyst for hydrogen generation from NaBH4 solution. *International Journal of Hydrogen Energy* 34: 2164-2173.

Chorkendorff, I. and J.W. Niemantsverdriet. 2017. *Concepts of Modern Catalysis and Kinetics.* John Wiley & Sons.

Crabtree, R.H., H. Felkin and G.E. Morris. 1976. Activation of molecular hydrogen by cationic iridium diene complexes. *Journal of the Chemical Society, Chemical Communications* 716-717.

Dalal, M.K. and N.R. Ram. 2001. Catalytic activity of polymer-bound Ru(III)–EDTA complex. *Bulletin of Materials Science* 24: 237-241.

Davies, J.A. and A. Sood. 1989. Polystyrene functionalized with sulfoxide groups as a macromolecular ligand for iridium(III) and palladium(II). *Asian Journal of Chemistry* 1: 1-6.

Fan, Q., C. Ren, C. Yeung, W. Hu and A.S.C. Chan. 1999. Highly effective soluble polymer-supported catalysts for asymmetric hydrogenation. *Journal of American Chemical Society* 121: 7407-7408.

Fan, Q.H., Y.M. Li and A.S.C. Chan. 2002. Recoverable catalysts for asymmetric organic synthesis. *Chemical Reviews* 102: 3385-3466.

Freire, C., C. Pereira and S. Rebelo. 2012. Green oxidation catalysis with metal complexes: From bulk to nano recyclable hybrid catalysts. *Catalysis* 24: 116-203.

Griffith, W.P. 1967. *The Chemistry of the Rare Platinum Metals (Os, Ru, Ir and Rh).* United Kingdom: Interscience.

Hao, S., J. Yang, P. Liu, J. Xu, C. Yang and F. Li. 2021. Linear-organic-polymer-supported iridium complex as a recyclable auto-tandem catalyst for the synthesis of quinazolinones via selective hydration/acceptorless dehydrogenative coupling from o-aminobenzonitriles. *Organic Letters* 23: 2553-2558.

Hoveyda, A.H., S.B. Garber, J.S. Kingsbury and B.L. Gray 2000. Efficient and recyclable monomeric and dendritic Ru-based metathesis catalysts. *Journal of American Chemical Society* 122: 8168-8179.

Hronec, M., Z. Cvengrošová, M. Králik, G. Palma and B. Corain. 1996. Hydrogenation of benzene to cyclohexene over polymer-supported ruthenium catalysts. *Journal of Molecular Catalysis A: Chemical* 105: 25-30.

Hwang, J., A.A. Dabbawala and D.K. Mishra. 2016. Selective hydrogenation of D-glucose using amine functionalized nanoporous polymer supported Ru nanoparticles based catalyst. *Catalysis Today* 265: 163-173.

Iwasa, S., A. Elfotoh, H.W. Chua, S. Murakami, K. Phomkeona and K. Shibatomi. 2013. A novel porous-polymer-supported Ru(II)-dm-Pheox catalyst and its application in highly efficient N-H insertion reactions. *Advanced Materials Research* 626: 411-414.

Iwatate, K., S.R. Dasgupta, R.L. Schneider, G.C. Smith and K.L. Watters. 1975. A study of substitution reactions in Rh6(CO)16 using simple and polymer bound phosphine ligands. *Inorganica Chimica Acta* 15: 191-195.

Lamarty, F., S. Varray, R. Lazaro and J. Martinez. 2003. New soluble-polymer bound ruthenium carbene catalysts: Synthesis, characterization, and application to ring-closing metathesis. *Organometallics* 22: 2426-2435.

Lau, T., Z. Hu, C. Leung, Y. Tsang, H. Du, H. Liang and Y. Qiua. 2011. A recyclable polymer-supported ruthenium catalyst for the oxidative degradation of bisphenol A in water using hydrogen peroxide. *New Journal Chemistry* 35: 149-155.

Lee, A.-L. and S.V. Ley. 2002. A polymer-supported iridium catalyst for the stereoselective isomerization double bond. *Synlett* 3: 516-518.

Li, C. and Y. Liu (eds.) 2014. *Bridging Heterogeneous and Homogeneous Catalysis: Concepts, Strategies, and Applications*. Weinheim: Wiley-VCH.

Li, Y., Z. Li, F. Li, Q. Wang and F. Tao. 2005. Preparation of polymer-supported Ru-TsDPEN catalysts and use for enantioselective synthesis of (S)-fluoxetine. *Organic & Biomolecular Chemistry* 3: 2513-2518.

Liang, Y., C. Watson, T. Malinski, J. Tepera and D.E. Bergbreiter. 2016. Soluble polymer supports for homogeneous catalysis in flow reactions. *Pure and Applied Chemistry* 88: 953-960.

Lu, J. and P.H. Toy. 2009. Organic polymer supports for synthesis and for reagent and catalyst immobilization. *Chemical Review* 109: 815-838.

McNamara, C.A., M.J. Dixon and M. Bradley. 2002. Recoverable catalysts and reagents using recyclable polystyrene-based supports. *Chemical Reviews* 102: 3275-3300.

McNamara, R. and S. Roller. 2004. *Immobilized Catalysts: Solid Phases, Immobilization and Applications* (A. Kirschning (ed.), 1. Berlin, Heidelberg: Springer Berlin Heidelberg.

Merrifield, R.B. 1963. Solid phase peptide synthesis. I. The synthesis of a tetrapeptide. *Journal of American Chemical Society* 85: 2149-2154.

Nolan, S.P., L. Jafarpour, M. Heck, C. Baylon, H.M. Lee and C. Mioskowski. 2002. Preparation and activity of recyclable polymer-supported ruthenium olefin metathesis catalysts. *Organometallics* 21: 671-679.

Park, K., S. Padmanaban, S. Kim, K. Jung and S. Yoon. 2021. NNN pincer-functionalized porous organic polymer supported Ru(III) as a heterogeneous catalyst for levulinic acid hydrogenation to γ-valerolactone. *ChemCatChem* 13: 695-703.

Phelps, J.C. and D.E. Bergbreiter. 1989. A polyepbylene-bound Ru(II) catalyst for inter- and intramolecular Kharasch reactions. *Tetrahedron Letters* 30: 3915-3918.

Pittman, C.U. and L.R. Smith. 1974. Sequéntial multistep reactions catalyzed by polymer-anchored homogeneous catalysts. *Journal of the American Chemical Society* 97: 1749-1754.

Rafalkjoo, J.J., S. Ibtob, R. Gates and G. Schrade. 1978. Polymer-bound tetranuclear iridium carbonyl catalyst. *Journal of the Chemical Society, Chemical Communications* 540-541.

Randl, S., N. Buschmann, S.J. Connon and S. Blechert. 2001. Highly efficient and recyclable polymer-bound catalyst for olefin metathesis reactions highly efficient and recyclable polymer-bound catalyst for olefin metathesis reactions. *Synlett* 10: 1547-1550.

Rase, H.F. 2016. *Handbook of Commercial Catalysts: Heterogeneous Catalysts*. CRC Press.

Regalbuto, J. (ed.) 2007. *Catalyst Preparation: Science and Engineering*. Boca Raton, FL: CRC Press, Taylor & Francis Group.

Ross, J.R.H. 2012. *Heterogeneous Catalysis: Fundamentals and Applications*. Germany: Elsevier Science.

Rossi, L.M., N.J.S. Costa, F.P. Silva and R. Wojcieszak. 2012. Magnetic nanomaterials in catalysis: Advanced catalysts for magnetic separation and beyond. *Green Chemistry* 16: 2906-2933.

Schürer, S.C., S. Gessler, N. Buschmann and S. Blechert. 2000. Synthesis and application of a permanently immobilized olefin-metathesis catalyst. *Angewandte Chemie International Edition* 39: 3898-3901.

Shao, X., X. Yang, J. Xu, S. Liu, S. Miao, X. Liu, X. Su, H. Duan, Y. Huang and T. Zhang. 2019. Iridium single-atom catalyst performing a quasi-homogeneous hydrogenation transformation of CO_2 to formate. *Chem* 5: 693-705.

Sharma, C., A.K. Srivastava, K.N. Sharma and R.K. Joshi. 2020. Half-sandwich (η5-Cp*) Rh(III) complexes of pyrazolated organo-sulfur/selenium/tellurium ligands: Efficient catalysts for base/solvent free C–N coupling of chloroarenes under aerobic conditions. *Organic & Biomolecular Chemistry* 18: 3599-3606.

Sharma, K.N., N. Satrawala, A.K. Srivastava, M. Ali and R.K. Joshi. 2019. Palladium(II) ligated with a selenated (Se, CNHC, N⁻)-type pincer ligand: An efficient catalyst for Mizoroki–Heck and Suzuki–Miyaura coupling in water. *Organic & Biomolecular Chemistry* 17: 8969-8976.

Somorjai, G.A. and C.J. Kliewer. 2009. Reaction selectivity in heterogeneous catalysis. *Reaction Kinetics and Catalysis Letters* 96(2): 191-208.

Stuntz, G.F. and J.R. Shapley. 1976. Direct synthesis of mono- and disubstituted phosphorus ligand derivatives of dodecacarbonyltetrairidium. *Inorganic Chemistry* 15: 1994-1997.

Sun, R., A. Kann, H. Hartmann, A. Besmehn, P.C.J. Hausoul and R. Palkovits. 2019. Direct synthesis of methyl formate from CO_2 with phosphine-based polymer-bound Ru catalysts. *ChemSusChem* 12: 3278-3285.

Thomas, G.L., C. Böhner, M. Ladlow and D.R. Spring. 2005. Polystyrene-supported triphenylarsines: Useful ligands in palladium-catalyzed aryl halide homocoupling reactions and a catalyst for alkene epoxidation using hydrogen peroxide. *Tetrahedron* 61: 12153.

Tomar, V., Y. Upadhyay, A.K. Srivastava, M. Nemiwal, R.K. Joshi and P. Mathur. 2021. Selenated NHC-Pd(II) catalyzed Suzuki-Miyaura coupling of ferrocene substituted β-chloro-cinnamaldehydes, acrylonitriles and malononitriles for the synthesis of novel ferrocene derivatives and their solvatochromic studies. *Journal of Organometallic Chemistry* 940: 121752.

Urakawa, A., Y.H. Ng, P. Sit and W.Y. Teoh. 2021. *Heterogeneous Catalysts: Advanced Design, Characterization, and Applications*, 2 Volumes. Germany: Wiley.

Vaska, L. 1968. Reversible activation of covalent molecules by transition-metal complexes. The role of the covalent molecule. *Accounts of Chemical Research* 1(11): 335-344.

Wang, D., J. Li, H. Liu, H. Zhu and W. Yao. 2021. Highly efficient and recyclable porous organic polymer supported iridium catalysts for dehydrogenation and borrowing hydrogen reactions in water. *ChemCatChem* 13: 4751-4758.

Xu, Z.-Y., Y. Luo, D.-W. Zhang, H. Wang, X.-W. Sun and Z.-T. Li. 2020. Iridium complex-linked porous organic polymers for recyclable, broad-scope photocatalysis of organic transformations. *Green Chemistry* 22: 136-143.

Yao, Q. 2000. A soluble polymer-bound ruthenium carbene complex: A robust and reusable catalyst for ring-closing olefin metathesis. *Angewandte Chemie International Edition* 39: 3896-3898.

Yu, X., J. Huang, W. Yu and C. Che. 2000. Polymer-supported ruthenium porphyrins: Versatile and robust epoxidation catalysts with unusual selectivity. *Journal of American Chemical Society* 122: 5337-5342.

Yusuff, M., S.C. Pearly and N. Sridevi. 2007. Characterization and catalytic activity of polymer supported ruthenium schiff base complexes towards catechol oxidation. *Journal of Applied Polymer Science* 105: 997-1002.

Palladium Containing Polymer-supported Catalyst

Yachana Upadhyay*, Avinash Kumar Srivastava and Raj Kumar Joshi*
Department of Chemistry, Malaviya National Institute of Technology Jaipur,
Jaipur - 302017, Rajasthan, India

1. Introduction

More than a century of research has contributed to the development of homogeneous catalysis. In every aspect of chemical synthesis, homogeneous catalysis has notable involvement. Organic molecules work as organocatalysts, or some organic molecules work as ligands channelizing the catalytic potential of transitions metal (List 2007, Eelkema et al. 2019, Lundgren et al. 2016). This control over the catalytic potential was so effective that researchers could perform regio-, stereo-, and enantio-selective reactions (Suib 1993). Because of this exceptional potential of catalysis, catalysts have become a critical parameter to optimize any reaction for commercial or societal benefits. In every aspect, we can say that a catalyst can make any reaction productive by reducing cost and effort. Consider a couple of parameters that influence any chemical reaction are temperature and time while the type of solvents and in several cases the pressure of gas participating in a reaction also controls the productivity of a reaction. It is observed in several instances that a suitable catalyst can overcome the use or requirement of any of these parameters. Suitable catalysts can reduce the required reaction time, temperature, and pressure of a gas used as a reactant. In addition, it has been observed that suitable catalysts can replace the need for co-catalysts or some particular reagents required to pursue any chemical transformation, in the words of Alwin Mittasch, a renowned chemical engineer. He did all the optimization of Haber's Process for commercial application (Dronsfield 2007).

*Corresponding authors: yachana02@gmail.com, rkjoshi.chy@mnit.ac.in

Mittasch said, *"Chemistry without catalysis would be a sword without a handle, a light without brilliance, a bell without sound"*, which appears very true in the present scenario. It is because all the major industrial processes are capable of fulfilling commercial and societal demands only because suitable catalysts are involved. Moreover, continuous research is ongoing to improve the quality of catalysts further.

At a laboratory scale synthesis, there are several aspects through which anyone can say the quality and efficiency of catalyst has been improved. For example, at small scale, the most important parameter that shows the quality and activity of the catalysts is the product yield. However, on a commercial scale the most important factor that ensures applicability of the catalyst is the economic viability. However, from last two decades considerable efforts have been invested to increase the applicability of 3d transition metal-based catalysis for commercial applications. Nevertheless, 4d transition elements Ru, Rh, Pd and Ag are the most versatile elements as catalysts. Especially these four elements as catalysts have the widest applications. The Nobel Prize was awarded to some eminent researchers in 2005 and 2010 for establishment of Ru and Pd based catalytic methodologies. These catalytic methodologies were economical and safe compared to the earlier known synthetic procedures.

Pd as a catalyst for variety of chemical transformations has excellent applicability. Especially for CC cross coupling and CN coupling reactions, these reactions are part of API, natural product synthesis, molecular material development, and pharmaceuticals. The pioneering works of Heck, Suzuki, Nigeshi, Casser, Sonogashira and Buchwald reactions are some of the most popular protocols. Compared to homogeneous catalysts, heterogeneous catalysts are always more economical. With the growing societal and industrial demands, there is always a need for new catalysts. However, usually introducing a novel catalyst and replacing an older and established catalyst requires a lot of studies and optimizations in every aspect. Since the idea of partial heterogenization over a solid support has been established, it has become a very well accepted alternate for catalysis. It is because a heterogeneous catalyst usually shows low catalytic activity compared to homogeneous catalysts. Storage, transportation, handling and lack of reusability are the common problems usually faced with the homogeneous catalysts.

Polymers are a wide class of chemical compounds or we can say a material. They are classified on the basis of nature of polymerization process, presence of cross linkers, morphology of the formed polymer – either cross linked or non-cross linked – and the chemical nature of used monomers or the polymers formed as a product. Large variety of polymers were known for their application as a support material. During the 1970s, it was discovered that catalyst support materials made up of polymers such as polyamides and polyurea were used as support material. The report by Zhang et al. in 1983 is one of the first report where the idea of polyurea-supported Pd catalyst originated. That was a two-step catalyst preparation in which co-monomers 2,4-toluenediisocynanate complexed with Pd acetate and 4,40-diamino-2,20-bipyriridine were condensed followed by the reduction of Pd(II) by LiAlH4, which produces the final catalyst for use. The idea of Pd containing monomer for polymerization shows an excellent outcome as a novel polymer-supported catalyst and in comparison to Pd/C, this new catalyst shows 40% less Pd loading and a greater life time for catalysis. However, other results show that this new catalyst was greatly affected by the polarity of the solvent.

As it appears, polymer-supported Pd catalysts have a very rich and diverse history, which can be a difficult task to converse in only one chapter. Therefore, in this chapter we are going to discuss the recent developments of past two decades. Even in these two decades, there are significant reports available with excellent outcomes.

2. Polystyrene-supported Pd catalyst

Styrene is one of the most commonly available polymers and its immense number of functionalization possibilities makes it extremely useful for the domestic and commercial applications (Wünsch 2012). This versatility of polystyrenes in it various forms attracts researchers worldwide to explore its application in catalysis. Merrifield resins are the cross-linked polystyrene which contains chloromethyl functionality. Merrifield resins are one of the most common examples and frequently used for the development of Pd supported catalysts. In year 2001, Buchwald and coworkers extensively explored the C–C and C–N coupling reactions using Pd catalysts; they came up with an innovative idea to use this Merrifield resin as catalyst support (Buchwald 2001). The reported methodology was uncomplicated and refined, which involved linking biphenylphosphine, a widely recognized ligand, to the polymer backbone through an ether linkage. A biphenyl framework is easy and a well-known synthetic protocol: 2-dicyclohexylphosphine-2-methoxybiphenyl was synthesized through a reaction of 2-methoxyphenyl magnesium bromide with benzyne and subsequent addition of chlorodicyclohexylphosphine. This is followed by deprotection of aryl methyl ether with boron tribromide, which produces the desired ligand. The formed ligand with hydroxyl group was reacted with Merrifield resin cross linked with 1% divinylbenzene, the reaction of NaH in DMF solvent produces resin bound dicyclohexylphosphine ligand which is further complexed by the Pd(OAc)$_2$. The formed catalyst was quite stable, active and reusable for Suzuki coupling and C-N coupling reactions. One of the limitation, which is observed here, was that the multistep synthetic protocol requires absolute care in inert gas environment that restricts its applicability for scalable applications (Scheme 1).

Technical simplicity is always the foremost requirement for any methodological advancement for its application in scalable synthesis. Wet impregnation is one of the most widely used method for the development of heterogeneous catalysts because of its simplicity, low cost and because it is a less waste producing method. Shekher and co-workers (2002) reported a polymer supported Pd catalyst for Suzuki coupling. The synthesis protocol involved a simple wet impregnation

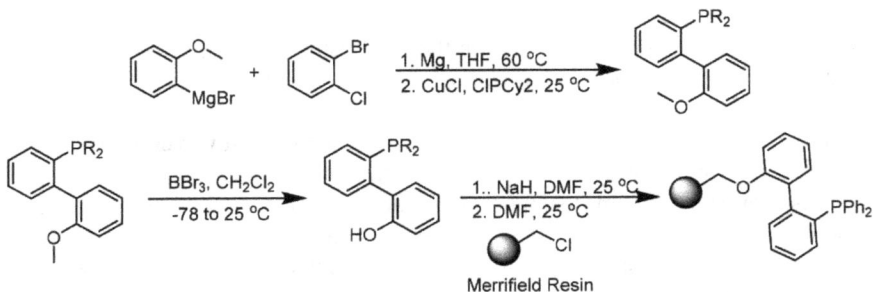

Scheme 1. Synthesis of Merrifield resin functionalized dicyclohexylphosphine ligand (Buchwald 2001)

for polymer-phosphine bound with PdCl$_2$ in ethanol. Polystyrene functionalized with triphenyl phosphine can be directly ligated with PdCl$_2$ or Pd(0) (Shekhar et al. 2002) (Scheme 2).

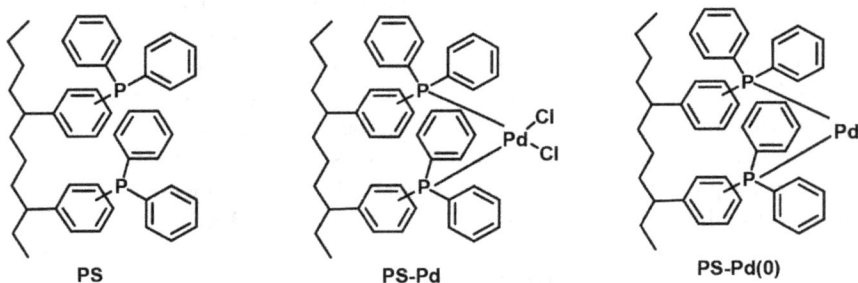

Scheme 2. Triphenylphosphine functionalized polystyrene (Shekhar et al. 2002)

Similar to the work of Buchwald and coworkers (2001), Ram and coworkers (2003) reported the use of chloromethylated poly(styrene-divinylbenzene) copolymers (PS-DVB) with 8–14% crosslinking, which was functionalized with Schiff base ligand based Pd(II) complex. Initially, chloromethylated polystyrene was converted to aldehyde functional group through a reaction with NaHCO$_3$ and DMSO. Formed PS-DVB with aldehyde was reacted with ethylenediamine or 2-aminopyridine to produce Schiff base – functionalized PS-DVB followed by a reaction with PdCl$_2$ in ethanol solvent to produce desired catalyst. This simple synthetic approach produces bidentate Schiff-base-ligand complexed with Pd for an efficient hydrogenation of olefins (Ram et al. 2003) (Scheme 3).

Compared to p-chloromethyl polystyrene, another approach was developed with p-anilinopolystyrene, which was reported by Mukherjee in 2003. In here, macropourous polystyrene cross-linked with divinylbenzene was used as a polymer backbone and a similar methodology of schiff base is used complex with Pd for catalytic applications. Some other alternates like nitropoly-N-vinyl carbazole are also reported in this work (Mukherjee 2003).

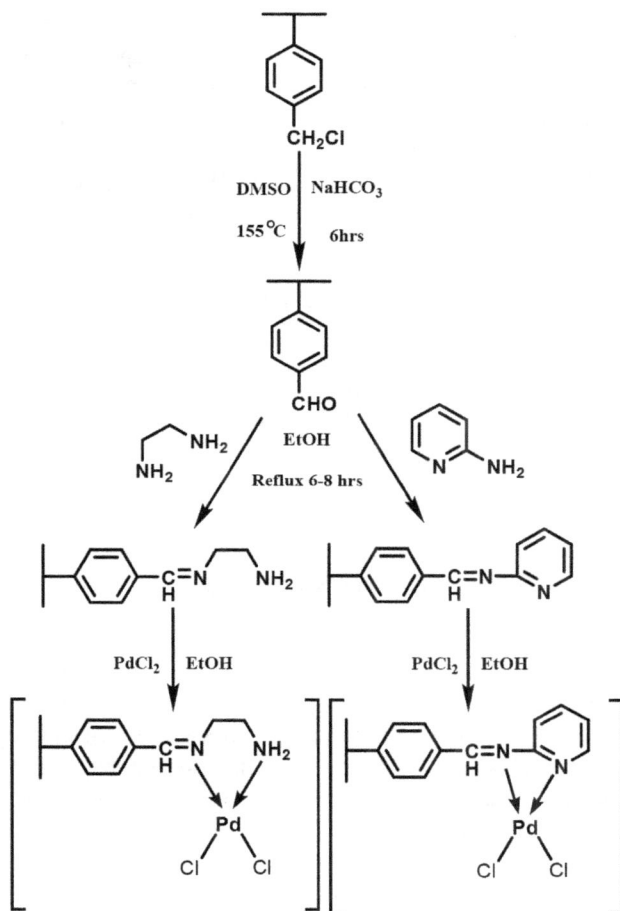

Scheme 3. Synthesis of Schiff base ligand bonded with chloromethylated poly(styrene-divinylbenzene) copolymers (PS-DVB) (Ram et al. 2003)

N-Heterocyclic carbene (NHC) has gained a considerable interest in the recent decades. These ligands are well known for the development of stable and active catalysts. It is because NHCs are conformationally stabilized ligand, and its strong σ–donor abilities makes these catalysts more active (Gonzalez 2016). Even in some recent reports, chalcogenated NHCs are gaining more interest due to multiple donor sites of different nature, which makes them more facile to interact with variety of metals and their oxidation states (Joshi et al. 2019, 2021, 2021, 2022). Lee and co-workers (2005) reported a novel methodology to develop a polymer-supported catalyst using a NHC ligand based Pd complex supported on Poly(1-methylimidazoliummethyl styrene)-*surface grafted*-poly(styrene) resin. A methodology known as suspension polymerization is used to produce the desired catalyst. Polymer-supported N-heterocyclic carbene precursor has been prepared by 1-methyl-3-(4-vinylbenzyl)imidazolium hexaflouorophosphate, which was used to synthesize and copolymerize with styrene and divinylbenzene in aqueous media. This idea was interesting research, which produces an extremely stable and reusable catalyst; most effectively, this methodology produces a catalyst, which has all the active sites embedded on the polymer surface (Lee et al. 2005) (Scheme 4).

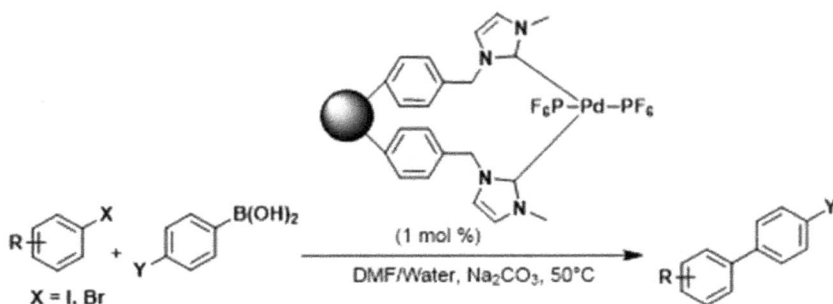

Scheme 4. Synthesis of NHC ligand based Pd complex supported on Poly(1-methylimidazoliummethyl styrene)-*surface grafted*-poly(styrene) resin (Lee et al. 2005)

Bakherad and coworkers (2009) reported a polymer-supported Pd(II) ethylenediamine complex as a highly active catalyst that is used for sonogashira coupling reaction. The synthesis protocol includes the use of chloromethylated polystyrene cross-linked by 2% divinylbenzene. As reported, this polymer is flexible and capable of interaction with metallic atom via ligands that are attached to the polymer beads. The merits of this prepared catalyst are important; this catalyst was prepared using commercially available reagents and the synthetic protocol was very convenient (Bakherad et al. 2009) (Scheme 5).

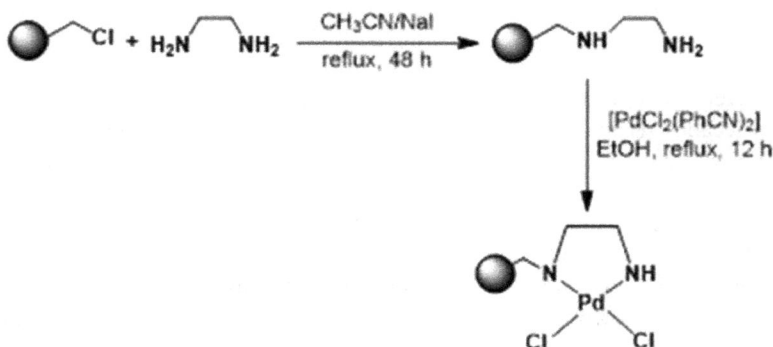

Scheme 5. Synthesis of Pd(II) ethylenediamine complex supported over chloromethylated polystyrene cross-linked by 2% divinylbenzene

A reusable, air-stable and polymer anchored Schiff base ligated Pd catalyst has been synthesized and reported by Mondal and coworkers in 2009. The remarkable advantages of this catalyst are that it is economical, easily accessible, and reusable. As discussed by the author, this is free from fire hazard or explosions and has shown notable activity towards variety of substrate with excellent yields of products (Mondal et al. 2009) (Scheme 6).

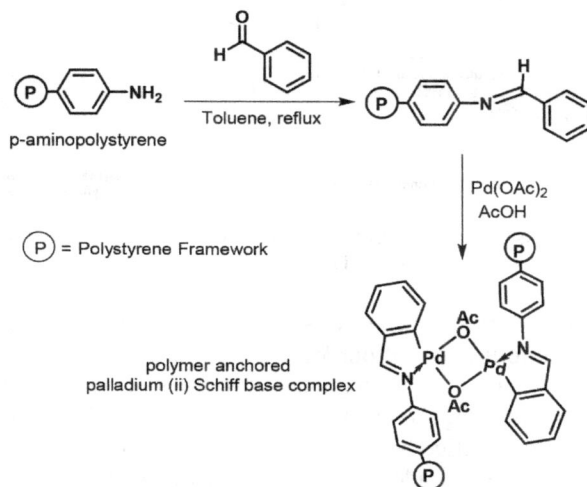

Scheme 6. Synthesis of polystyrene anchored Schiff base ligated Pd complex (Mondal et al. 2009)

Udaykumar and coworkers (2010) showed the used of Pd imidazole complex supported over chloromethylated polystyrene for the hydrogenation of benzylideneaniline. In here, polystyrene cross-linked with 6.5% divinyl benzene is used as a catalyst support. Functionalization of PS-DVB with imidazole was done via a simple reaction performed under a mixture of toluene and acetonitrile on a hot water bath for 60 h. The imidazole-functionalized catalyst was separated through soxhlet extraction followed by a reaction of PdCl$_2$ in ethanol that produces desired product (Udaykumar et al. 2010) (Scheme 7).

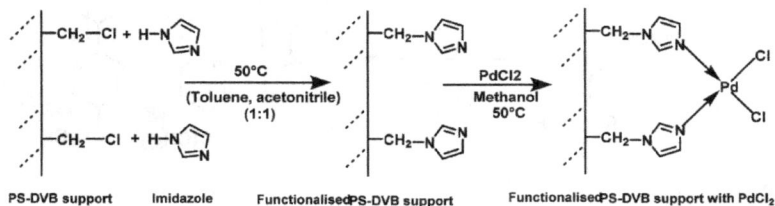

Scheme 7. Synthesis of Pd imidazole complex supported over chloromethylated polystyrene (Udaykumar et al. 2010)

The same group of Udaykumar and coworkers in 2011 reported another catalyst with slight modification in the ligand. Polymer supported Pd-2-methylimidazole was synthesized and reported for the catalytic activity. There is no reasonable explanation that has been provided for the effect of methylimidazole on the catalytic reaction (Udaykumar et al. 2010) (Scheme 8).

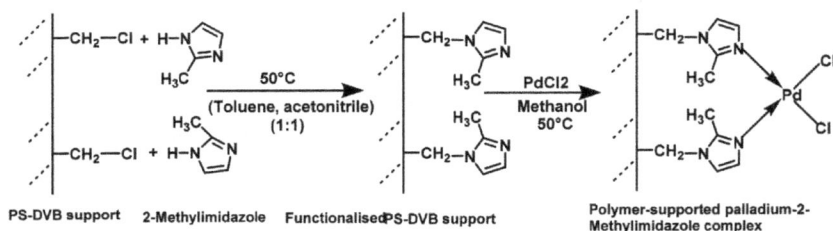

PS-DVB support   2-Methylimidazole   Functionalised PS-DVB support   Polymer-supported palladium-2-Methylimidazole complex

Scheme 8. Synthesis of Pd methylimidazole complex supported over chloromethylated polystyrene (Udaykumar et al. 2011)

In the previous decade, the use of styrene as catalyst support for Pd catalyst has been reduced; to the best of our knowledge, only two reports are available. Extensive work by various research groups has been done on the soluble polymer-supported catalysts. Bergbreiter and coworkers in 2015 reported synthesis strategies for the more effective soluble ligands for phosphine ligand based Pd complex. RAFT (Reversible addition-fragmentation chain transfer) polymerization is used to prepare an alkane soluble polyl(4-alkylstyrene)-bound phosphine ligands. In here, 4-dodecylstyrene and 4-tert-butylstyrene were copolymerized with 7% of 4-diphenylphosphine styrene or 4-chloromethylstyrene using RAFT chemistry and produced poly(tert-butylstyrene-co-4-dodecylstyrene) copolymers. The formed polymers with chloromethylene functionality react with the phenolic group of dicyclohexylbiarylphosphine ligand. This work includes the development of supported polymer catalyst using a soluble polymer (Bergbreiter et al. 2015) (Schemes 9, 10 and 11).

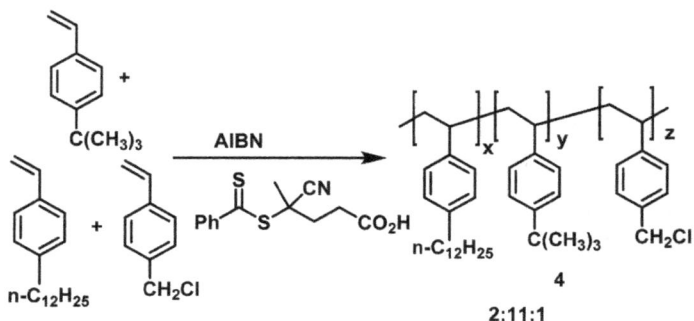

Scheme 9. RAFT polymerisation of 4-dodecylstyrene and 4-tert-butylstyrene were copolymerized with 4-chloromethylstyrene (Bergbreiter et al. 2015)

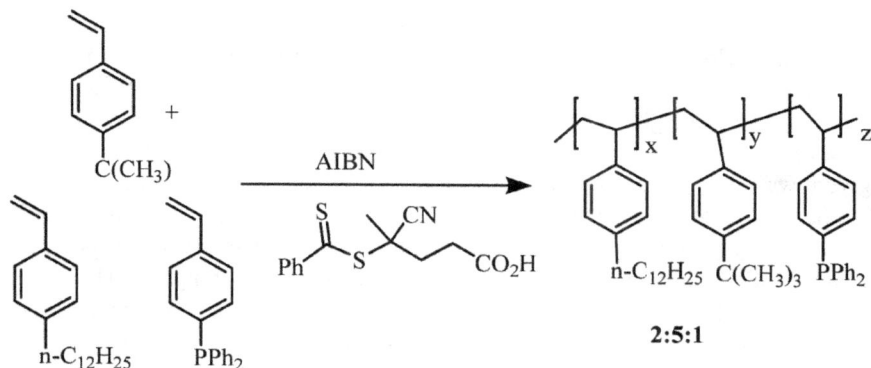

Scheme 10. RAFT polymerisation of 4-dodecylstyrene and 4-tert-butylstyrene were copolymerized with 4-diphenylphosphine styrene (Bergbreiter et al. 2015)

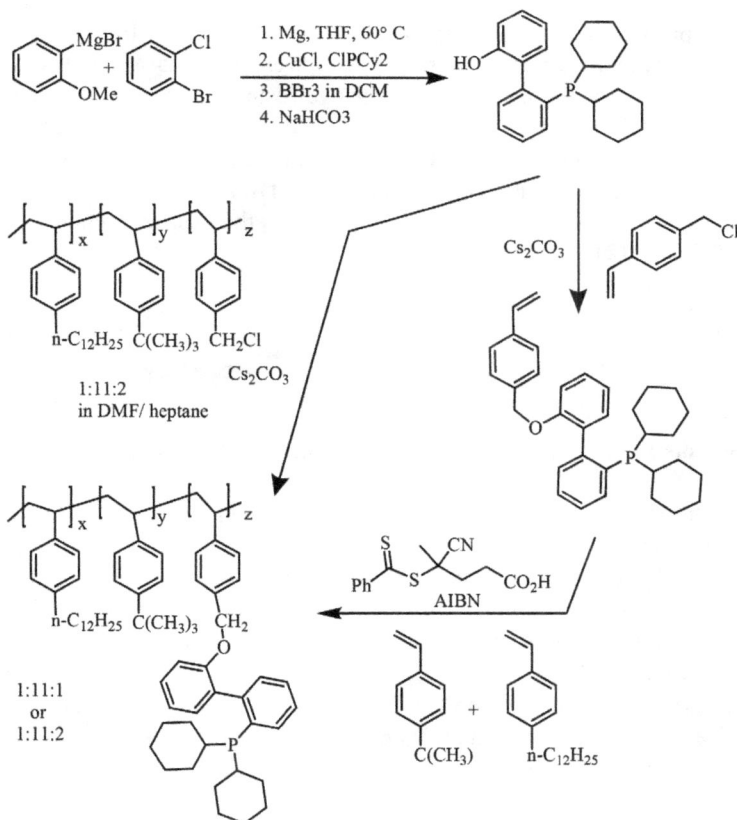

Scheme 11. Functionalisation of poly(tert-butylstyrene-co-4-dodecylstyrene) copolymers dicyclohexylbiarylphosphine ligand (Bergbreiter et al. 2015)

Merrifield resin is one of the most common form of polystyrene which is commercially available. In one of the most recent report by Sreekumar and coworkers (2020), they had shown the use of dendromerized form of Merrifield resin with ethylene glycol. The synthesis procedure was relatively easy compared to several previous reports. The merrfield resin polymer supported by ethylene glycol was allowed to swell in the DMF and a methanolic solution of Pd(OAc)$_2$ is charged to get the desired catalyst. Compared to the earlier reports, this methodology produces symmetrical dendrons bonded on to the linear polymer chain, resulting in a rigid and permanent polymer chain having Pd. This is an intresting phenomenon which produces a site isolation to the peripheral active sites that are available for catalysis (Sreekumar et al. 2020) (Figure 1).

3. Polymer support other than polystyrene

As we have discussed earlier, polymers are a diverse class of materials with excellent properties to withstand harsh conditions. Apart from polystyrene, in previous decades several other polymers were used as a catalyst support. Amberlyst is one of the neutral ion exchange resin and bears a basic chemical nature; it is an inexpensive and commercially available polymer. Ye and coworkers (2008) reported the use of amberlyst A-21 as catalyst support. The synthesis protocol is quite reasonable: Pd(OPf)$_2$ and amberlyst A-21 were directly refluxed in acetone for 24 h that produces the desired product (Ye et al. 2008) (Scheme 12).

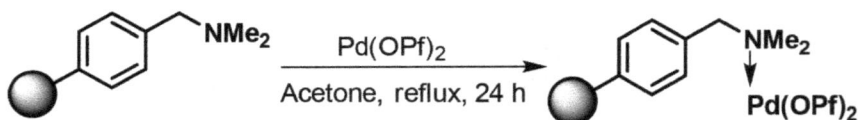

Scheme 12. Amberlyst A-21 as catalyst support for Pd(OPf)$_2$ (Ye et al. 2008)

Another report by Li and coworkers (2011) reported the use of a most commonly available biopolymer, cellulose. Cellulose is one of the best material as a catalyst support because it is naturally available or it can be easily synthesized and can have wide applications. A slightly modified cellulose, functionalized with OPPh$_2$ is an attractive option as a catalyst support. Nano Pd is an active catalyst but it is difficult to bind with the ligands and polymer supports; also, nano Pd are sensitive to air and moisture. In this report, Cellulose – OPPh$_2$ – Pd^0 was synthesized and reported for catalytic application. The two-step synthesis protocol was not difficult and easily produces the desired product. Initially, cellulose was reacted with PPh$_2$Cl, which produces an intermediate Cellulose – PPh$_2$, which is further reacted by PdCl2 in ethanol solution, and affords desired catalyst (Li et al. 2011) (Scheme 13).

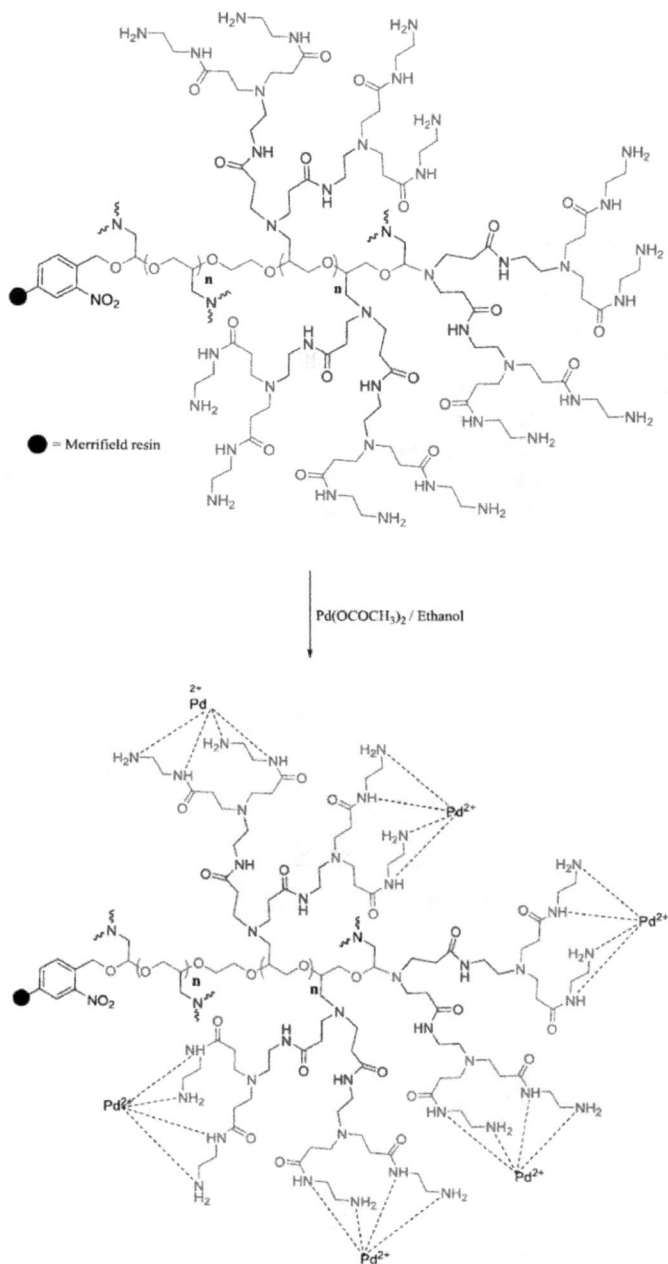

Figure 1. Pd ligation to the dendromerized form of Merrifield resin with ethylene glycol. With permission from Wiley (Sreekumar et al. 2020)

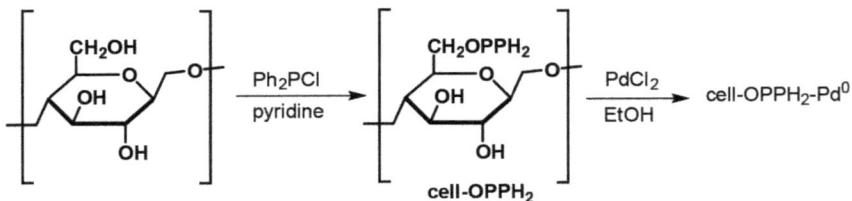

Scheme 13. Synthesis of Cellulose – $OPPh_2$ – Pd^0 catalyst (Li et al. 2011)

Shou and coworkers (2012) reported a porous polymer synthesized by copolymerization of divinylbenzene and methylacrylate monomers, which were reacted in presence of AIBN; after stirring for 3 h, the mixture was transferred to autoclave for 24 h at 100°C, which produces the solid product. The formed product contains carboxyl group, which upon treatment with HCl converted to acyl group. The formed acyl group was further reacted with acylamidation reaction with 1,6-hexadiamine followed by a reaction with pyridine-2-aldehyde to produce Schiff base. The formed product is a polymer-supported ligand, which can be complexed with $Pd(OAc)_2$ (Shou et al. 2012) (Scheme 14).

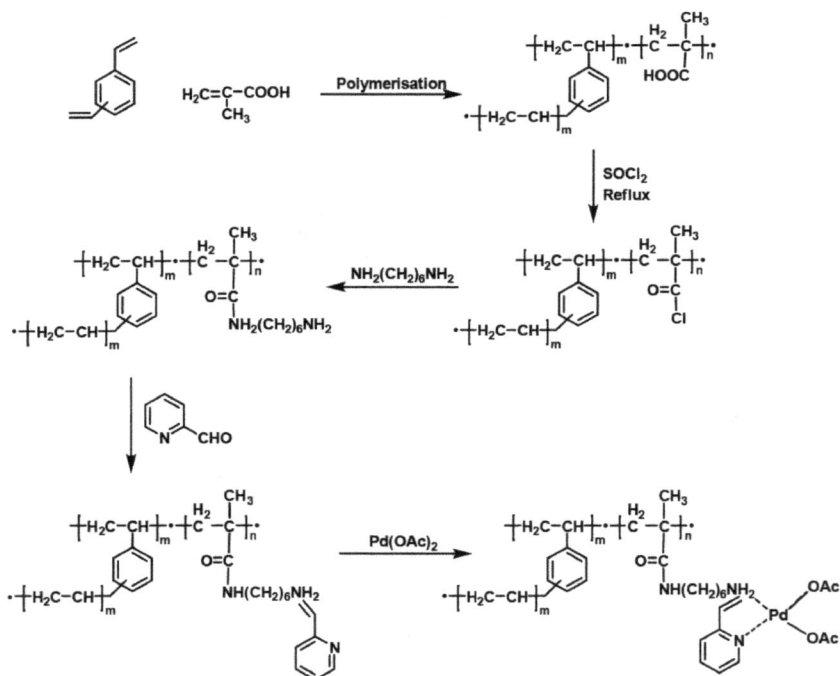

Scheme 14. Synthesis of Schiff base functionalized porous polymer synthesized by copolymerization of divinylbenzene and methylacrylate monomers (Shou et al. 2012)

In the last few decades, MOFs (Metal Organic Framework) are extensively studied for their applications in catalysis and they are well suitable for various other applications. The reasons for the applications of MOF, particularly in the area of catalysis, are the porosity and metal bonding capacities as well as their multichemical functionalities. However, it is observed that MOF have very poor hydrothermal stability and it is not much stable in certain organic solvents in the longer duration reactions (Zhou et al. 2012, Gao et al. 2012, Wu et al. 2012, Kreno et al. 2012, Gu et al. 2011, Aijaz et al. 2012). While looking for potential alternatives, researchers came up with an idea by making a composite of MOF with microporous polymers, which are relatively stable because of covalent bonds.

Recently, covalent organic polymers (COP's) were synthesized and reported by Xiang and coworkers (Xiang et al. 2012). This novel polymer is quite porous and extremely stable under hydrothermal conditions. Therefore, COPs are well suited as catalyst support. Zhou and coworkers (2013) reported an initial report of a Pd catalyst supported by COP-4. COP-4 was synthesized via a reaction of 1,5-cyclooctadiene and 2,4,6-tris-(4-bromo-phenyl)-[1,3,5-] triazine in presence of bis(1,5-dicyclooctadiene) Ni (0) catalyst in DMF. Further, the Pd nitrate was impregnated to the prepared COP-4 in DMF or aqueous media (Zhou et al. 2013). This work shows a high dispersion of superfine particles, and hence the high catalytic activity and stability (Figure 2).

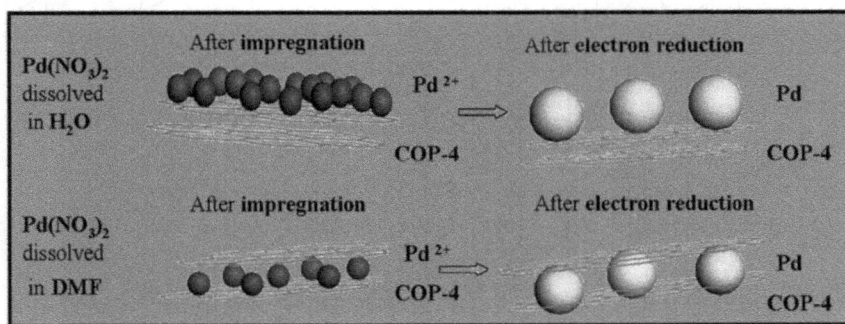

Figure 2. Showing the covalent organic polymers (COPs) as polymer support. With permission from ACS (Zhou et al. 2013)

Schiff base ligands or Schiff base functionalized polymer supports are excellent in terms of the stability they impart to the metal complexes. Numerous reports are available showing applications of these catalysts. The reasons for the privilege these ligands are receiving are because of the multidentate chelating capabilities and ability to withstand under strong oxidizing and reducing conditions, and most importantly, the ease of availability of precursors and synthesis of desired ligands. Taking these merits into account, Zhao and coworkers (2014) reported the use of polyallylamine polymer as catalyst support. In addition, the novel part of the work is that they included phosphine functionality to the polymer support, which brings an interesting outcome that it provides multiple coordination sites for the Pd acetate. The synthetic protocol was relatively easy, the cross-linked polymer of polyallylamine and epichlorohydrin were obtained under basic reaction conditions and converted to carbonate by an ion exchange reaction with NaHCO$_3$. Further, a reaction with 3-(diphenylphosphanyl)benzaldehyde produces a Schiff base functionalized polyallylamine polymer, followed by a reaction with Pd acetate that afforded the desired catalyst (Zhao et al. 2014) (Scheme 15).

Sajiki and coworkers have reported a slight modification of the work of Zhao and coworkers in 2019. The authors suggest that polyethyleneimine polymers, polystyrene-divinylbenzene and polymethacrylate polymers have similar Pd scavenging capabilities. It is also mentioned that due to the similar affinities to Pd, these polymer supports show similar type of selectivity and catalytic efficiency. However, polyethyleneimine modified polymers show better catalytic activities

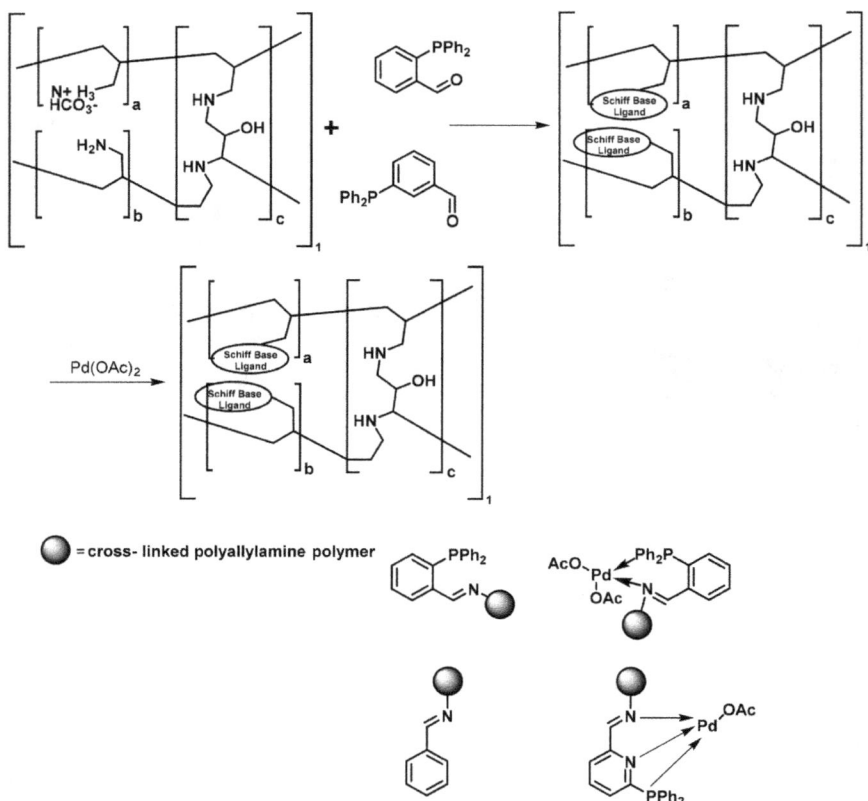

Scheme 15. Phosphine functionalized Polyallylamine polymer as catalyst support (Zhao et al. 2014)

towards the chemoselective hydrogenations. In this report, a comparative study has been performed to show the ability to chelate with the residual or leached Pd present in the sonogashira reaction (Sajiki et al. 2019) (Figure 3).

Porous polymer monoliths are a new class of polymer, which is recently discovered. In a very small duration of its discovery, monolith polymers had proved their presence in various areas of application, most importantly in the production of ultrapure water (Murayama et al. 2011). Moreover, these polymers were studied for their applications in the flow chemistry. Because most of the polymers swell and show different physical properties in the organic solvents, it is an important drawback which restricts their applications in the flow and microchannel reactors. Sajiki and coworkers (2017) reported the use of polystyrene-divinylbenzene based monolith polymer as a catalyst support. They prepared an unfunctionalized polymer via two-step successive polymerization of styrene and divinylbenzene using sorbitan fatty acid as emulsifier and radical

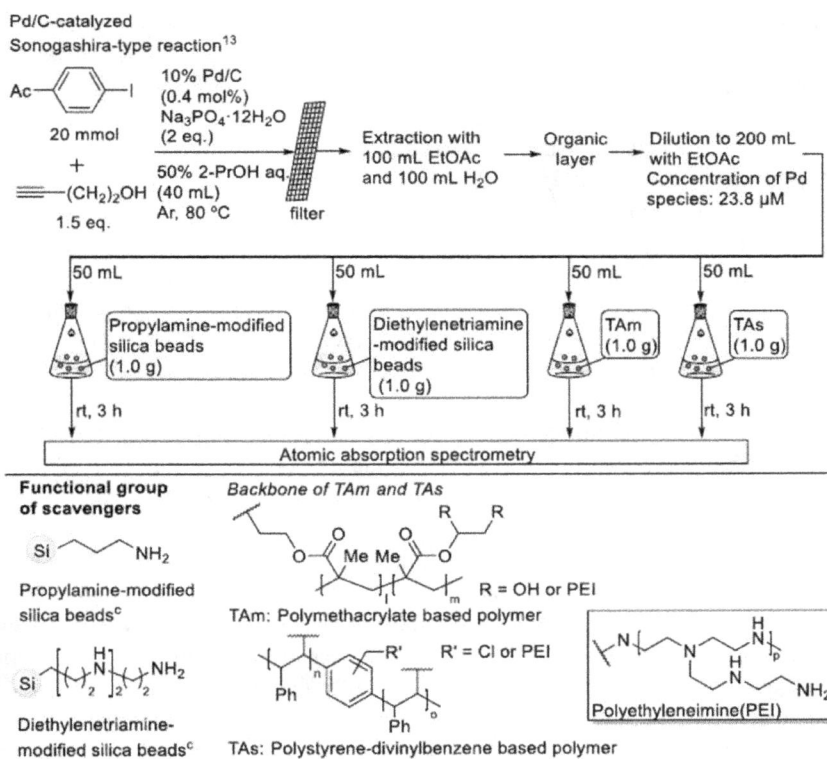

Figure 3. Schematic showing the comparative study of chelating ability of different ligands functionalized over polymer support polyethyleneimine polymers, polystyrene-divinylbenzene and polymethacrylate polymers. With permission from ACS Omega (Sajiki et al. 2019)

indicator AIBN, followed by the sulphonation of monolithic polymer using an acidic cation exchange resin. Finally, this polymer has been treated with the methanolic solution of Pd acetate to obtain the desired catalyst (Sajik et al. 2017) (Scheme 16).

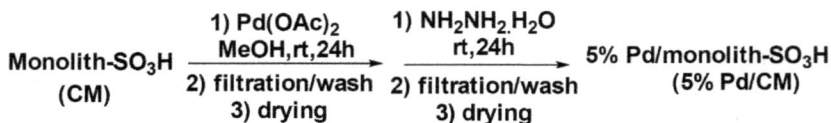

$$\text{Monolith-SO}_3\text{H} \xrightarrow[\substack{2)\text{ filtration/wash} \\ 3)\text{ drying}}]{\substack{1)\text{ Pd(OAc)}_2 \\ \text{MeOH,rt,24h}}} \xrightarrow[\substack{2)\text{ filtration/wash} \\ 3)\text{ drying}}]{\substack{1)\text{ NH}_2\text{NH}_2.\text{H}_2\text{O} \\ \text{rt,24h}}} \text{5\% Pd/monolith-SO}_3\text{H}$$

Monolith-SO$_3$H (CM)   →   5% Pd/monolith-SO$_3$H (5% Pd/CM)

Scheme 16. Showing the use of porous polymer monoliths as a catalyst support (Sajik et al. 2017)

An alternate to the porous polymer monoliths, porous organic polymers (POPs) are emerging as a versatile material for a catalyst support. After introduction of COFs (Covalent Organic Frameworks), these porous polymers can be synthesized by two different methods via two-component condensation or by two-component coupling. After establishment of these synthesis protocols, because of their ease of synthesis, these polymers are gaining substantial interest. In addition, due to their large surface area, and flexible binding properties, hierarchical porosity provides a dispersion medium to produce a single atom catalyst also. Recently in 2018, Zhu and coworkers reported the use of 1,3,5-TDTB (1,3,5-triformylphloroglucinol) based POP and used as a catalyst support. Using Duff formylation reaction, POP were synthesized, followed by the reaction with Pd acetate, which produces the desired catalyst (Zhu et al. 2018) (Figure 4), while in another report by Ding and coworkers (2020), a PPh$_3$ functionalized POP as catalyst support was used (Ding et al. 2020) (Figure 5).

Anas and coworkers showed the use of polyacrylonitrile as a catalyst support. The novelty of the work is situated in the linear structure of this molecule, and because it is thermoplastic it has a high melting point above 300 °C. In addition, this polymeric material provides the diversity to produce copolymers with different acrylonitriles. Based on this perspective, two new catalysts have been reported. These two modifications were based on the functionalization of the polyacrylonitriles with ethylene diamine and mono-ethanolamine (Anas et al. 2021, 2022) (Schemes 17 and 18).

Nevertheless, it is a new decade coming up with new targets and new horizons are yet to be achieved. In the previous year, an excellent report by Li and coworkers (2021) showed an innovative idea of polymer supported catalysis. As we know, catalyst deactivations and aggregations are the most common problems, and these problems are yet to be addressed, specially in heterogeneous catalysts, and even in the polymer-supported catalysis. This report includes an exceptional idea by developing a hyper crosslinked polymer integrated NHC (N-Heterocyclic carbenes) as a polymer support for Pd catalyst. This polymeric support has potential to recapture the aggregated or leached Pd (Li et al. 2021) (Figure 6).

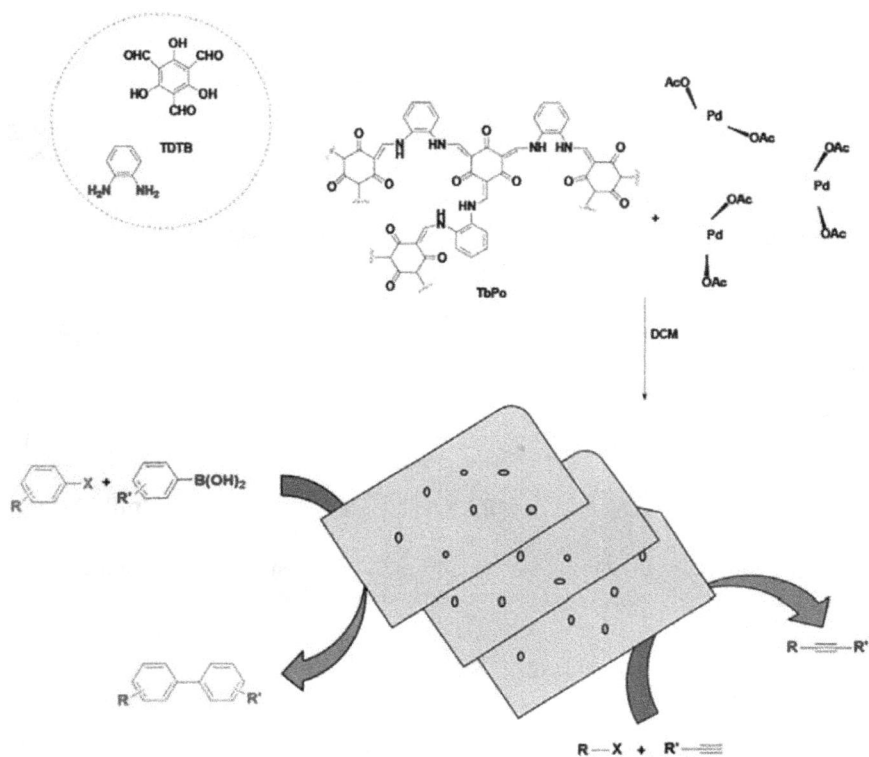

Figure 4. Synthesis of 1,3,5-TDTB (1,3,5-triformylphloroglucinol) based POP for catalyst support (Ding et al. 2020)

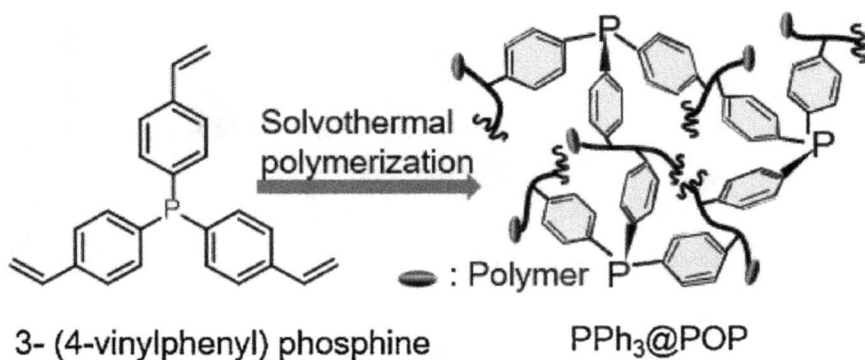

Figure 5. Synthesis of PPh3 functionalized POP. With permission from Elsevier (Ding et al. 2020)

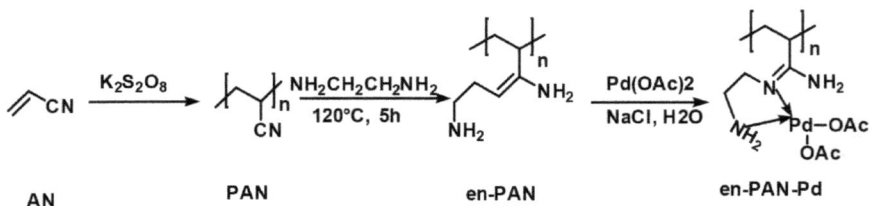

Scheme 17. Synthesis of polyacrylonitriles functionalised with ethylene diamine
(Anas et al. 2021)

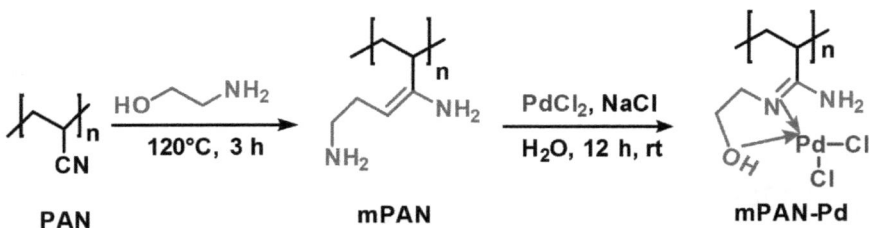

Scheme 18. Synthesis of polyacrylonitriles with ethylene diamine and mono-
ethanolamine (Anas et al. 2022).

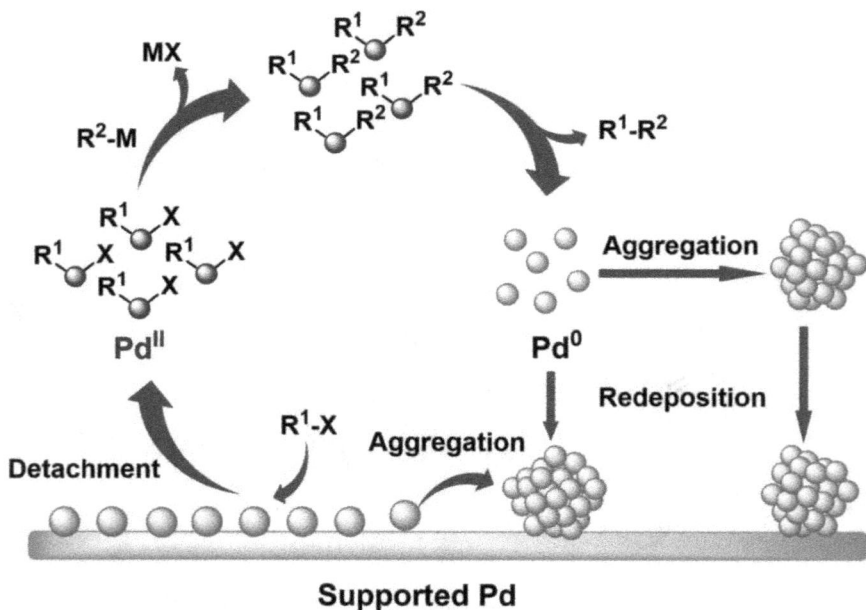

Figure 6. Hyper crosslinked Kapton polymer integrated NHC (N-Heterocyclic carbenes);
reaction cycle shows the detachment of Pd in the reaction media and redepostion of free
Pd back to the polymer surface (Li et al. 2021)

4. Conclusion

Heterogeneous catalysis is unquestionably a field of fascination, which is continuously growing because of the efforts of excellent research groups. Due to this, extensive research is leading to the development of more dependable and credible methods for catalyst production, characterization, and catalyst applications.

Catalysis systems, which are chemically relevant and interesting, and also effective in terms of catalytic applications with potential reusability and reduced metal leaching, will always, remain in demand. Even though there are substantial and effective reports available, very few are getting attention from the industries for commercial applications. The most common reason is the lack of innovations and modifications in the existing polymeric supports. Moreover, a number of new polymers are being developed by the researchers, which have considerable potential as catalyst support but their application is yet limited.

Pd based catalysts have significant applications in laboratory and industrial scale applications. At present, to make synthetic protocols, energy efficient industries are shifting towards the flow and microchannel reactors and these reaction techniques are well suited with the homogeneous catalysts but with homogeneous catalysis, the question of separation of catalyst is a major problem. Therefore, there is a robust requirement to develop heterogeneous catalysts suitable for these types of reactor techniques.

References

Aijaz, A., A. Karkamkar, Y.J. Choi, N. Tsumori, E. Ronnebro, T. Autrey, H. Shioyama and Q. Xu. 2012. Immobilizing highly catalytically active Pt nanoparticles inside the pores of metal–organic framework: A double solvents approach. *Journal of the American Chemical Society* 134: 13926-13929.

Antony, R., G.L. Tembe, M. Ravindranathan and R.N. Ram. 2003. Polymer-supported palladium (II) complexes and their catalytic study. *Journal of Applied Polymer Science* 90: 370-378.

Bakherad, M., A. Keivanloo, B. Bahramian and T.A. Kamali. 2009. Synthesis of novel 6-(substituted benzyl)imidazo[2,1-b][1,3]thiazole catalyzed by polystyrene-supported palladium(II) ethylenediamine complex. *Journal of the Brazilian Chemical Society* 20(5): 907-912.

Bi, Y., J. Chen, Y. Dong, W. Guo, W. Zhu and T. Li. 2020. Polymer-supported palladium (II) containing N_2O_2: An efficient and robust heterogeneous catalyst for C–C coupling reactions. *Journal of Polymer Science Part A: Polymer Chemistry* 56: 2344-2353.

Chen, M., X. Mou, S. Wang, X. Chen, X. Tan, M. Chen, Z. Zhao, C. Huang, W. Yang, R. Lin and Y. Ding. 2020. Porous organic polymer-supported palladium catalyst for hydroesterification of olefins. *Molecular Catalysis* 498: 111239.

Diez-Gonzalez, S. 2016. *N-Heterocyclic Carbenes: From Laboratory Curiosities to Efficient Synthetic Tools.* United Kingdom: Royal Society of Chemistry.

Dronsfield, A. 2007. *Who Really Discovered the Haber Process?* RSC Education in Chemistry.

Du, Q. and Y. Li. 2011. Air-stable, recyclable and time-efficient diphenylphosphinite cellulose-supported palladium nanoparticles as a catalyst for Suzuki–Miyaura reactions. *Beilstein Journal of Organic Chemistry* 7: 378-385.

Gao, W.Y., W.M. Yan, R. Cai, K. Williams, A. Salas, L. Wojtas, X.D. Shi and S.Q. Ma. 2012. A pillared metal–organic framework incorporated with 1,2,3-triazole moieties exhibiting remarkable enhancement of CO_2 uptake. *Chemical Communications* 48: 8898-8900.

George, S. and K. Sreekumar. 2020. Heterogeneous palladium (II)-complexed dendronized polymer: A rare palladium catalyst for the one-pot synthesis of 2-arylbenzoxazoles. *Applied Organometallic Chemistry* e6083.

Gu, X.J., Z.H. Lu, H.L. Jiang, T. Akita and Q. Xu. 2011. Synergistic catalysis of metal–organic framework-immobilized Au–Pd nanoparticles in dehydrogenation of formic acid for chemical hydrogen storage. *Journal of the American Chemical Society* 133: 11822-11825.

Inoue, H., H. Takada and M. Murayama. 2011. High efficiency decomposition of hydrogen peroxide by palladium-supported monolithic ion exchange resin. *Kobunshi Ronbunshu* 68: 320.

Islam, S.M., P. Mondal, A.S. Roy, S. Mondal and M. Mobarak. 2009. An efficient and reusable polymer-supported palladium catalyst for the Suzuki cross-coupling reactions of aryl halides. *Journal of Chemical Research* 756-760.

Joshi, R.K., K.N. Sharma, N. Satrawala, A.K. Srivastava and M. Ali. 2019. Palladium (II) ligated with a selenated (Se, CNHC, N−)-type pincer ligand: An efficient catalyst for Mizoroki–Heck and Suzuki–Miyaura coupling in water. *Organic & Biomolecular Chemistry* 17: 8969-8976.

Joshi, R.K., V. Tomar, Y. Upadhyay, A.K. Srivastava, M. Nemiwal and P. Mathur. 2021. Selenated NHC-Pd(II) catalyzed Suzuki-Miyaura coupling of ferrocene substituted β-chloro-cinnamaldehydes, acrylonitriles and malononitriles for the synthesis of novel ferrocene derivatives and their solvatochromic studies. *Journal of Organometallic Chemistry* 940: 121752.

Joshi, R.K., S. Kumari, C. Sharma, A.K. Srivastava, N. Satrawala and K.N. Sharma. 2021. Half-sandwich (η6-benzene)ruthenium(II) complex of picolyl functionalized N-heterocyclic carbene as an efficient catalyst for thioether directed C–H alkenylation of arenes. *European Journal of Inorganic Chemistry* 3648-3653.

Joshi, R.K., S. Kumari, C. Sharma, N. Satrawala, A.K. Srivastava and K.N. Sharma. 2022. Selenium-directed ortho C–H activation of benzyl selenide by a selenated NHC– Half-Pincer ruthenium(II) complex. *Organometallics* 41: 1403-1411.

Khamatnurova, T.V., D. Zhang, J. Suriboot, H.S. Bazzi and D.E. Bergbreiter. 2015. Soluble polymer-supported hindered phosphine ligands for palladium-catalyzed aryl amination. *Catalysis Science & Technology* 5: 2378-2383.

Kim, J.-H., J.-W. Kim, M. Shokouhimehr and Y.S. Lee. 2005. Polymer-supported N-heterocyclic carbene-palladium complex for heterogeneous Suzuki cross-coupling reaction. *Journal of Organic Chemistry* 70: 6714-6720.

Kreno, L.E., L. Kirsty, O.K. Farha, M. Allendorf, R.P.V. Duyne and J.T. Hupp. 2012. Metal–organic framework materials as chemical sensors. *Chemical Reviews* 112: 1105-1125.

Li, J.R., J. Sculley and H.C. Zhou. 2012. Metal–organic frameworks for separations. *Chemical Reviews* 112: 869-932.

List, B. 2007. Introduction: Organocatalysis. *Chemical Reviews* 107: 5413-5415.

Liu, X., X. Zhao and M. Lu. 2014. Novel polymer supported iminopyridylphosphine palladium(II) complexes: An efficient catalyst for Suzukie-Miyaura and Heck cross-coupling reactions. *Journal of Organometallic Chemistry* 768: 23-27.

Monguchi, Y., F. Wakayama, S. Ueda, R. Ito, H. Takada, H. Inoue, A. Nakamura, Y. Sawama and H. Sajiki. 2017. Amphipathic monolith-supported palladium catalysts for chemoselective hydrogenation and cross-coupling reactions. *RSC Advances* 7: 1833-1840.

Mukherjee, D. 2003. Polymer supported palladium(II) complexes as hydrogenation catalysts. *Indian Journal of Chemistry*, 42B: 346-352.

Parrish, C.A. and S.L. Buchwald. 2001. Use of polymer-supported dialkylphosphinobiphenyl ligands for palladium-catalyzed amination and Suzuki reactions. *Journal of Organic Chemistry* 66: 3820-3827.

Qi, S., Z. LongFeng, S. ZhenHua, M. XiangJu and X. Feng-Shou. 2012. Porous polymer supported palladium catalyst for cross coupling reactions with high activity and recyclability. *Science China Chemistry* 55: 2095-2103.

Saranya, T.V., P.R. Sruthi, N. Ayana and S. Anas. 2021. An efficient polymer supported palladium catalyst for ortho selective C–H olefination of anilides. *Chemistry Select* 6: 2615-2620.

Shieh, W.-C., R. Shekhar, T. Blacklock and A. Tedesco. 2002. A simple, recyclable, polymer-supported palladium catalyst for suzuki coupling—An effective way to minimize palladium contamination. *Synthetic Communications* 32: 1059-1067.

Sruthi, P.R., P.P. Nimmi, S.S. Babu and S. Anas. 2022. Highly efficient and reusable polymer supported palladium catalyst for copper free sonogashira reaction in water. *Chemistry Select* 7: e202104273

Stradiotto, M. and R.J. Lundgren. 2016. *Ligand Design in Metal Chemistry: Reactivity and Catalysis*. Germany: Wiley.

Suib, S.L. 1993. Selectivity in catalysis: An overview. *ACS Symposium Series* (Chapter 1), 517: 1-19.

Udayakumar, V., S. Alexander, V. Gayathri, Shivakumaraiah, K.R. Patil and B. Viswanathan. 2010. Polymer-supported palladium-imidazole complex catalyst for hydrogenation of substituted benzylideneanilines. *Journal of Molecular Catalysis A: Chemical* 317: 111-117.

van der Helm, M.P., B. Klemm and R. Eelkema. 2019. Organocatalysis in aqueous media. *Nature Reviews Chemistry* 3: 491-508.

Velu, U., A. Stanislaus, A. Virupaiah, Shivakumaraiah, K.R. Patil and V. Balasubramanian. 2011. Synthesis, characterization of polymer-supported palladium-2-methylimidazole complex catalyst for the hydrogenation of aromatic nitro compounds. *Journal of Chemical Research* 35(2): 112-115.

Wu, D., Q.Y. Yang, C.L. Zhong, D.H. Liu, H.L. Huang, W.J. Zhang and G. Maurin. 2012. Revealing the structure–property relationships of metal–organic frameworks for CO_2 capture from flue gas. *Langmuir* 28: 12094-12099.

Wünsch, J.R. 2012. *Polystyrene: Synthesis, Production and Applications*. Smithers Rapra Publishing.

Xiang, Z.H. and D.P. Cao. 2012. Synthesis of luminescent covalent–organic polymers for detecting nitroaromatic explosives and small organic molecules. *Macromolecular Rapid Communications* 33: 1184-1190.

Xiang, Z.H., D.P. Cao, W.C. Wang, W.T. Yang, B.Y. Han and J.M. Lu 2012. Postsynthetic lithium modification of covalent-organic polymers for enhancing hydrogen and carbon dioxide storage. *Journal of Physical Chemistry C* 116: 5974-5980.

Yamada, T., Y. Kobayashi, N. Ito, T. Ichikawa, K. Park, K. Kunishima, S. Ueda, M. Mizuno, T. Adachi, Y. Sawama, Y. Monguchi and H. Sajiki. 2019. Polyethyleneimine-modified polymer as an efficient palladium scavenger and effective catalyst support for a functional heterogeneous palladium catalyst. *ACS Omega* 4: 10243-10251.

Ye, Z.-W. and W.-B. Yi. 2008. Polymer-supported palladium perfluorooctanesulfonate [Pd(OPf)$_2$]: A recyclable and ligand-free palladium catalyst for copper-free Sonogashira coupling reaction in water under aerobic conditions. *Journal of Fluorine Chemistry* 129: 1124-1128.

Yue, C., Q. Xing, P. Sun, Z. Zhao, H. Lv and F. Li. 2021. Enhancing stability by trapping palladium inside N-heterocyclic carbene-functionalized hypercrosslinked polymers for heterogeneous C–C bond formations. *Nature Communications* 12: 1875.

Zhang, K. and D.C. Neckers. 1983. Diaminobipyridine-TDI polyureas: Synthesis, metal complexes, and catalytic activity. *J. Polym. Sci. Polym. Chem. Ed.* 21: 3115.

Zhou, H.C., J. Park, J.R. Li, E.C. Sanudo and D.Q. Yuan 2012. A porous metal–organic framework with helical chain building units exhibiting facile transition from micro- to meso-porosity. *Chemical Communications* 48: 883-885.

Zhou, Y., Z. Xiang, D. Cao and C. Liu. 2013. Covalent organic polymer supported palladium catalysts for CO oxidation. *Chemical Communications* 49: 5633-5635.

Cobalt, Copper, and Rhodium Containing Polymer-supported Catalyst

Gitanjali Arora[1], Nirmala Kumari Jangid[1*], Anamika Srivastava[1], Navjeet Kaur[2] and Jaya Dwivedi[1]

[1] Department of Chemistry, Banasthali Vidyapith, Banasthali - 304022, Rajasthan, India
[2] Department of Chemistry & Division of Research and Development,
Lovely Professional University, Phagwara - 144411, Punjab, India

1. Introduction

Water is one of the most vital elements for life on Earth. Nowadays, it's necessary to have access to clean water, which is crucial for maintaining excellent health (Anderson et al. 2019). One of today's most significant and potentially fatal problems is environmental contamination, which demands a specific focus from researchers. Environmentally hazardous toxins in nature are rising quickly as a result of human activities like industrialization (Bengtsson et al. 2019). The advancements made in science have both positive and negative effects on the environment. Poor industrial effluent discharge practices into soil and water bodies, surface runoff from agricultural lands, municipal waste disposal that is left untreated, and mining activities are all responsible for this pollution, which negatively affects health and life and degrades our environment (Delkash et al. 2018). Due to their refractory properties, heavy metals, a naturally occurring component of the Earth's crust, and an economically significant mineral cause heavy metal pollution that degrades water quality and has an impact on the food chain. To do this, numerous researchers have carried out studies to identify effective techniques for wastewater remediation (Munir et al. 2021). Consequently, a lot of study has focused on designing and developing strategies

---

*Corresponding author: nirmalajangid.111@gmail.com

to cope with these contaminants. Among many other approaches, the creation of an active catalyst has drawn a lot of interest in addressing the potentially harmful contaminants present in aqueous environments (Farhan et al. 2022).

Numerous research from both the economic and environmental aspects has been conducted on the catalytic reactions utilizing catalysts based on transition metals. There have been numerous research initiatives over the past 15 years to develop new, environmentally-friendly catalytic systems for reactant transformations; replacing traditional processes with newer catalytic conversions using different metallic catalysts under cleaner, greener conditions has also continued (Javaid et al. 2019). These catalytic systems are frequently homogeneous, making it impossible to separate or extract the catalysts from the reaction. Due to their renewability, reusability, ease of handling, and insolubility in solvents, heterogeneous catalytic systems have recently received a lot of attention from researchers to overcome the shortcomings of their homogeneous counterparts (Nasrollahzadeh et al. 2021a).

Copper (Cu), arsenic (As), rhodium (Rh), lead (Pb), mercury (Hg), cobalt (Co), nickel (Ni), cobalt (Co), zinc (Zn), cadmium (Cd), and chromium (Cr) are examples of heavy metals with specific densities of around 5 g/mL or above and higher mass numbers (Qasem et al. 2021). These substances are naturally present in the Earth's crust and are regarded as being incredibly dangerous even at low concentrations. They can enter the human body by the consumption of contaminated food, water, or air and cannot biodegrade (Tabelin et al. 2018). However, the human body's metabolism needs just a little amount of heavy metals as trace elements. They pose a serious threat to all living organisms as their level rises. In addition, domestic and agricultural runoff, industrial waste, and other sources of heavy metals are the main causes of water contamination. As a result, heavy metal ion removal from polluted streams is crucial (Rathi et al. 2021).

Applications of polymer-based materials for adsorption and catalysis have grown significantly over the last few years. Novel polymeric composites and nanocomposites have been created to address the issue of the growing scale of water contamination by various mono- and multivalent heavy metal ions in recent years from an adsorption perspective (Eskandari et al. 2020). The polymer can be coupled with a variety of organic/inorganic, natural, or synthetic materials to create novel sorbents that have enhanced sorption characteristics for heavy metal ions and higher removal efficiencies due to its remarkable properties like the ease of production and biocompatibility (Zhang et al. 2021a). The polymer-supported catalyst has shown its ability as a suitable platform for hydrogenizing the catalyst of various catalytic reactions, operating either as a protective material or as a catalytic component (Hou et al. 2020).

Polymer-based adsorbents were used in optics and medication administration in addition to wastewater treatment. Physical characteristics of polymer-supported adsorbents include mechanical stability, large surface area, size distribution, and regenerative, flexible, controllable surface chemistry with high surface area and tiny inter-fibrous pore size (Pereao et al. 2019).

Recently, metal catalysts based on chitosan and its derivatives have drawn a lot of interest. The benefits of heterogeneous processes include better product isolation, reusable catalysts, enhanced steric control of reaction intermediates, and more (Wang et al. 2022). These advantages led scientists to immobilize a homogenous catalytic site on a variety of supports, including polymers, zeolites, silica, and magnetic materials. Because it plays a vital role in transition metal-catalyzed processes and is environmentally friendly, chitosan (CS) has recently attracted a lot of attention (Dohendou et al. 2021). For C–C bond-forming procedures such as the Suzuki cross-coupling reaction (CS-supported Pd catalyst) (Chen et al. 2022), Henry reaction (CS-supported Ti catalyst) (Dhakshinamoorthy et al. 2021), hydroformylation process (CS-supported Rh catalyst) (Madkour et al. 2021), and C-N bond-forming methods such as click reaction (Patureau et al. 2019), chitosan-supported metal complexes have been used as catalysts. There have been reports from several groups about these chemicals' catalytic activity (Nasrollahzadeh et al. 2021b).

Numerous studies have shown how chitosan and its derivatives successfully function to adsorb metal ions like lead, cadmium, copper, cobalt, rhodium, and nickel as well as oxyanions and complexed metal ions (Adeyemi et al. 2022). Chitosan is a common adsorbent for extracting metal ions from aqueous solutions and is one of the most used adsorbents in waste treatment. The primary factor influencing chitosan's ability to bind metal ions is its amine groups ($-NH_2$), which can serve as coordination sites for different metals. The degree of deacetylation, the kind of metal ion, and solution characteristics like pH all have an impact on how frequently metal is adsorbable (Wang et al. 2019).

The removal of heavy metals from contaminated water has recently drawn attention to several adsorbents with high selectivity and metal-binding capabilities (Kaur et al. 2022). To improve the sorption selectivity and adsorption capability of chitosan for metal ions, several chitosan derivatives have been made by grafting extra functional groups such as succinic anhydride, heparin, histidine, and N, O-carboxymethyl through a cross-linked chitosan backbone (Gim et al. 2019). Chitosan has also undergone chemical changes to increase its selectivity and capacity toward heavy metal ions. To facilitate chitosan's capacity for heavy metal adsorption, substituted chitosan was synthesized using a simple and effective technique (Zhang et al. 2021b).

The concentration of metal ions ultimately dropped as the immersion time was further extended. The maximum adsorption efficiency was achieved when unmodified chitosan was present. The best efficiency is 99.4% and 99.8%, respectively, when Co (II) and Cu (II) metal ions are present in an alkaline medium (pH 9) (Kalak et al. 2021). In a neutral medium, the adsorption efficiency is 47.2% and 73.9%, respectively (pH 7). The adsorption effectiveness of glycine-chitosan modified biosorbents dropped to 94.1% and 95.4%, respectively, in the presence of Co (II) and Cu (II) in an alkaline medium (pH 9) (Wang et al. 2018). In a neutral solution (pH 7), the adsorption effectiveness of Co (II) and Cu (II) metal ions were reduced by 34.4 % and 65.8 %, respectively. The adsorption effectiveness of the biopolymer treated with chloroacetic acid followed a similar pattern for Co (II) and Cu (II) metal ions (Emamy et al. 2021). In an alkaline medium, Co (II) and Cu (II) had adsorption efficiencies of 65,3% and 78,2%, respectively. However, the adsorption efficiency dropped to 26.6 and 34.4 %, respectively, when the medium's pH was adjusted to neutral (pH 7). In an alkaline medium, the process of metal ion adsorption is more efficient than it is in a neutral one. Additionally, cobalt ions are less likely to adsorb than copper ions (Kwak et al. 2019).

2.   Chitosan-supported catalysts

The development of biomaterial-supported catalysts for green chemical synthesis has drawn more attention in recent years. Polysaccharides are frequently utilized as substrates for biocatalysts in the synthesis of complex compounds due to their outstanding properties, including non-toxicity, renewability, availability, biocompatibility, and biodegradability (Bilal et al. 2019). The N-deacetylated derivative of chitin, chitosan, offers an excellent substitute to other biomaterials for the immobilization of catalytic metals due to its availability of amine and hydroxyl chelating functional groups, physical moldability (to provide porous open framework hydrogels), stability of metal ions, and chemical flexibility due to the chemical change of amino groups (Rafiee et al. 2019).

Chitosan is a chitin derivative that is created by deacetylating the chitin polymer. It is a biodegradable, biocompatible, non-toxic, and environmentally friendly biopolymer (Peter et al. 2021). Although chitosan has diverse applications, its inability to dissolve in water limits the scope of those uses. To make chitosan soluble in aqueous fluids, it is therefore subjected to various changes. Numerous carboxymethyl chitosan derivatives that are water-soluble have been created (Journot et al. 2020).

Chitosan has the potential to function as a useful stabilizing polymer for metal nanoparticles, reducing metal ions to a zero-valent oxidation state while managing nanoparticle development. This biopolymer's significant affinity for interacting with the metal nanoparticles resulted in stability and prevented metal leaching (Ken et al. 2020).

To create coordination sites, chitosan has been functionalized chitosan-metal complexes have created heterogeneous catalysts for oxidation, cycloaddition, oxidative carboxylation of olefins, acetalization, hydrogenation, cyclopropanation of olefins, and cross-coupling mechanisms (Rafiee et al. 2018).

2.1 Chitosan-supported catalyst containing cobalt

Several highly effective and magnetically recyclable cobalt catalytic systems (Co-ligand@MNPs/Ch) were produced by integrating magnetic chitosan with a few stable and easily accessible organic compounds (Hajipour et al. 2021). CHNS, FT-IR, SEM, TGA, XRD, TEM, VSM, and ICP-OES are a few of the physicochemical methods that were used to confirm the structure of these nanocomposites (Ouardi et al. 2021). In moderate and environmentally friendly reaction conditions, these nanocomposites demonstrate excellent catalytic efficiency for Suzuki and Heck cross-coupling reactions. Our catalysts' advantages include easy accessibility to raw materials, the ability to function in the presence of air, and environmentally benign circumstances (Honk et al. 2020).

Electrostatic interactions and the density functional theory (DFT) model of the molecular approach were also used to characterize and gain insight into the function and impact of the ligands present in these catalysts (Sellaoui et al. 2019).

First, a model reaction involving the reactions of bromobenzene and phenylacetylene was designed, and the effects of various variables such as catalyst concentrations, base, solvent, and temperature were examined (Munusamy et al. 2019). The produced catalyst's catalytic activity was also tested in a control experiment under identical reaction circumstances, along with that of magnetic chitosan and cobalt-free magnetic nanofiber (Amrutha et al. 2022). Finally, it was determined that K_3PO_4 and polyethylene glycol (PEG-200) were the best bases and solvents, respectively. The other ideal parameters were chosen to be 60 degrees with 5 milligrams of catalyst. After the reaction conditions were improved, the range of the catalyst Co-DMM@MNPs/Cs in the Sonogashira coupling reaction was broadened to include the reaction of aryl halides and phenylacetylene with varied electron-withdrawing and electron-donating substituents as shown in Scheme 1 (Nair et al. 2021).

The coupling products had high yields due to the electron-withdrawing and electron-donating substituents of aryl halides (62%–82%). Aryl chlorides remained inactive under these circumstances, but the reaction of sterically inhibited aryl halides likewise produced a high yield of the appropriate product (Babij et al. 2019).

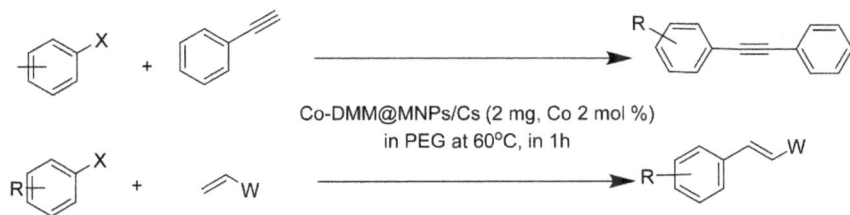

Scheme 1. Sonogashira coupling reactions of Co-DMM@MNPs/Cs are shown, and the conditions are optimized

Using an external magnetic bar, it is simple to remove the produced catalyst Co-DMM@MNPs/Cs from the reaction media and re-use it for additional reaction runs. Six recycling of Co-DMM@MNPs/Cs is possible without significantly lowering coupling yields. ICP-AES was used to determine the precise quantity of cobalt used in the final catalytic process, and it was discovered that only 15% of cobalt was lost (Mohajer et al. 2021).

Pure cobalt nanoparticle (Co-NPs) and a complex of cobalt N-heterocyclic ligands supported by multi-walled carbon nanotubes (Co-NHC@MWCNTs) were synthesized as the catalyst (Han et al. 2019). In green solvents, this catalyst produced very high yields of derivatives of propargylamine. The catalytic system is simply recoverable and may be utilized seven times without losing any of its functionality (Darroudi et al. 2021). To compare the two catalytic systems, the Sonogashira reaction of phenylacetylene and iodobenzene was carried out in their presence (Scheme 2 and Scheme 3). Four equivalents of KOH in EtOH/H_2O

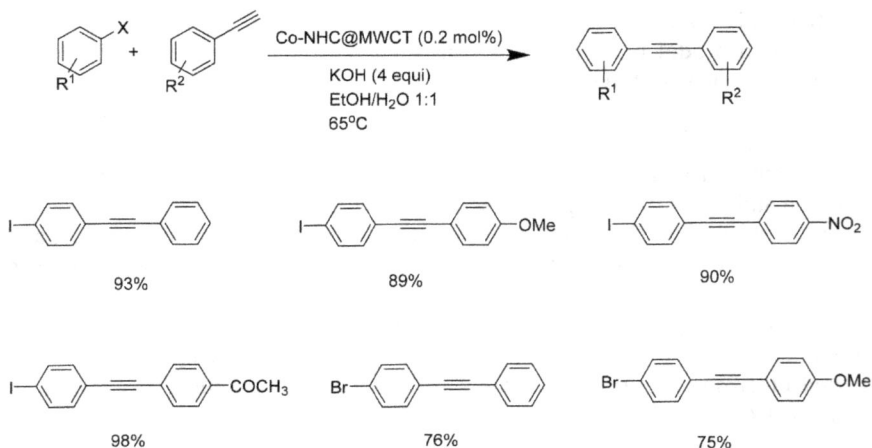

Scheme 2. Sonogashira's work on the cross-coupling of aryl halides with terminal acetylenes while using Co-NHC@MWCNTs

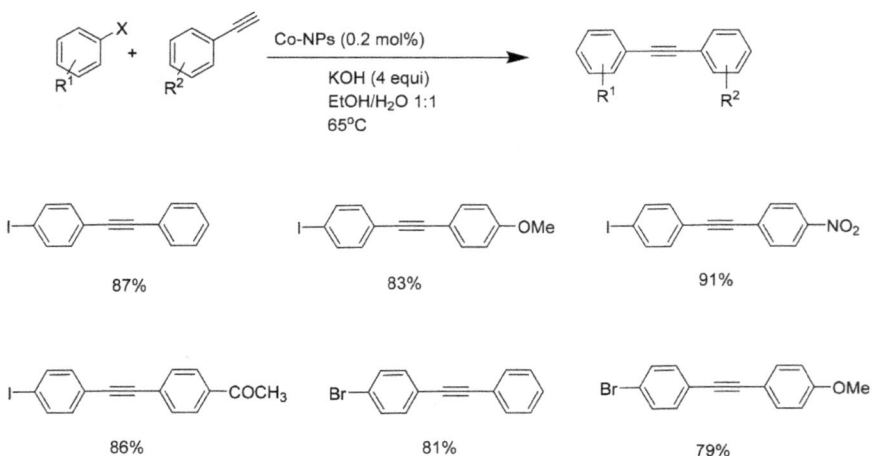

Scheme 3. Aryl halide cross-coupling with terminal acetylenes by Sonogashira in the presence of Co-NPs

at 65°C were used in the process. The necessary compounds were produced in acceptable yields by effectively coupling aryl chlorides, which were less reactive and showed greater reactivity, with aryl iodides (Cook et al. 2021).

The use of both catalysts resulted in excellent yields of up to 98 percent for the reaction catalyzed by Co-NPs and a still-good yield of up to 91 percent for the procedure catalyzed by Co-NHC@MWCNTs. As demonstrated, the Co-N-heterocyclic ligand molecule exhibits higher activity (Cheng et al. 2020).

Another example is the Suzuki cross-coupling reaction which is one of the most often used techniques for creating carbon-carbon bonds in organic synthesis. Commercially, this reaction is used to produce a variety of compounds, including active medicinal components (Salih et al. 2019). Despite having strong catalytic activity, palladium metal's employment in Suzuki cross-coupling reactions is undesired because of its availability, high cost, and toxicity. Pharmaceutical products have standard limits of palladium traces that are typically established at ppm levels (Diyali et al. 2022). As a result, there is an increasing desire to change it for better alternatives that make use of the Earth's abundant metals. First-row transition metals, such as Suzuki cross-coupling reactions with nickel and iron catalysts, are particularly attractive in this respect (Nassar et al. 2021).

These catalytic systems have several benefits, including cost, accessibility, and efficiency, but they also have significant drawbacks, including the inability to be reused. It is necessary to use a unique catalytic mechanism to get around this constraint (Sestelo et al. 2020). The cross-coupling reaction of aryl halides with phenylboronic acid has been described here, and heterogeneous cobalt catalysts synthesized from reliable, non-toxic, and easily available commercial precursors have been used. We are unaware of any reports of Suzuki reactions with a heterogonous cobalt catalyst (Sobhani et al. 2020).

One of the other most important reactions in organic synthesis is the Heck cross-coupling reaction, which is used in the agrochemical, pharmaceutical, and fine chemical industries (Devendar et al. 2018). Palladium is a common component in Heck couplings' conventional catalytic mechanism. However, the use of palladium-free catalytic systems is more interesting in the context of current organic synthesis due to its high cost and toxicity (Zhu et al. 2019). Nevertheless, a variety of synthetic processes using nickel-based palladium-free catalysts are available. Heck reactions have been seen for copper, iron, and cobalt. Cobalt stands out among them because it is easily accessible, non-toxic, inexpensive, stable, and has potent catalytic properties (Goetzke et al. 2021). There are a few studies on the Heck coupling reaction's use of cobalt catalysts. For instance, cobalt nanoparticles were reported in 2009 by Qi et al. as a catalyst for the Heck cross-coupling process (Bankar et al. 2020). It is noteworthy that the class of aryl halides was not described, despite the challenging reaction conditions. Recently, Zhu et al. employed uniform Co-B amorphous alloy nanoparticles as a catalyst in the Heck process as shown in Scheme 4 (Wang et al. 2020).

Nanoparticle-based catalytic systems frequently exhibit more activity than comparable bulk materials because of their high surface-to-volume ratio as shown in Table 1. However, the catalytic activity of the nanoparticles is diminished when they group to create bigger bulk particles. Utilizing the right stabilizers or protective chemicals will help you prevent it (Hajipour et al. 2020).

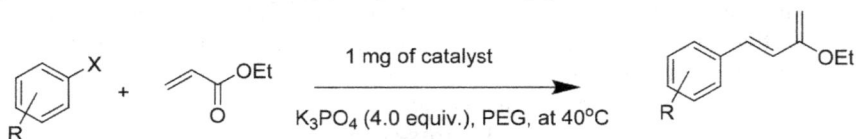

Scheme 4. Pd/Cu-Free Cobalt-catalyzed Suzuki and Heck using green Bio-Magnetic Hybrid and DFT-based Theoretical study

Table 1. Variation of Substrates of Scheme 4 to produce a better yield

Entry	X	R	Yield (%)
1	I	2-NO$_2$	84
2	I	H	93
3	Br	4-OMe	92
4	Br	H	88
5	Br	4-CHO	96
6	Br	3-NO$_2$	95
7	Br	4-CN	95
8	Cl	4-NO$_2$	83
9	Cl	2-Cl	73
10	Cl	4-CHO	71

Aryl halide (1.0 mmol), methyl acrylate (1.1 mmol), K$_3$PO$_4$ (4.0 equiv.), and 1 mg of catalyst (2.4 mol percent of Co) were used in the reaction, which was conducted at 40°C for one hour (Lee et al. 2021).

2.2 Chitosan-supported catalyst containing copper

It was possible to make a very effective and easily recoverable heterogeneous Cu catalyst for the production of aryl sulfones by simply swirling an aqueous solution of chitosan in water containing copper ions (Zare et al. 2020). The coupling of aryl halides with sodium sulfinates to easily produce the corresponding sulfones in good to excellent yields was catalyzed by the chitosan@copper catalyst as shown in Scheme 5 and with varying substrates and reaction conditions shown in Table 2. Numerous re-uses of the highly active catalyst are possible without losing any of its catalytic power. Additionally, this process made it simple to create the commercially available medication zolimidine (antiulcer) as shown in Scheme 6 (Reddy et al. 2021).

The CuSO$_4$, CuI, and Cu(OAc)$_2$ CS-supported copper salts are often synthesized using the following method: to create the catalyst, chitosan was immersed in an aqueous medium of CuSO$_4$, CuI, or Cu(OAc)$_2$ for three hours under neutral circumstances at 50°C (Chang et al. 2020). After copper adsorption, the material was thoroughly washed to remove any remaining Cu compounds. Then, it was dried underneath the vacuum at 50°C for a whole night to generate the chitosan@copper catalyst (Narayanan et al. 2020).

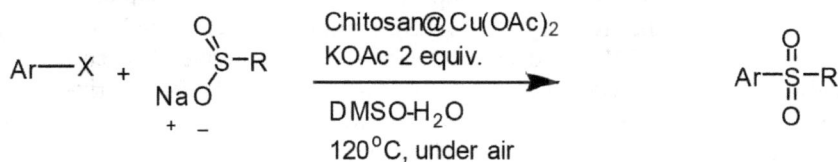

Scheme 5. The action of sodium sulfinate with aryl halides

Table 2. Showing Variations of Substrates and Reaction conditions for Scheme 5

S. No.	Ar	R	X	Yield (%)
1	Ar = p-MeO-C_6H_4	Ph	I	91
2	Ar = p-Me-C_6H_4	Ph	I	90
3	Ar = p-Cl-C_6H_4	Ph	I	82
4	Ar = p-NO_2-C_6H_4	Ph	I	83
5	Ar = p-CF_3-C_6H_4	Ph	I	80
6	Ar = Ph	Ph	I	91
7	Ar = o-COOMe-C_6H_4	Ph	I	65
8	Ar = p-C_6H_4-C_6H_4	Ph	I	72
9	Ar=Ph	Me	I	95
10	Ar = p-MeO-C_6H_4	Me	I	91
11	Ar=Ph	p-Me-C_6H_4	I	90
12	Ar = p-Me-C_6H_4	p-Me-C_6H_4	I	87

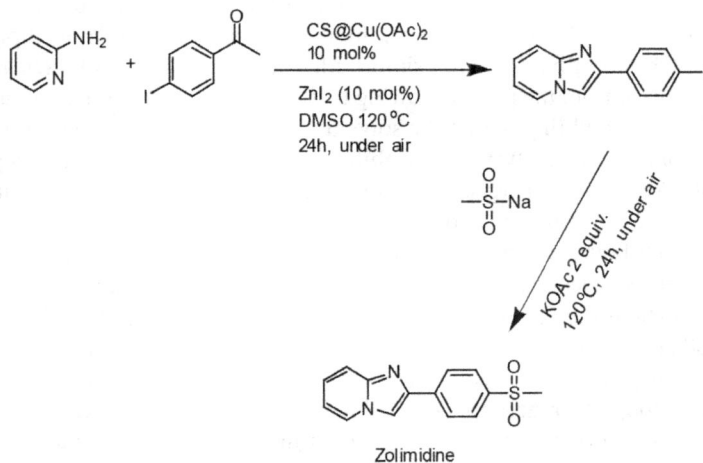

Zolimidine

Scheme 6. Application of the one-pot synthesis of zolimidine using the CS@Cu(OAc)$_2$-catalyzed C-S coupling

The material was extensively cleaned to remove any stray Cu compounds following the extraction of copper ions from wastewater using a chitosan membrane, and then it was dried under a vacuum at 50°C overnight to form the chitosan@copper catalyst (Lu et al. 2022).

Transmission electron microscopy (TEM), inductively coupled plasma-atomic emission spectrometry (ICP-AES), thermogravimetric analysis (TGA), X-ray diffraction (XRD), and infrared analysis (FT-IR) have all been used to analyze catalysts (Niakan et al. 2020). According to FT-IR characterization of the chitosan catalysts, the pure chitosan powder displayed unique absorption bands of O–H and N–H stretching vibrations of amine groups at 3440 cm^{-1}. When compared to chitosan, the peak at 3426 cm^{-1} for the CS@copper catalysts is sharper and stronger. The peaks at about 1085, 1384, 1606, and 2877–2925 cm^{-1} are corresponding to the chitosan molecules' C–OH, C–N, N–H, and C–H stretching models, respectively (Hu et al. 2021). The copper catalysts' FT-IR results showed that the metal Cu coordinates with the molecules -NH$_2$ and -OH. However, the thermal stability is relatively low than that of chitosan due to the drop in the number of principal amino groups on further coordination with the metal Cu (Huang et al. 20201). Importantly, the TG demonstrated the durability of these catalysts up to 200 °C, indicating that they are compatible with the majority of organic reactions due to their high thermal stability. The copper metal's signals were not detected by XRD because of its chelation with chitosan or even its low percentage level (Xu et al. 2021). The results of the TEM investigation revealed that the dispersion of the copper nanoparticles was excellent and that their average diameter was between 3 and 8 nm. ICP-AES was used to measure the copper loading levels of the catalysts, and the values were 1.95, 1.50, and 1.46 mmol g^{-1} (Wei et al. 2021).

The recovery and recycling of the catalyst is a major issue with a transition-metal catalyzed process for sustainability, both economically and environmentally concerned. This is due to the frequent high cost and toxicity of transition metal catalysts (Dhawa et al. 2021). A simple filtration was adequate to separate the catalyst solution from the products owing to the high solubility of the products and the catalyst's insolubility within the solvent. The C–S coupling reaction was then used to examine the catalyst's recyclability (Caliskan et al. 2021). The separated CS@Cu(OAc)$_2$ was recharged in the recycling experiment with a new substrate for the subsequent run under the same reaction conditions. It was remarkable that even after being utilized five times, the catalyst was still catalytically active. The required product was produced via the C–S coupling reaction with yields of 87% and 85% in the fourth and fifth runs, respectively (Ding et al. 2018).

Another example of an ion-containing chitosan-supported-catalyst is numerous C–N coupling reactions, including Buchwald-Hartwig amination processes, Chan-Lam coupling reactions, and Ullmann reactions, which are described in the literature (Cortes et al. 2019). Various metal catalysts, including nickel, palladium, and copper, are used in all of these reactions together with various coordinating ligands, including bisamines, amino acids, DPPF, phosphoramidites, and oxime-phosphine oxides as shown by Table 3 (Zahmatkesh et al. 2019). The toxicity carried on by the presence of metals in the products is a severe drawback of these reactions, even though these catalysts and ligands are utilized at minimal loadings. Han et al. described N-arylation of amines/amino alcohols with aryl halides utilizing free CuCl as a catalyst as shown in Scheme 7 (Palacios et al. 2021). This reaction has a longer reaction time (8 h), but it also results in substantial metal contamination of the end products. Therefore, it is highly essential to create novel approaches that take into account a decrease in reaction time, minimal metal contamination of products, simple reaction conditions, simple product and catalyst isolation, and catalyst renewability. Due to their low cost and low toxicity, applications of eco-friendly polymer-supported metal catalysts are particularly promising in this aspect (Miceli et al. 2021).

Scheme 7. Reaction yields of chitosan@copper-catalyzed C–N cross-couplings between aryl halides and diamines/amino alcohol in acetonitrile at 1 equiv and 2 equiv, respectively

Scanning electron microscopy (SEM), TGA, powder XRD, and IR analyses were used to characterize the produced chitosan@copper catalyst. To determine the size and shape of the manufactured copper catalyst with chitosan incorporated, a SEM analysis at a 15 kV accelerating voltage is used (Jayaramudu et al. 2020).

2.3 Chitosan-supported catalyst containing rhodium

Transition metals have a variety of uses as catalysts, whether in their elemental state or as the central core of metal complexes. They may be in several oxidation states and readily switch between them, which allows them to lend or take electrons from various reagents, which is the major factor in their remarkable performance (Favre et al. 2022). As a result, transition metals work by combining with the reagent to produce complexes. There are just a few practical uses for transition metal complexes (TMCs), even though several research organizations have concentrated on their synthesis. This is because homogenous complexes have several serious drawbacks, including high cost, particular handling needs, challenges with recycling and reuse, and a non-contentious mode of operation (Liu et al. 2021).

Numerous additional metals, including nickel, iron, copper, and rhodium, were successfully used for the Suzuki reaction after the development of the first

Table 3. Showing variations of substrates and reaction conditions for Scheme 7

Entry	Aryl Halide (1)	Diamines/ Alcohol (2)	Product (3)	Catalyst load (mg)	Reaction time (h)	Yield (%)
1				100	3	94
2				120	5	88
3				120	4	88
4				150	6	87
5				120	5	92
6				100	6	89

palladium catalyst (Arora et al. 2021). Rhodium-based catalysts were shown to be beneficial for the Suzuki reaction with halobenzens that retain both aldehydes and nitriles on their aromatic ring when used as nano- and micro-particles on diverse supports, providing efficient heterogeneous catalysts (Leviev et al. 2020). We finally introduced a simple and effective method for immobilizing TMCs in polysaccharide matrices. An aqueous solution of polysaccharides like Carrageenan types lambda (λ), iota (i), xanthan-gum (x), and kappa (κ) was combined with an aqueous solution of rhodium or palladium salts with sodium triphenylphosphine trisulfonate (TPPTS) as a ligand to produce the heterogeneous catalysts (Boukhatem et al. 2021).

FTIR analysis showed that the resulting xerogels, or "sponge-like" solid catalyst systems, allowed the recycling of the catalyst without the complex being leached. This was only achieved by forming a new linkage between the hydroxyl groups on the polysaccharide and the sodium sulfonic acid salt on the TPPTS ligand (Leviev et al. 2019). Due to their renewable nature, biodegradability, and low toxicity, polysaccharides derived from natural sources, such as carrageenans, and chitosan, cellulose, have subsequently found several uses in the biomedical product, pharmaceutical, food, cosmetic, and building block sectors. A variety of polysaccharides were also utilized as substrates for metal catalysts, notably palladium and rhodium catalysts for the Suzuki process, resulting in heterogeneous and recyclable systems (Wolfson et al. 2020).

The homogeneous reaction with rhodium chloride saw a poor conversion rate when a polysaccharide was added, in contrast to the reactions with palladium (Bauer et al. 2020). This led to the hypothesis that rhodium, unlike palladium, would interact with the ligands on the polymer backbone and, as a result, might be attached to the polysaccharide even in the lack of a ligand. In light of these earlier results, the goal of this work is to clarify the interaction between rhodium and polysaccharides and assess the viability of using it as a heterogeneous catalyst for the Suzuki process (Villemin et al. 2019).

The heterogenization of several polysaccharides which are derived from natural sources takes place on a relatively simple, ligand-free rhodium chloride medium. The Suzuki cross-coupling of halobenzenes with phenylboronic acid and sodium carbonate in ethanol utilizing the novel heterogeneous catalysts is shown in Scheme 8 (Ontman et al. 2021).

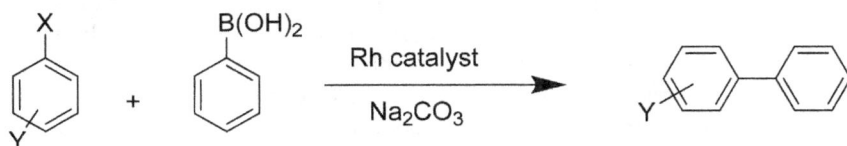

Scheme 8. Suzuki cross-coupling of halobenzene and phenylboronic acid

To sum things up, $RhCl3$ was successfully immobilized to a straightforward procedure, and it was successfully used in a Suzuki cross-coupling process. The salt did not seep into the reaction mixture, and the catalyst was successfully regenerated with a rise in activity during the first and second recycles (Kadib et al. 2020). In addition, the evaluation of the lyophilized -$RhCl3$ system showed that the salt is integrated into the support even though FTIR indicated that there is probably no new link between the salt and the polymer (Peng et al. 2022). Additionally, SEM images show that the addition of $RhCl_3$ produced a structure that was more porous and structured than the native one, pointing to a potential interaction between the two (Loreto et al. 2021). Additionally, XPS research revealed a typical Rh^{+3} binding energy when the heterogeneous catalyst's Rh:Cl atomic ratio was about 3:1 (Bai et al. 2020). Finally, viscosity studies showed that RhCl$_3$ interacts with polysaccharides since adding RhCl$_3$ to an aquatic solution caused the viscosity to rise. It was therefore proposed that the rhodium salt was not entrapped inside the polysaccharide matrix but rather was immobilized by -carrageenan coordination to RhCl$_3$ via the sulphate groups in their backbone (Hammi et al. 2020).

Another example is hydroformylation, which has been extensively employed in industry to produce aldehydes from alkenes. The OXEA method, which has been generating 8.0×10^5 tons of C4 and C5 aldehydes from propene or butene yearly since 1984, is a significant case in point (Sole et al. 2020). The catalyst used in this procedure is Rh/P(C$_6$H$_4$SO$_3$Na)$_3$ (TPPTS). Aldehydes are the raw material used to create a variety of beneficial secondary products, including specialized chemicals (which are relevant to the organic synthesis of fragrances and complex natural products) and alcohols (used to make detergents). Production capacity grew to 6.6×10^6 tons in 1995 (Mika et al. 2018). The creation of highly active and selective catalysts for the hydroformylation process utilizing various transition metals has received a lot of attention during the past several decades. Due to their high activity and selectivity under more favorable circumstances, Rh complex-based catalysts are the most often utilized for this reaction (Motokura et al. 2021).

However, issues with the separation of the catalyst/product combination have restricted the industry's use of homogeneous hydroformylation systems in practice. Distillation separation is also a time-consuming, energy-intensive, and equipment-corrosive process (Shende et al. 2019). As a result, several techniques have been used to address this issue, including supported aqueous phase, ionic liquids, aqueous biphasic, supported liquid phase, supercritical fluids, and supported ionic liquid-phase catalysts (Jin et al. 2019). Despite solving the separation problem, these methods frequently lead to metal leaching and poor aldehyde product regioselectivity. Alternatives include immobilizing homogeneous catalysts on solid supports such as mesoporous materials, metal oxides, polymers, dendritic scaffolds, and various forms of carbon (Cho et al. 2020).

Researchers have lately looked to biopolymers as supports for transition metal catalysts due to their apparent renewability, abundance in nature, biodegradability, and non-toxicity. Several biopolymers, namely starch, cellulose, gelatin, chitosan, and alignate, have been suggested in this context. These activities do result in more environmentally friendly and sustainable chemistry (Carrion et al. 2021).

It is simple to alter the amine groups in chitosan to provide ligand donor sites for efficient and reliable metal coordination. The creation of fresh chitosaniminophosphine Pd catalysts for carbon-carbon cross-coupling processes was recently reported (Goff et al. 2020). The catalysts provided yields that were on a level with or better than those made using a homogeneous catalyst of a related kind. They also exhibited excellent activity. As part of our continuous efforts to expand the utilization of the chitosan-Schiff base ligands, the first occurrences of chitosan-iminopyridyl and -iminophosphine Rh complexes have been created. An unmodified chitosan-Rh catalyst was reportedly employed for 1-hexene hydroformylation processes in the past, but no research on metal leaching or catalyst recyclability was published (Burkart et al. 2019).

We now present the preparation, characterization, and analysis of the new supported catalysts used in the hydroformylation of 1-octene. The utilization of biodegradable and non-toxic biopolymer support and the atom economy are two green chemistry concepts that are used in this study (Raghuvanshi et al. 2020). The hydroformylation activity of model mononuclear Rh homogeneous catalysts has also been produced and investigated in comparison to that of their heterogenized counterparts (Siangwata et al. 2019).

By treating chitosan-Schiff base ligands (1 and 2) with [RhCl(CO)$_2$], chitosan-supported Rh catalysts (3 and 4) were quickly created as shown above in Scheme 9. As a result, dry acetone was heated to room temperature while the appropriate chitosan-Schiff base ligand (loading values: 0.12 mmol g^{-1} (1) and 0.13 mmol g^{-1} (2)) was mixed with too much [RhCl(CO)$_2$]$_2$. The supported catalysts (3 and 4) were successfully manufactured into stable light-orange and purple solids with good yields (Garba et al. 2021).

Microanalysis, FT-IR, UV-vis, solid state ^{31}P and ^{13}C NMR spectroscopy, ICP-MS, PXRD, and TEM have all been used to describe them. These techniques have been used to confirm that the reported structure of the chitosan-supported Rh catalysts (3 and 4) and the predicted structure of the chitosan Schiff base ligands (1 and 2) are accurate (Yazdanseta et al. 2022).

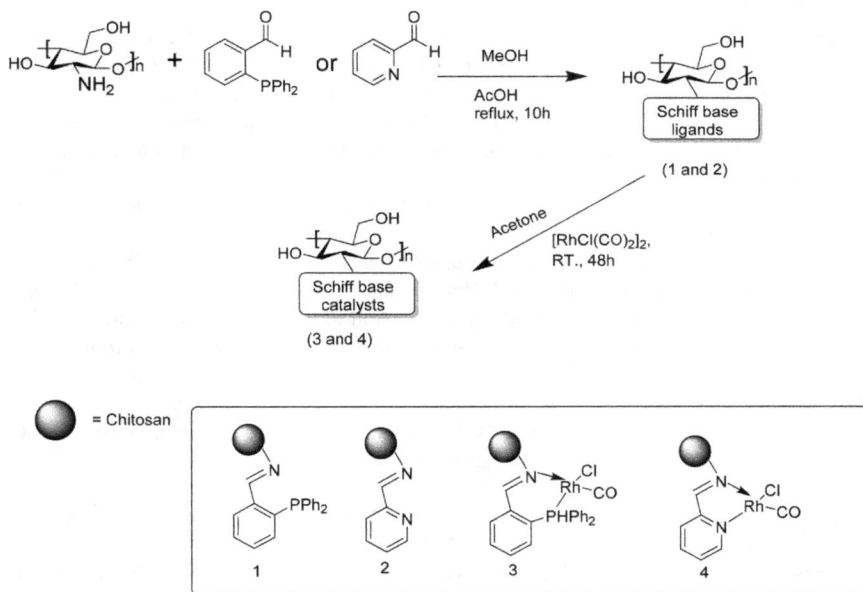

Scheme 9. Schematic for the synthesis of supported catalysts (3 and 4), Schiff base ligands (1 and 2)

3. Conclusion

Compared to the typical heating approach, this catalyst demonstrated outstanding activity with a slower reaction time and much less catalyst. When the catalyst's reusability was tested, it was discovered that it could be reused up to seven times. Tests for mercury toxicity and leaching also revealed that the catalyst was heterogeneous. As a result of its benefits, including ease of setup, quick reaction times, high reaction yields, the lack of solvents, high selectivity with low catalyst amounts, long lifetimes, high thermal stability, high TON and TOF values, renewability, and the presence of oxygen and moisture insensitivities, we can anticipate that this new efficient synthesis technique with a novel catalyst will find use in a wide range of applications in various fields.

References

Adeyemi, S.A., P. Kumar, V. Pillay and Y.E. Choonara. 2022. Environmentally sustainable and safe production of nanomedicines. *Sustainable Nanotechnology: Strategies, Products, and Applications* 329-354.

Amrutha, S., S. Radhika and G. Anilkumar. 2022. Recent developments and trends in the iron-and cobalt-catalyzed Sonogashira reactions. *Beilstein Journal of Organic Chemistry* 18(1): 262-285.

Anderson, E.P., S. Jackson, R.E. Tharme, M. Douglas, J.E. Flotemersch, M. Zwarteveen, C. Lokgariwar, M. Montoya, A. Wali, G.T. Tipa, T.D. Jardine, J.D. Olden, L. Cheng, J. Conallin, B. Cosens, C. Dickens, D. Garrick, D. Groenfeldt, J. Kabogo, D.J. Roux, A. Ruhi and A.H. Arthington. 2019. Understanding rivers and their social relations: A critical step to advance environmental water management. *Wiley Interdisciplinary Reviews: Wate*, 6(6): e1381.

Arora, A., S. Singh, P. Oswal, D. Nautiyal, G.K. Rao, S. Kumar and A. Kumar. 2021. Preformed molecular complexes of metals with organoselenium ligands: Syntheses and applications in catalysis. *Coordination Chemistry Reviews* 438: 213885.

Bai, S., F. Liu, B. Huang, F. Li, H. Lin, T. Wu, M. Sun, J. Wu, Q. Shao, Y. Xu and X. Huang. 2020. High-efficiency direct methane conversion to oxygenates on a cerium dioxide nanowires supported rhodium single-atom catalyst. *Nature Communications* 11(1): 1-9.

Bankar, D.B., K.G. Kanade, R.R. Hawaldar, S.S. Arbuj, M.D. Shinde, S.P. Takle, D.P. Amalnerkar and S.T. Shinde. 2020. Facile synthesis of nanostructured Ni-Co/ZnO material: An efficient and inexpensive catalyst for Heck reactions under ligand-free conditions. *Arabian Journal of Chemistry* 13(12): 9005-9018.

Bauer, E.B. 2020. Transition metal catalyzed glycosylation reactions – An overview. *Organic & Biomolecular Chemistry* 18(45): 9160-9180.

Bengtsson, J., J.M. Bullock, B. Egoh, C. Everson, T. Everson, T. O'Connor, P.J. O'Farrell, H.G. Smith and R. Lindborg. 2019. Grasslands – More important for ecosystem services than you might think. *Ecosphere* 10(2): e02582.

Bilal, M. and H.M. Iqbal. 2019. Naturally-derived biopolymers: Potential platforms for enzyme immobilization. *International Journal of Biological Macromolecules* 130: 462-482.

Boukhatem, A., K. Bouarab and A. Yahia. 2021. Kappa (κ)-carrageenan as a novel viscosity-modifying admixture for cement-based materials – Effect on rheology, stability, and strength development. *Cement and Concrete Composites* 124: 104221.

Burkart, M.D., N. Hazari, C.L. Tway and E.L. Zeitler. 2019. Opportunities and challenges for catalysis in carbon dioxide utilization. *ACS Catalysis* 9(9): 7937-7956.

Çalışkan, M. and T. Baran. 2021. Decorated palladium nanoparticles on chitosan/ δ-FeOOH microspheres: A highly active and recyclable catalyst for Suzuki coupling reaction and cyanation of aryl halides. *International Journal of Biological Macromolecules* 174: 120-133.

Carrion, C.C., M. Nasrollahzadeh, M. Sajjadi, B. Jaleh, G.J. Soufi and S. Iravani. 2021. Lignin, lipid, protein, hyaluronic acid, starch, cellulose, gum, pectin, alginate and chitosan-based nanomaterials for cancer nanotherapy: Challenges and opportunities. *International Journal of Biological Macromolecules* 178: 193-228.

Chang, T.L. 2020. *In-Situ Formation of Metal/Alloy Nanoparticles and Characterization* (Doctoral dissertation), Stevens Institute of Technology.

Chen, Y., S. Yang, T. Zhang, M. Xu, J. Zhao, M. Zeng, K. Sun, R. Feng, Z. Yang, P. Zhang, B. Wang and X. Cao. 2022. Positron annihilation study of chitosan and its derived carbon/pillared montmorillonite clay stabilized Pd species nanocomposites. *Polymer Testing* 107689.

Cheng, S., W. Wei, X. Zhang, H. Yu, M. Huang and M. Kazemnejadi. 2020. A new approach to large scale production of dimethyl sulfone: A promising and strong recyclable solvent for ligand-free Cu-catalyzed C–C cross-coupling reactions. *Green Chemistry* 22(6): 2069-2076.

Cho, I.H., D.H. Kim and S. Park. 2020. Electrochemical biosensors: Perspective on functional nanomaterials for on-site analysis. *Biomaterials Research* 24(1): 1-12.

Cook, X.A., L.R. Pantaine, D.C. Blakemore, I.B. Moses, N.W. Sach, A. Shavnya and M.C. Willis. 2021. Base-activated latent heteroaromatic sulfinates as nucleophilic coupling partners in palladium-catalyzed cross-coupling reactions. *Angewandte Chemie International Edition* 60(41): 22461-22468.

Darroudi, M., H. Rouh, M. Hasanzadeh and N. Shadjou. 2021. Cu/SiO2-Pr-NH-Benz as a novel nanocatalyst for the efficient synthesis of 1,4-disubstituted triazoles and propargyl amine derivatives in an aqueous solution. *Heliyon* 7(4): e06766.

Delkash, M., F.A. Al-Faraj and M. Scholz. 2018. Impacts of anthropogenic land use changes on nutrient concentrations in surface waterbodies: A review. *CLEAN – Soil, Air, Water* 46(5): 1800051.

Devendar, P., R.Y. Qu, W.M. Kang, B. He and G.F. Yang. 2018. Palladium-catalyzed cross-coupling reactions: A powerful tool for the synthesis of agrochemicals. *Journal of Agricultural and Food Chemistry* 66(34): 8914-8934.

Dhakshinamoorthy, A., M. Jacob, N.S. Vignesh and P. Varalakshmi. 2021. Pristine and modified chitosan as solid catalysts for catalysis and biodiesel production:

A mini review. *International Journal of Biological Macromolecules* 167: 807-833.

Dhawa, U., N. Kaplaneris and L. Ackermann. 2021. Green strategies for transition metal-catalyzed C–H activation in molecular syntheses. *Organic Chemistry Frontiers* 8(17): 4886-4913.

Ding, K., A. He, D. Zhong, L. Fan, S. Liu, Y. Wang, Y. Liu, P. Chen, H. Lei and R. Ruan. 2018. Improving hydrocarbon yield via catalytic fast co-pyrolysis of biomass and plastic over ceria and HZSM-5: An analytical pyrolyzer analysis. *Bioresource Technology* 268: 1-8.

Diyali, N., S. Rasaily and B. Biswas. 2022. Metal–organic framework: An emergent catalyst in C–N cross-coupling reactions. *Coordination Chemistry Reviews* 469: 214667.

Dohendou, M., K. Pakzad, Z. Nezafat, M. Nasrollahzadeh and M.G. Dekamin. 2021. Progresses in chitin, chitosan, starch, cellulose, pectin, alginate, gelatin and gum based (nano) catalysts for the Heck coupling reactions: A review. *International Journal of Biological Macromolecules* 192: 771-819.

El Kadib, A. 2020. Green and functional aerogels by macromolecular and textural engineering of chitosan microspheres. *The Chemical Record* 20(8): 753-772.

El Ouardi, Y., A. Giove, M. Laatikainen, C. Branger and K. Laatikainen. 2021. Benefit of ion imprinting technique in solid-phase extraction of heavy metals, special focus on the last decade. *Journal of Environmental Chemical Engineering* 9(6): 106548.

Emamy, F.H., A. Bumajdad and J.P. Lukaszewicz. 2021. Adsorption of hexavalent chromium and divalent lead ions on the nitrogen-enriched chitosan-based activated carbon. *Nanomaterials* 11(8): 1907.

Eskandari, E., M. Kosari, M.H.D.A. Farahani, N.D. Khiavi, M. Saeedikhani, R. Katal and M. Zarinejad. 2020. A review on polyaniline-based materials applications in heavy metals removal and catalytic processes. *Separation and Purification Technology* 231: 115901.

Farhan, A., E.U. Rashid, M. Waqas, H. Ahmad, S. Nawaz, J. Munawar, A. Rahdar, S. Varjani and M. Bilal. 2022. Multifunctional graphene-based nanocomposites and nanohybrids for the abatement of agro-industrial pollutants in aqueous environments—A review. *Environmental Pollution* 119557.

Favre, D., C.E. Bobst, S.J. Eyles, H. Murakami, D.C. Crans and I.A. Kaltashov. 2022. Solution- and gas-phase behavior of decavanadate: Implications for mass spectrometric analysis of redox-active polyoxidometalates. *Inorganic Chemistry Frontiers* 9(7): 1556-1564.

Forero-Cortés, P.A. and A.M. Haydl. 2019. The 25th anniversary of the Buchwald–Hartwig amination: Development, applications, and outlook. *Organic Process Research & Development* 23(8): 1478-1483.

Garba, H.W., M.S. Abdullahi, M.S.S. Jamil and N.A. Endot. 2021. Efficient catalytic reduction of 4-nitrophenol using copper (II) complexes with N, O-chelating schiff base ligands. *Molecules* 26(19): 5876.

Goetzke, F.W., A.M. Hell, L. van Dijk and S.P. Fletcher. 2021. A catalytic asymmetric cross-coupling approach to the synthesis of cyclobutanes. *Nature Chemistry* 13(9): 880-886.

Hajipour, A.R., Z. Khorsandi, Z. Abeshtiani and S. Zakeri. 2020. Pd/Cu-free Heck and C–N coupling reactions using two modified magnetic chitosan cobalt catalysts: Efficient, inexpensive and green heterogeneous catalysts. *Journal of Inorganic and Organometallic Polymers and Materials* 30(6): 2163-2171.

Hajipour, A.R., Z. Khorsandi, M. Ahmadi, H. Jouypazadeh, B. Mohammadi and H. Farrokhpour. 2021. Pd/Cu-free cobalt-catalyzed Suzuki and Heck using green bio-magnetic hybrid and DFT-based theoretical study. *Catalysis Letters* 151(10): 2842-2850.

Hammi, N., S. Chen, F. Dumeignil, S. Royer and A. El Kadib. 2020. Chitosan as a sustainable precursor for nitrogen-containing carbon nanomaterials: Synthesis and uses. *Materials Today Sustainability* 10: 100053.

Han, B., P. Ma, X. Cong, H. Chen and X. Zeng. 2019. Chromium- and cobalt-catalyzed, regiocontrolled hydrogenation of polycyclic aromatic hydrocarbons: A combined experimental and theoretical study. *Journal of the American Chemical Society* 141(22): 9018-9026.

Hou, H., X. Zeng and X. Zhang. 2020. Production of hydrogen peroxide by photocatalytic processes. *Angewandte Chemie International Edition* 59(40): 17356-17376.

Hu, K., Z. Li, L. Bai, F. Yang, X. Chu, J. Bian, Z. Zhang, H. Xu and L. Jing. 2021. Synergetic subnano Ni- and Mn-oxo clusters anchored by chitosan oligomers on 2D g-C3N4 boost photocatalytic CO_2 reduction. *Solar Rrl* 5(6): 2000472.

Huang, T., W. Hao, B. Jin, J. Zhang, J. Guo, L. Luo, Q. Zhang and R. Peng. 2021. Novel energetic coordination compound [Cu (AT) 4] Cl2 for catalytic thermal decomposition of ammonium perchlorate. *Journal of Solid State Chemistry* 304: 122622.

Javaid, R. and U.Y. Qazi. 2019. Catalytic oxidation process for the degradation of synthetic dyes: An overview. *International Journal of Environmental Research and Public Health* 16(11): 2066.

Jayaramudu, T., K. Varaprasad, K.K. Reddy, R.D. Pyarasani, A. Akbari-Fakhrabadi and J. Amalraj. 2020. Chitosan-pluronic based Cu nanocomposite hydrogels for prototype antimicrobial applications. *International Journal of Biological Macromolecules* 143: 825-832.

Jin, X., J. Feng, S. Li, H. Song, C., Yu, K. Zhao and F. Kong. 2019. A novel homogeneous catalysis – liquid/solid separation system for highly effective recycling of homogeneous catalyst based on a phosphine-functionalized polyether guanidinium ionic liquid. *Molecular Catalysis* 475: 110503.

Kalak, T., R. Cierpiszewski and M. Ulewicz. 2021. High efficiency of the removal process of Pb (II) and Cu (II) ions with the use of fly ash from incineration of sunflower and wood waste using the CFBC technology. *Energies* 14(6): 1771.

Kaur, J., P. Sengupta and S. Mukhopadhyay. 2022. Critical review of bioadsorption on modified cellulose and removal of divalent heavy metals (Cd, Pb, and Cu). *Industrial & Engineering Chemistry Research* 61(5): 1921-1954.

Ken, D.S. and A. Sinha. 2020. Recent developments in surface modification of nano zero-valent iron (nZVI): Remediation, toxicity and environmental impacts. *Environmental Nanotechnology, Monitoring & Management* 14: 100344.

Kohler, D.G. 2018. *I. Palladium-catalyzed anti-Markovnikov oxidative amination of olefins II. Catalytic deracemization of axially chiral diols III. Three component carboamination of electron deficient alkenes* (Doctoral dissertation), University of Illinois at Urbana-Champaign.

Kumar, P.S., R. Gayathri and B.S. Rathi. 2021. A review on adsorptive separation of toxic metals from aquatic system using biochar produced from agro-waste. *Chemosphere* 285: 131438.

Kwak, J.H., M.S. Islam, S. Wang, S.A. Messele, M.A. Naeth, M.G. El-Din and S.X. Chang. 2019. Biochar properties and lead (II) adsorption capacity depend on feedstock type, pyrolysis temperature, and steam activation. *Chemosphere* 231: 393-404.

Le Goff, R., O. Mahé, R. Le Coz-Botrel, S. Malo, J.M. Goupil, J.F. Brière and I. Dez. 2020. Insight in chitosan aerogels derivatives – Application in catalysis. *Reactive and Functional Polymers* 146: 104393.

Lee, N.R. 2021. *Expanding the Palette of Organic Synthesis in Water: I. Carbonyl Iron Powder as a Reagent for Nitro Group Reduction. II. B-Alkyl Suzuki-Miyaura Couplings in Water. III. Development of a Low-Foaming Surfactant for Organic Synthesis in Water. IV. "ppm" Tsuji-Trost Allylations in Water* (Doctoral dissertation), University of California, Santa Barbara.

Leviev, S., A. Wolfson and O. Levy-Ontman. 2019. RhCl (TPPTS) 3 supported on iota-carrageenan as recyclable catalysts for Suzuki cross-coupling. *Journal of Applied Polymer Science* 136(45): 48200.

Leviev, S., A. Wolfson and O. Levy-Ontman. 2020. Novel iota carrageenan-based RhCl3 as an efficient and recyclable catalyst in Suzuki cross coupling. *Molecular Catalysis* 486: 110841.

Levy-Ontman, O., E. Arbit, S. Leviev and A. Wolfson. 2021. Effect of reactor configurations on the Suzuki cross-coupling reaction using a carrageenan-based RhCl 3 catalyst. *Organic Communications* 14(4).

Liu, R., A. Zhou, X. Zhang, J. Mu, H. Che, Y. Wang, Q. Zhang and Z. Kou. 2021. Fundamentals, advances and challenges of transition metal compounds-based supercapacitors. *Chemical Engineering Journal* 412: 128611.

Loreto, D. and A. Merlino. 2021. The interaction of rhodium compounds with proteins: A structural overview. *Coordination Chemistry Reviews* 442: 213999.

Lu, L., H. Li and A. Lei. 2022. Oxidative cross-coupling reactions between two nucleophiles. *Chinese Journal of Chemistry* 40(2): 256-266.

Madkour, M., K.D. Khalil and F.A. Al-Sagheer. 2021. Heterogeneous hybrid nanocomposite based on chitosan/magnesia hybrid films: Ecofriendly and recyclable solid catalysts for organic reactions. *Polymers* 13(20): 3583.

Miceli, M., P. Frontera, A. Macario and A. Malara. 2021. Recovery/reuse of heterogeneous supported spent catalysts. *Catalysts* 11(5): 591.

Mika, L.T. and I.T. Horváth. 2018. Fluorous catalysis. *Green Techniques for Organic Synthesis and Medicinal Chemistry* 219-268.

Mohajer, F., M.M. Heravi, V. Zadsirjan and N. Poormohammad. 2021. Copper-free Sonogashira cross-coupling reactions: an overview. *RSC Advances* 11(12): 6885-6925.

Motokura, K., S. Ding, K. Usui and Y. Kong. 2021. Enhanced catalysis based on the surface environment of the silica-supported metal complex. *ACS Catalysis* 11(19): 11985-12018.

Munir, N., M. Jahangeer, A. Bouyahya, N. El Omari, R. Ghchime, A. Balahbib, S. Aboulaghras, Z. Mahmood, M. Akram, S.M.A. Shah, I.N. Mikolaychik, M. Derkho, M. Rebezov, B. Venkidasamy, M. Thiruvengadam and M.A. Shariati. 2021. Heavy metal contamination of natural foods is a serious health issue: A review. *Sustainability* 14(1): 161.

Munusamy, S., P. Muniyappan and V. Galmari. 2019. Synthesis and structural characterization of palladium (II) 2-(arylazo) naphtholate complexes and their catalytic activity in Suzuki and Sonogashira coupling reactions. *Journal of Coordination Chemistry* 72(11): 1910-1921.

Nair, P.P., R.M. Philip and G. Anilkumar. 2021. Nickel catalysts in Sonogashira coupling reactions. *Organic & Biomolecular Chemistry* 19(19): 4228-4242.

Narayanan, V.S., P.V. Prasath, K. Ravichandran, D. Easwaramoorthy, Z. Shahnavaz, F. Mohammad, H.A. Al-Lohedan, S. Paiman, W.C. Oh and S. Sagadevan. 2020. Schiff-base derived chitosan impregnated copper oxide nanoparticles: An effective photocatalyst in direct sunlight. *Materials Science in Semiconductor Processing* 119: 105238.

Nasrollahzadeh, M., N.S.S. Bidgoli, Z. Nezafat and N. Shafiei. 2021. Catalytic applications of biopolymer-based metal nanoparticles. *Biopolymer-based Metal Nanoparticle Chemistry for Sustainable Applications* 423-516. Elsevier.

Nasrollahzadeh, M., N. Motahharifar, Z. Nezafat and M. Shokouhimehr. 2021. Chitosan supported 1-phenyl-1H-tetrazole-5-thiol ionic liquid copper (II) complex as an efficient catalyst for the synthesis of arylaminotetrazoles. *Journal of Molecular Liquids* 341: 117398.

Nassar, Y., F. Rodier, V. Ferey and J. Cossy. 2021. Cross-coupling of ketone enolates with grignard and zinc reagents with first-row transition metal catalysts. *ACS Catalysis* 11(9): 5736-5761.

Niakan, M., Z. Asadi and M. Masteri-Farahani. 2020. Fe (III)-salen complex supported on dendrimer functionalized magnetite nanoparticles as a highly active and selective catalyst for the green oxidation of sulfides. *Journal of Physics and Chemistry of Solids* 147: 109642.

Palacios, F., A.M.O. de Retana and M. Jesús. 2021. Copper-catalyzed synthesis of aziridines. *Copper in N-Heterocyclic Chemistry* 1-48. Elsevier.

Patureau, F.W. 2019. The phenol-phenothiazine coupling: An oxidative click concept. *ChemCatChem* 11(21): 5227-5231.

Peng, K., W. Wang, J. Zhang, Y. Ma, L. Lin, Q. Gan, Y. Chen and C. Feng. 2022. Preparation of chitosan/sodium alginate conductive hydrogels with high salt contents and their application in flexible supercapacitors. *Carbohydrate Polymers* 278: 118927.

Pereao, O., C. Bode-Aluko, K. Laatikainen, A. Nechaev and L. Petrik. 2019. Morphology, modification and characterisation of electrospun polymer nanofiber adsorbent material used in metal ion removal. *Journal of Polymers and the Environment* 27(9): 1843-1860.

Pérez Sestelo, J. and L.A. Sarandeses. 2020. Advances in cross-coupling reactions. *Molecules* 25(19): 4500.

Peter, S., N. Lyczko, D. Gopakumar, H.J. Maria, A. Nzihou and S. Thomas. 2021. Chitin and chitosan based composites for energy and environmental applications: A review. *Waste and Biomass Valorization* 12(9): 4777-4804.

Qasem, N.A., R.H. Mohammed and D.U. Lawal. 2021. Removal of heavy metal ions from wastewater: A comprehensive and critical review. *NPJ Clean Water* 4(1): 1-15.

Rafiee, F. 2019. Recent advances in the application of chitosan and chitosan derivatives as bio supported catalyst in the cross coupling reactions. *Current Organic Chemistry* 23(4): 390-408.

Rafiee, F. and S.A. Hosseini. 2018. CNC pincer palladium complex supported on magnetic chitosan as highly efficient and recyclable nanocatalyst in C—C coupling reactions. *Applied Organometallic Chemistry* 32(11): e4519.

Raghuvanshi, K., C. Zhu, M. Ramezani, S. Menegatti, E.E. Santiso, D. Mason, J. Rodgers, M.E. Janka and M. Abolhasani. 2020. Highly efficient 1-octene hydroformylation at low syngas pressure: From single-droplet screening to continuous flow synthesis. *ACS Catalysis* 10(14): 7535-7542.

Reddy, R.J. and A.H. Kumari. 2021. Synthesis and applications of sodium sulfinates (RSO 2 Na): A powerful building block for the synthesis of organosulfur compounds. *RSC Advances* 11(16): 9130-9221.

Salih, K.S. and Y. Baqi. 2019. Microwave-assisted palladium-catalyzed cross-coupling reactions: Generation of carbon–carbon bond. *Catalysts* 10(1): 4.

Sellaoui, L., D.I. Mendoza-Castillo, H.E. Reynel-Ávila, B.A. Ávila-Camacho, L.L. Díaz-Muñoz, H. Ghalla, A.B. Petriciolet and A.B. Lamine. 2019. Understanding the adsorption of Pb^{2+}, Hg^{2+} and Zn^{2+} from aqueous solution on a lignocellulosic biomass char using advanced statistical physics models and density functional theory simulations. *Chemical Engineering Journal* 365: 305-316.

Shende, V.S., V.B. Saptal and B.M. Bhanage. 2019. Recent advances utilized in the recycling of homogeneous catalysis. *The Chemical Record* 19(9): 2022-2043.

Siangwata, S., N.C. Breckwoldt, N.J. Goosen and G.S. Smith. 2019. Olefin hydroformylation and kinetic studies using mono- and trinuclear N, O-chelate rhodium (I)-aryl ether precatalysts. *Applied Catalysis A: General* 585: 117179.

Sobhani, S., H.H. Moghadam, J. Skibsted and J.M. Sansano. 2020. A hydrophilic heterogeneous cobalt catalyst for fluoride-free Hiyama, Suzuki, Heck and Hirao cross-coupling reactions in water. *Green Chemistry* 22(4): 1353-1365.

Sole, R. 2020. Synthesis and characterization of biopolymers and transition metal complexes for the valorizaion of biomasses. 8(1).

Tabelin, C.B., T. Igarashi, M. Villacorte-Tabelin, I. Park, E.M. Opiso, M. Ito and N. Hiroyoshi. 2018. Arsenic, selenium, boron, lead, cadmium, copper, and zinc in naturally contaminated rocks: A review of their sources, modes of enrichment, mechanisms of release, and mitigation strategies. *Science of the Total Environment* 645: 1522-1553.

Villemin, E., Y.C. Ong, C.M. Thomas and G. Gasser. 2019. Polymer encapsulation of ruthenium complexes for biological and medicinal applications. *Nature Reviews Chemistry* 3(4): 261-282.

Wang, F., Y. Zhang, Y. Luo, Y. Wang and H. Zhu. 2020. Preparation of dandelion-like Co–Mo–P/CNTs-Ni foam catalyst and its performance in hydrogen production by alcoholysis of sodium borohydride. *International Journal of Hydrogen Energy* 45(55): 30443-30454.

Wang, J.S., C. Li, J. Ying, T. Xu, W. Lu, C.Y. Li and X.F. Wu. 2022. Activated carbon fibers supported palladium as efficient and easy-separable catalyst for carbonylative cyclization of o-alkynylphenols with nitroarenes: Facile construction of benzofuran-3-carboxamides. *Journal of Catalysis* 413: 713-719.

Wang, R., R. Liang, T. Dai, J. Chen, X. Shuai and C. Liu. 2019. Pectin-based adsorbents for heavy metal ions: A review. *Trends in Food Science & Technology* 91: 319-329.

Wang, T., Y. Cao, G. Qu, Q. Sun, T. Xia, X. Guo, H. Jia and L. Zhu. 2018. Novel Cu (II)–EDTA decomplexation by discharge plasma oxidation and coupled Cu removal by alkaline precipitation: Underneath mechanisms. *Environmental Science & Technology* 52(14): 7884-7891.

Wei, Y., K.A. Salih, K. Rabie, K.Z. Elwakeel, Y.E. Zayed, M.F. Hamza and E. Guibal. 2021. Development of phosphoryl-functionalized algal-PEI beads for the sorption of Nd (III) and Mo (VI) from aqueous solutions –Application for rare earth recovery from acid leachates. *Chemical Engineering Journal* 412: 127399.

Wolfson, A. and O. Levy-Ontman. 2020. Development and application of palladium nanoparticles on renewable polysaccharides as catalysts for the Suzuki cross-coupling of halobenzenes and phenylboronic acids. *Molecular Catalysis* 493: 111048.

Xu, S., Q. Gao, C. Zhou, J. Li, L. Shen and H. Lin. 2021. Improved thermal stability and heat-aging resistance of silicone rubber via incorporation of UiO-66-NH2. *Materials Chemistry and Physics* 274: 125182.

Yang, Q., N.R. Babij and S. Good. 2019. Potential safety hazards associated with Pd-catalyzed cross-coupling reactions. *Organic Process Research & Development* 23(12): 2608-2626.

Yazdanseta, S., K. Yasin, M. Setoodehkhah, M. Ghanbari and E. Fadaee. 2022. Anchoring Cu (II) on $Fe_3O_4@ SiO_2$/Schiff base: A green, recyclable, and extremely efficient magnetic nanocatalyst for the synthesis of 2-amino-4H-chromene derivatives. *Research on Chemical Intermediates* 1-22.

Zahmatkesh, S., M. Esmaeilpour and A. Mollaiy Poli. 2019. Ligand complex of copper (II) supported on superparamagnetic $Fe_3O_4@ SiO_2$ nanoparticles: An efficient and magnetically separable catalyst for N-arylation of nitrogen-containing heterocycles with aryl halides. *Inorganic and Nano-Metal Chemistry* 49(10): 323-334.

Zhang, T., W. Wang, Y. Zhao, H. Bai, T. Wen, S. Kang, G. Song, S. Song and S. Komarneni. 2021. Removal of heavy metals and dyes by clay-based adsorbents: From natural clays to 1D and 2D nano-composites. *Chemical Engineering Journal* 420: 127574.

Zhang, Y., M. Zhao, Q. Cheng, C. Wang, H. Li, X. Han, Z. Fan, G. Su, D. Pan and Z. Li. 2021. Research progress of adsorption and removal of heavy metals by chitosan and its derivatives: A review. *Chemosphere* 279: 130927.

Zhu, J. and V.N. Lindsay. 2019. Benzimidazolyl palladium complexes as highly active and general bifunctional catalysts in sustainable cross-coupling reactions. *ACS Catalysis* 9(8): 6993-6998.

Index

For Product Safety Concerns and Information please contact our EU
representative GPSR@taylorandfrancis.com
Taylor & Francis Verlag GmbH, Kaufingerstraße 24, 80331 München, Germany